Strategies for Accelerating Cleanup at Toxic Waste Sites
Fast-tracking Environmental Actions and Decision Making

Strategies for Accelerating Cleanup at Toxic Waste Sites
Fast-tracking Environmental Actions and Decision Making

Scott M. Payne

Lewis Publishers
Boca Raton   Boston   London   New York   Washington, D.C.

Library of Congress Cataloging-in-Publication Data

Payne, Scott M.
    Accelerating cleanup at toxic waste sites : fast-tracking environmental actions and decision making / Scott M. Payne.
        p. cm.
    Includes bibliographical references and index.
    ISBN 1-56670-237-2 (alk. paper)
    1. Hazardous waste site remediation--United States. 2. Government productivity--United States. 3. Strategic planning--United States. I. Title.
TD1040.P39   1997
363.738'4--dc21
                                                                                                             97-37138
                                                                                                             CIP

    This book contains information obtained from authentic and highly regarded sources. Reprinted material is quoted with permission, and sources are indicated. A wide variety of references are listed. Reasonable efforts have been made to publish reliable data and information, but the author and the publisher cannot assume responsibility for the validity of all materials or for the consequences of their use.
    Neither this book nor any part may be reproduced or transmitted in any form or by any means, electronic or mechanical, including photocopying, microfilming, and recording, or by any information storage or retrieval system, without prior permission in writing from the publisher.
    The consent of CRC Press LLC does not extend to copying for general distribution, for promotion, for creating new works, or for resale. Specific permission must be obtained in writing from CRC Press LLC for such copying.
    Direct all inquiries to CRC Press LLC, 2000 Corporate Blvd., N.W., Boca Raton, Florida 33431.

© 1998 by CRC Press LLC
Lewis Publishers is an imprint of CRC Press LLC

No claim to original U.S. Government works
International Standard Book Number 1-56670-237-2
Library of Congress Card Number 97-37138
Printed in the United States of America 1 2 3 4 5 6 7 8 9 0
Printed on acid-free paper

To my wife, son, and daughter —
who are my reasons for being.

To my teachers —
who are my reasons for knowing.

ACKNOWLEDGMENTS

I want to thank the many people who helped with the preparation of this book. Cheryl Schmidt helped with word processing. Jim Dushin prepared the graphical displays, many of which were interpreted from cryptic handwritten scratches on paper. Steve MacNeill provided the primary technical review of this book and acted as a sounding board for new and innovative approaches. In addition, I want to thank the many private companies, professional organizations, and government agencies referenced in this book for their support and help.

Lastly, I want to thank my wife Ann. Her editorial skills and the extra help caring for our family while I wrote this book are most appreciated.

PREFACE

Accelerating cleanup, streamlining decision making, and reducing environmental costs at contaminated sites are essential principles which the environmental industry must embrace. This book links together the elements of these principles, including the technology and science, innovative approaches, environmental leadership, communication, regulatory processes, and other elements that drive increased environmental performance. Currently, the quality and service offered by the environmental industry is perceived as unsatisfactory. A clear message is being communicated requesting that *we* do more cleanup with less resources at our nation's contaminated sites. "We" in this case refers to both private and government entities working *together* to achieve these goals.

In the last 5 to 10 years, there have been numerous papers and guidance developed to help fast-track and streamline environmental response actions. More recently, environmental laws have become more relaxed in favor of fast-tracking *protective* environmental response actions. This book brings together many of these concepts, ideas, and guidance under one cover. This book will hopefully be a valuable resource to environmental professionals for developing strategic plans to more efficiently and effectively investigate, clean up, and close environmental sites — no matter who they work for or what regulatory process they are following.

This book also links together the *technical* and *nontechnical* concepts and approaches of accelerated cleanup and streamlined decision making. Topics such as environmental leadership, trust, communication, teamwork, and cooperation are also discussed. As a scientific- and engineering-based industry, these elements are sometimes overlooked, which can impede progress. Linking these elements with the technical elements, such as innovative technologies, provides a basis for developing comprehensive strategies to accelerate, streamline, and reduce costs of environmental actions at toxic waste sites.

Scott M. Payne

Scott M. Payne is an employee of Tetra Tech EM, Inc. The positions and opinions presented in this book are those of the author and do not necessarily represent the positions and opinions of Tetra Tech EM, Inc.

The Author

Mr. Payne is a senior environmental consultant with Tetra Tech EM Inc. in Helena, Montana and is a registered geologist in the State of California and a professional geologist in the State of Wyoming. Mr. Payne has over 11 years of professional environmental consulting experience in numerous areas including investigation, remediation, decision making, site closeout, and program and project management. Mr. Payne has a B.S. in environmental sciences from Northland College (1985) and an M.S. in geology with a computer modeling emphasis from the University of Montana (1989). He is the author of two books, and numerous articles, decision documents, engineering studies, and professional consulting reports related to all aspects of site characterization, environmental modeling, cleanup, and organizational leadership. He has extensive experience characterizing contaminants in both soil and groundwater, evaluating groundwater movement and aquifer systems, interpreting geochemical results, and numerical and analytical modeling related to groundwater flow and solute fate and transport. Other experience includes numerous soil remediation projects and evaluating and design of water treatment and disposal systems, oversight at Superfund sites, and experience with abandoned mine reclamation. Mr. Payne has focused his career on accelerating and streamlining the environmental process at contaminated sites. Mr. Payne is also the Chief Operating Officer and on the Board of Directors for JVI, a professional software development corporation and business management consulting group focused on improving workplace leadership, communication, trust, and teamwork. Mr. Payne is also an officer and director for the non-profit center Potters Farm, Inc., a business and spiritual retreat center located in Washburn, Wisconsin.

Contents

1 Introduction 1
   1.1 Reasons for Accelerating Cleanup and Streamlining Decision Making 2
   1.2 Who Should Read This Book 4
       1.2.1 What Kind of Book is This? 5
   1.3 Fundamental Principles 6
       1.3.1 Risk and Decision Making 8
       1.3.2 The Scientific Method 9
       1.3.3 Strategy and Approach 10
       1.3.4 Investigation and Cleanup Technology 12
       1.3.5 The Human Element 12
   1.4 Book Organization 12

2 U.S. Environmental Protection Agency 15
   2.1 EPA'S Information Transfer and Technology Demonstration Programs 16
       2.1.1 Remedial Technologies 17
       2.1.2 Accelerated Investigation 20
   2.2 CERCLA Early Actions and Accelerated Cleanup 24
       2.2.1 Removal Actions 25
       2.2.2 Interim Remedial Actions 33
       2.2.3 Superfund Accelerated Cleanup Model 33
       2.2.4 Integrated Investigations 38
       2.2.5 Presumptive Remedies and Plug-In Remedies 42
       2.2.6 The Record of Decision 48
       2.2.7 Superfund Case Study for Accelerated Cleanup 55
   2.3 RCRA and Other EPA Programs 58
       2.3.1 Resource Conservation and Recovery Act 59
       2.3.2 RCRA Corrective Actions 59
       2.3.3 USTs and RCRA 69
       2.3.4 RCRA Waste Minimization 70
       2.3.5 Integration of Environmental Programs 71
       2.3.6 Evaluating the Impracticability of Groundwater Cleanup 74
       2.3.7 Additional EPA Information 76
   2.4 References 77

3 U.S. Department of Defense and Military Branches 79
   3.1 Department of Defense 79
       3.1.1 The Environmental Security Program 80
       3.1.2 Lead Agency Responsibility 82
       3.1.3 Budgeting as a Means of Fast-tracking Cleanup 83
       3.1.4 DoD's Aggressive Approach for Site Cleanup 85
       3.1.5 Base Realignment and Closure 86

	3.1.6	Establishing Partnerships and Enhancing Community Involvement 96
	3.1.7	Community Outreach Programs 98
	3.1.8	Determination of Land Use 98
	3.1.9	Measuring Progress at DoD Installations 99
	3.1.10	Road to ROD 102
3.2	U.S. Air Force 106	
	3.2.1	The Air Force's Installation Restoration Program and Accelerated Cleanup 106
	3.2.2	Air Force Center For Environmental Excellence 113
3.3	U.S. Navy 128	
	3.3.1	Navy Environmental Leadership Program 129
	3.3.2	Integrated Project Schedules 130
	3.3.3	Developing a Comprehensive Scope of Work 139
	3.3.4	Accelerated Scheduling 141
3.4	U.S. Army 142	
3.5	References 143	

4 U.S. Department of Energy 145
   4.1 Office of Environmental Management 147
   4.2 OEM Environmental Restoration Program 148
       4.2.1 OEM's Environmental Restoration Strategy and Accomplishments 148
   4.3 Office of Health and Environmental Research 161
       4.3.1 Overview of the NABIR Program 161
   4.4 References 171

5 State and Local Environmental Programs 173
   5.1 Research Programs 174
   5.2 Voluntary Cleanup and Negotiating Actions under State Programs 178
       5.2.1 Addressing Voluntary Actions from a Cultural and Flexibility Perspective 179
   5.3 State Underground Storage Tank Programs 182
       5.3.1 Acceleration through Preventing Future UST Releases 183
       5.3.2 Considerations for Pilot Cleanup Technologies and Actions during UST Removals 185
   5.4 Western Governors Association 194
       5.4.1 Interstate Technology and Regulatory Cooperation Working Group Findings 194
   5.5 State Abandoned Mine Reclamation Programs 198
       5.5.1 Case Study – Montana's Approach for Abandoned Mine Reclamation 199
   5.6 Developing Flexible Record of Decisions under CERCLA 201
       5.6.1 Case Study – Streamside Tailings Project 201

	5.7	California Base Closure Environmental Committee	202
	5.8	References	204

6 Private Sector, Professional Organizations, and Concepts of Innovative Solutions 205
   6.1 Observational Approach 206
      6.1.1 Observational Method 206
      6.1.2 Sequential Risk Mitigation 214
   6.2 Risk-Based Cleanup Approach 217
      6.2.1 Other ASTM Standards 220
   6.3 Computer Modeling 221
      6.3.1 Modeling Techniques and Decision Making 222
   6.4 Geophysical Tools and Applications 230
      6.4.1 Selecting and Planning Geophysical Applications 231
      6.4.2 Case Study – Electromagnetic Offset Logging 239
   6.5 Innovative Technologies and Approaches 246
      6.5.1 General Concepts of Environmental Innovation 246
      6.5.2 Concepts of Innovative Technologies 248
      6.5.3 Concepts of Innovative Approaches 250
   6.6 References 253

7 Communication, Teamwork, Leadership, and Trust 255
   7.1 General Concepts 255
      7.1.1 Evidence Supporting the Need to Value People in the Environmental Profession 256
      7.1.2 Science-Based Information 259
   7.2 Communication 259
      7.2.1 The Learning Process 260
      7.2.2 Understanding Basic Dialogue 261
      7.2.3 Communicating Success and Failure 264
      7.2.4 I and Thou Concepts 264
      7.2.5 Behavior Styles and Gender Differences 265
   7.3 Teamwork 267
      7.3.1 Mission and Vision 267
      7.3.2 Partnership Models 269
      7.3.3 Empowerment and Accountability 272
   7.4 Leadership 273
      7.4.1 Leadership Skills 275
   7.5 Trust 279
   7.6 People Training and Intervention 280
   7.7 References 281

8 Strategic Planning and Lowering Environmental Costs 283
   8.1 Strategic Planning 284
      8.1.1 External Strategy Development and Implementation Process 286

        8.1.2   Large-Scale Strategy Elements for Accelerating and Streamlining Environmental Projects 287
        8.1.3   Small-Scale Strategy for Accelerating and Streamlining Environmental Projects 297
        8.1.4   Strategic Plan 301
        8.1.5   Brownfields and Environmental Strategic Planning 302
  8.2   Lowering Environmental Costs 316
        8.2.1   Reducing Remediation Costs 319
        8.2.2   Remedial Performance Evaluations 322
        8.2.3   Property Tax Reduction 324
        8.2.4   Increasing Efficiency by Understanding What Motivates People 327
  8.3   References 328

Appendix A   Environmental Resource Guide 329

Appendix B   Completed U.S. EPA Site Measuring and Monitoring Technologies Evaluation Projects As of 1994 351

Appendix C   Remediation Technology Matrix 355

Index 421

CHAPTER 1

Introduction

The diesel engine begins to roar as the blade lowers and the operator scrapes off the top 3 inches of soil at a National Priorities List (NPL) site. The project engineer scratches her neck through a protective suit and then begins to take copious notes as she keeps one eye on the operator and another eye on the front-end loader gearing up to load contaminated soil into roll-off bins for on-site soil washing treatment. After 10 years of investigation, public comment, arguing the question of who is responsible, designing the remedial system, and bidding the work, cleanup has finally begun. Perhaps the wait will be over soon, and in several years, the site will be safe for public use again.

Does it always have to take so long before action can begin? Was there, perhaps, a more efficient way to evaluate the site and more quickly protect human health and the environment? Is the overall cost of projects higher if the environmental process is accelerated? Can cleanup at all sites be accelerated? This book presents principles related to answering these, and many other, questions.

Overall, this book is about accelerating cleanup, streamlining decision making, and lowering environmental investigation and remediation costs. This book compiles a lengthy list of principles, techniques, approaches, strategies, models, and to some degree, *values,* that environmental professionals and government agencies have developed in order to accelerate cleanup, reduce costs, fast-track investigation, and shorten project schedules. This book is also about the interactions between people working in the environmental industry who are responsible for and make decisions that ultimately accelerate, or slow down, protection of human health and the environment. Concepts and principles of cooperation, communication, teamwork, leadership, and trust are presented in this book as fundamental components that are essential to running efficient and effective environmental programs and projects. Lastly, this book also describes concepts and principles of strategic planning, which ties together accelerated, streamlined, and cost-effective environmental approaches that can be applied to investigate and clean up contaminated sites in order to reach a final *destination.*

1.1 REASONS FOR ACCELERATING CLEANUP AND STREAMLINING DECISION MAKING

There are many reasons why it is important to accelerate cleanup and streamline decision making. Many of these reasons are site-specific, such as when extremely hazardous conditions are encountered and must be responded to in order to quickly protect human health and the environment, or perhaps, when site conditions clearly show that no action is protective of human health and the environment and that spending any further resources on the project would be inappropriate, implying the closure process should be fast-tracked. Many other site-specific, or "project-scale reasons," may necessitate accelerated or streamlined response actions. From a large- or grand-scale perspective, there are also important reasons for accelerating cleanup and decision making. For some, these reasons may seem less clear. However, they are just as important as the site-specific reasons. Following are five fundamental, grand-scale reasons that support accelerated cleanup and streamlined decision making:

Society is demanding faster service and closure at lower costs. Environmental firms and agencies have been given a clear message from the public and Congress, who directly and indirectly subsidize environmental investigation and restoration. The message clearly demands that the environmental industry must accelerate the overall environmental process used to remediate sites and accomplish this task more efficiently with fewer dollars. The reduced 1995 federal funding of environmental agencies and departments, including the U.S. Environmental Protection Agency (EPA), helps substantiate that more has to be accomplished with *less*. Many state and federal environmental agencies are faced with improving their image and ability to expedite cleanup, while continuing to protect human health and the environment. Those government and private organizations that embrace accelerated site investigation and cleanup, make good faith efforts to minimize the cost of actions through reasonable decision making, and adequately protect human health and the environment will remain viable in the future. Those organizations that do not will be out of business or, in the case of government agencies, will have their funding substantially cut back.

Application of innovative technologies is taking too long. The methods and techniques used to investigate and clean up environmental sites are continually being improved and expanded. In addition, innovative remedial and investigative technologies are constantly being successfully demonstrated and are potential alternatives that can be used to expedite response actions at many environmental sites. Applying these types of technologies is important in order to maintain growth and stabilize the environmental industry. Keeping up with innovative technologies requires that the process used to move projects into and through the decision-making phases must be efficient and effective. However, only since the early 1990s have accelerated and streamlined processes been seriously considered by environmental professionals; even though the opportunity to move through regulatory processes has always been present, a slower process has been followed. Historically, the time it takes to make remedial decisions, for example, finalizing the Record of Decision (ROD) for Superfund sites, is too long. Accelerating the regulatory and decision-making process is essential in order to increase the use of innovative technologies that improve performance. Approaches must be implemented that streamline the environmental process so that cleanup can begin in a reasonable amount of time, and so innovative technologies can be applied in order to accelerate response actions.

Past application and interpretation of environmental regulations have not met the intended spirit of environmental laws. No one intended that the current environmental Acts and regulations should hinder the cleanup process. Having discussed this perception with former U.S. EPA employees who wrote some of the regulations of Superfund, it was farthest from EPA's intention to implement a slow and sometimes ineffective process. Furthermore, if one takes the time to closely examine, for example, the National Oil and Hazardous Substances Contingency plan (NCP) for Superfund, it becomes clear that the regulations governing Superfund do not have to be interpreted as slow or inefficient. In reality, most environmental regulations such as Superfund are relatively flexible. Contrary to what many environmental professionals say, the regulations are not the *primary* reason why there is inaction in the environmental industry. More often, this can be attributed to the inability to get along with counterparts, poor application and interpretation of environmental regulations, technical inadequacy, and a "follow-the-leader mentality" which has led both the private sector and government to this state of perceived environmental inaction. Accelerating cleanup and streamlining decision-making is consistent with the existing environmental regulations. Revising the laws to more clearly define their intent may be in order, and in fact, pending changes will likely do so. However, if one takes a close look at the existing regulations, there are *many* opportunities, which some environmental professionals term "regulatory opportunities," that can be used to quickly implement actions and streamline decision making.

The environmental industry must promote environmental excellence, maintain values, and meet the needs and expectations of society and customers. For the environmental industry, environmental excellence stems from maintaining values and meeting the needs and expectations of society and customers. In this case, quality work and service are aligned with how we, as professionals, protect human health and the environment, meet the level of satisfaction society and customers dictates as excellent quality and service, maintain our individual values, and meet the spirit of environmental laws. Based on the current level of public satisfaction, accelerating cleanup and streamlining decision making are essential components required to achieve environmental excellence. In theory, protecting human health and the environment is the primary reason why environmental professionals are employed. The more *we* expedite protection of human health and the environment, and use reasonable and cost-effective approaches that are *protective*, the more satisfied society will be. In this case, "we" refers to all persons working in the environmental industry, including government and private-sector professionals. Historically, society as a whole and government spending authorities have not been highly satisfied with the environmental industry's efforts, and the industry is far from being judged as promoting environmental excellence. Likewise, the industry has historically not viewed government agencies as willing partners, eager to solve our nations contamination problems. The downsizing of the environmental industry in the mid-1990s is, in part, a result of unreasonable costs associated with environmental programs, inability to close sites early, and a lack of success stories. However, spills and releases periodically occur, and response and cleanup will be necessary for as long as chemicals and metals are used in our society. Likewise, undiscovered sites will be identified in the future, and other types of contamination (such as unexploded ordnance) will become an environmental focus and will need to be addressed. Those private and government organizations maintaining a slow pace in cleaning up sites will likely face serious challenges to viability in the future. More importantly, those private and government organizations who embrace environmental excellence, and specifically make acceleration, streamlining, and cost-reducing alternatives priorities, will be the *environmental leaders* of tomorrow.

A world economy and environmental leadership. In the U.S., the environmental industry has currently experienced a slowdown; however, environmental awareness and activity around the world has continued to grow during this same period. Our ability in the U.S. to accelerate and streamline is critical in order to help other countries efficiently and effectively remedy their environmental problems. Furthermore, successful accelerated investigation and cleanup programs are essential in order for the U.S. to remain the world's leader in environmental response actions and technology. As an analogy, consider the potential reasons why the Japanese enjoy a significant share of the automobile market in the U.S. Most would agree that the Japanese enjoy a significant share of the automobile market because they make good cars. In reality, they took advantage of a situation in the 1970s when American cars could have been better. In order for the U.S. to remain a leader in the environmental industry, and enjoy the market share of the *world* environmental industry, U.S. environmental firms and regulatory agencies must also be not only good at what we do, but also be the best at accelerating cleanup, streamlining decision making, reducing costs, and identifying reasonable solutions for our environmental problems. If we do not accomplish this feat, some other nation will become the world environmental leader of tomorrow.

1.2 WHO SHOULD READ THIS BOOK

The assumption is made that those who read this book are well versed in one or more of the environmental sciences/engineering disciplines, and have direct environmental experience from a private, state, or federal perspective. From a regulatory perspective, information in this book is related to the Resource Conservation and Recovery Act (RCRA); the Comprehensive Environmental Response, Compensation and Liability Act (CERCLA) as amended by the Superfund Amendments and Reauthorization Act (SARA) of 1986; and the equivalent state regulatory compliance programs. Other environmental programs also apply, such as EPA's Brownfields Initiative, the Toxic Substances Control Act (TSCA), state underground storage tank (UST) programs, military environmental programs, the U.S. Department of Energy (DOE) environmental programs, pollution prevention programs, and several other programs. Those navigating these regulatory programs and processes can gain knowledge from this book on how to accelerate and streamline environmental actions. Furthermore, this book is written so that readers having varied environmental backgrounds can learn about alternative cleanup technologies, cleanup models, regulatory processes, investigative strategies, etc. This information is useful in order to consider potentially more efficient or effective approaches when developing a site-specific or program-level strategic plan. As a result, the most efficient and effective environmental strategy can be developed and implemented at sites by using strategies and approaches developed under many regulatory problems.

The science and models presented in this book hold true under the majority of circumstances, regardless of the regulatory framework; however, the regulatory language and nomenclature may vary. For example, the concepts of using CERCLA operable units (OU) may be applied to other environmental programs, such as state UST programs, in order to separate extensive soil and groundwater corrective action efforts. In this case, the "OU" designation may be labeled something different, yet the concept of segregating media or location for cleanup would remain the same. In order to learn more about the basic regulatory background and cleanup technologies used in this book, other resources may be necessary. For example, an excellent background book is *Hazardous Waste Site Remediation Source Control* (1993) by Domenic Grasso. If you understand the information presented in Grasso's text on the regulatory framework and cleanup technologies, then many of the concepts and information in this book will be more easily understood. Other resources, such as those related to investigation and environmental decision making, are also important resources for understanding regulatory and technical background information.

1.2.1 What Kind of Book is This?

Strategies for Accelerating Cleanup at Toxic Waste Sites is certainly a technical book; however, most of the technical discussion is implicit rather than explicit. The technical and engineering information presented, including basic and advanced subjects, are generally gained on the job and less often through formal education. For this reason, this is a relatively *advanced* level book, because the concepts and principles are applied in nature vs. in theory. There are currently many technical books available on environmental engineering, investigation, and remediation, that support the technological concepts, case studies, models, approaches, and examples contained in this book. In some cases, more in-depth technical discussions are presented in order to illustrate certain less obvious principles related to the overall subject of acceleration and streamlining. For this reason, the reader should understand fundamental technical elements of the environmental processes and technologies in order to more easily understand the strategies and information presented in this book. In turn, the information presented in this book can be applied to site-specific and program-oriented objectives.

This book is in some ways a regulatory book because it contains various regulatory information related to common federal and state environmental programs. Specific and general information related to regulations are used to support ideas and concepts of acceleration and streamlining, which are in turn also linked to government guidance. In general, the regulatory and guidance information presented is necessary in regard to consideration of acceleration and streamlining; however, this book should not be considered a "purely regulatory book." In addition, this book is not a substitute for comprehensive regulatory resources, and most importantly, the actual regulations and guidance quoted.

This book also has a nontechnical, or "softer side," that focuses on the interactions between the people working in the environmental industry. When considering how to streamline and accelerate efforts, in no matter what industry, one element is *always* a factor — the human factor. In this case "always" is not used loosely. Cooperation, teamwork, leadership, trust, and communication are included in this book because they are factors that strongly influence *technical* decisions. It is essential to remember that we cannot separate or divorce ourselves from being human. For this very reason, the technical and regulatory information in this text is integrated with the human side of the industry in order to provide a truly comprehensive book, as opposed to a purely regulatory or technically focused book on acceleration, streamlining, and cost savings. Other nontechnical subjects described in this book include contracting strategies and strategic plan development, for example.

Lastly, this book contains little theoretical information and primarily focuses on practical application of the approaches and methods presented. In some cases, environmental industry *values,* from both private and government sector perspectives, are implicit or explicit in the information presented. Furthermore, examples and case studies that are presented come from many different perspectives. In an attempt to remain objective and open minded, the methods and approaches included in this book vary from "conservative" to "nonconservative" in terms of acceleration and streamlining. The realization that values are part of the information presented in this book helps project teams understand that they will have to use their *own* judgment and consider arguments for and against the various approaches and methods in order to make a final decision on the most effective strategy based on *site-specific conditions*. More importantly, the values of working together, cooperation, and taking a proactive stance on the need for environmental acceleration, streamlining, and cost reduction are inherent in all of the methods and approaches described in this book.

1.3 FUNDAMENTAL PRINCIPLES

Fundamental principles upon which this book is based are (1) cooperation and teamwork with all team members, (2) arrival at a common and/or predetermined project *destination*, and (3) movement away from the traditional linear thinking of how investigation and cleanup should be completed at contaminated sites. Technical ability is certainly a fundamental concept; however, for the purposes of this book, strong technical ability is considered a *prerequisite* in order to successfully accelerate cleanup and streamline decision making.

Cooperation and Teamwork

The concepts of cooperation and teamwork are implied whenever the phrase "project team" is used. Unless otherwise specified, the project team should include all counterparts, including the lead and support regulatory agencies, responsible parties (if any), the public, and other stakeholders. This understanding is important because it moves away from the "us and them" attitude that has plagued the environmental industry for years. Acceleration and streamlining are possible when people work together toward common goals and objectives.

Destination

In terms of a predetermined project destination, it is essential to define the desired end result(s) at the beginning of the project in order to successfully plan and implement the methods and strategies needed to get there. For example, as part of site evaluation, the desired end result should be clearly defined from a reuse standpoint, if applicable, and agreed upon by the project team ahead of time. Far too often, several investigative stages are completed before the project team steps back and truly considers what the final destination of the site will be. For example, sometimes environmental professionals become intoxicated with collecting more and more field samples to "fully characterize sites." However, no matter how much data is collected, there will remain some level of uncertainty, requiring the need to go back for more information and somehow fully characterize sites. These individuals suffer "technical tunnel vision" which decreases their ability to see the big picture. These individuals tend to maximize data collection and, unfortunately, minimize consideration of what drives decision making at sites. As a result, they become locked into the seemingly endless loop of collecting more data vs. stepping back to take a holistic look at sites in order to expedite closure. When clients, or the public, realize that substantial amounts of money are being spent and the site has not been cleaned up yet, problems arise. A direct result of this has been the downsizing of the environmental industry in the mid 1990s.

To avoid "technical tunnel vision," the desired destination should be defined as soon as possible, and redefined throughout a project's life. This is a fundamental concept of acceleration and streamlining (Figure 1-1). For example, under CERCLA, residential or industrial reuse generally requires that a comprehensive human health and ecological risk assessment be completed prior to establishing final cleanup levels. In this case, a decision to use industrial cleanup action levels should not be determined near the end of projects. Instead, consideration of land use, for example, should be integrated into the project planning and evaluation process in order to evaluate what the most *reasonable* land use may be as early as possible, whether industrial, recreation, residential, etc. The methods for investigating the site, and the overall approach used to clean up or close sites, should be compared with the regulatory framework worked in. Accelerating clean up and closure begins in the site discovery stage. Characterizing site conditions before deciding what a site will be used for generally postpones early identification of alternatives for site cleanup. Hindsight is 20/20, unless you take the initiative to look forward in the beginning.

Nontraditional Methods

Moving away from traditional methods means considering new or innovative approaches that bypass the traditional linear investigation and cleanup approaches that have been used in the past. Historically, project teams go step by step through the discovery, assessment, evaluation, and design stages in order to finally reach the action stage years later. While this linear approach eventually leads the project team through the decision making and the site restoration process, it can be extremely slow from a community standpoint, and extremely expensive from a funding standpoint. From a professional perspective, there are often ways to bypass the linear approach and initiate cleanup or other protective measures, such as institutional control, that may be more efficient than utilizing traditional approaches. The risk of failure may be greater in some cases, and therefore should be considered. The need

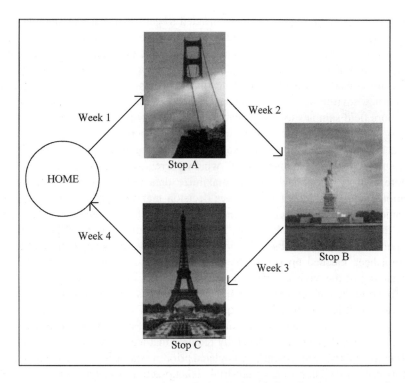

Figure 1-1 Tourists generally plan their main destinations before they leave on vacation, and especially their final destination, which is back home again. Knowing your main stops and how you will get back home again usually makes vacations more enjoyable and predictable. Environmental projects are similar. Knowing and planning your main, interim, and final destinations gives direction and focus that is needed to smoothly and efficiently execute environmental projects.

to pay attention to details is essential, along with project communication and agreement, in order to move forward when employing alternatives that may have a higher potential for failure. For example, a multitude of investigation and cleanup actions may be underway at the same time for one project in order to accelerate response actions. Tracking details and the overall inherent potential for failure with proactive approaches and innovative technologies is important in cases such as these in order to ensure the overall project goals and objectives are met and disaster avoided. The concepts of nonlinear thinking must integrate the elements of potential success *and* failure, which is further described below.

1.3.1 Risk and Decision Making

One of the questions in the opening paragraph asks whether accelerating cleanup is possible at all sites. The answer to this question is, unfortunately, no. Accelerating cleanup is possible at most sites, and quite often at least in specific areas or media of all sites, as long as the project team *wants* to accelerate cleanup. Without a desire

INTRODUCTION

Figure 1-2 Rock climber Jim Wilson risks a "lead fall" on steep rock, which if happens is *twice* the distance of the last piece of protection placed in the rock (bottom left). Risking a lead fall to many rock climbers is well worth it in order to make it to the top of a climb. For environmental projects, the potential of "falling," or failure to meet goals and objectives, is also present. The potential for failure must be evaluated, agreed upon, and integrated into environmental strategies in order to maximize performance, yet not get too far above our "protection" and risk a huge "lead fall" on environmental projects. (Photograph by Scott Payne.)

to accelerate and streamline, it will not happen. In addition, there is an inherent risk associated with accelerating cleanup (Figure 1-2). Those in the environmental industry who have a keen sense of the level of risk they are undertaking by proposing a proactive investigation or remedial decision are generally well versed in most earth science disciplines, understand engineering concepts, know the regulatory framework required for the job, have a strong technical background in cleanup technologies, and are able to work well with others. Most often, coworkers also provide background information needed for making project decisions. The inherent risk can be substantial if an unexperienced person blindly uses some of the techniques in this book for the sake of accelerating cleanup. Prior to doing a "belly flop" into the potentially cold waters of accelerated cleanup and streamlined decision making, the primary technical, regulatory, and engineering principles must be sufficiently covered in order to ensure that what the team proposes will actually have a relatively high probability of succeeding. The entire project team must understand, and acknowledge, the potential for failure of actions proposed and undertaken. In those cases where there is an exceptionally high potential for failure, a closer evaluation may be necessary in order to evaluate all alternatives and compare the potential avenues of action.

1.3.2 The Scientific Method

Proactive and accelerated environmental actions rely heavily on the science that is used to develop and to some degree govern the environmental industry. When aggressively pursuing accelerated investigation and cleanup, there may be individuals with an opposite viewpoint, and who will be anything but enthusiastic about

Figure 1-3 The Wright brothers testing their hypothesis that humans can fly. Indeed, they were correct and changed humankind forever. Following the scientific method, it is possible to develop a hypothesis and collect data to demonstrate that an idea or concept has merit and is worth trying or, in fact, works. (Source: The Bettmann Archive, with permission.)

"new and innovative" approaches for site investigation, cleanup, or streamlining the environmental process. For them, the risk of failure far outweighs the benefits of being proactive. In order to address this type of negative input from counterparts, supervisors, etc., it is essential to rely on the technical merit, background, work experience, cost comparison of alternatives, and case studies that lend support to the acceleration and/or streamlining proposal. Most importantly, the scientific method cannot be undermined when proposing proactive investigation and cleanup strategies (Figure 1-3). In other words — do not be *too* proactive! The environmental industry is a science-based industry. The human factor can be the catalyst for either slowing the process down or speeding it up. Using a scientifically based approach, coupled with a good understanding of the people and the politics at hand, helps generate an optimal approach for reaching consensus — depending on the level of trust shared on the project team. Failing to integrate a scientific element into acceleration proposals risks professional integrity, and the ability to reach consensus on "risky alternatives." The scientific elements of the environmental industry are its foundation and cannot be ignored when accelerating cleanup and decision making.

1.3.3 Strategy and Approach

Throughout this book, approaches and strategies are introduced that are used to accelerate investigation, cleanup, and site closure. Based on this information, site-specific, program, or strategic plans can be developed in order to plan and model acceleration and streamlining approaches. In some cases, a single strategy, such as

Figure 1-4 To fully grasp how something works, two vantage points are helpful and include (A) up close and (B) far away. A monitor screen up close outlines the individual pixels that make up the entire picture. Far away, the monitor shows the picture. From either vantage point, however, the other cannot be seen or easily understood.

the observational method, can be used to address site conditions. In other cases, a combination of several approaches, or methods, are the best alternative. This text is *not* a "cook book," or a complete anthology, outlining how to accelerate cleanup and address environmental compliance issues. Rather, this book presents general principles and methodology for acceleration, streamlining, lowering costs, and presents case studies that can be used singularly, or in unison, for improving site-specific environmental performance. These principles and methodologies can only be used as long as they complement the site conceptual model, work within the regulatory framework, and meet the project goals and objectives. As outlined in Chapter 8, a brief and concise strategic plan, primarily for larger projects, can be prepared that summarizes the strategy and approach that will be used to reach a final project destination, such as site closure under an early action scenario. In some cases, several strategies/approaches may have to be considered in order to meet project goals and objectives, or the strategic plan cannot be developed until there is sufficient information in order to make prudent decisions. Strategic plans should be reviewed periodically and revised, if necessary, based on new information and data. In this sense, the environmental strategy may be considered a living, breathing entity of projects that requires review in order to remain effective. A large portion of this book is about strategy, innovative approaches, and developing a strategic plan in order to expedite investigation, cleanup, and closure. Since there are many types of contaminated sites and regulatory programs, the case studies and examples presented in this book illustrate how strategies and approaches work. It may be necessary to step back, reread the section, and ponder the text for a while in order to fully grasp how the approach or strategy can be applied to other sites (Figure 1-4).

1.3.4 Investigation and Cleanup Technology

There is a disproportionate amount of technical discussion on investigation and cleanup technologies in this book compared to other topics related to acceleration and streamlining concepts and processes. In general, there is an absolute plethora of books on the market related to investigation and cleanup, ranging from soil vapor extraction, bioremediation, thermal processes, chemical destruction, and solidification and stabilization, to vadose zone monitoring, field analytical methods, geophysics, and many more. This book presents investigation and cleanup technologies that are both current and innovative, technologies that speed up the time it takes to characterize a site or clean it up, and the approaches that can be used for initiating actions early in the environmental process in order to fast-track response actions. The way in which investigation and cleanup technologies work is not presented in detail; however, knowledge of these general concepts is helpful when reading this book. Many detailed resources are available on these subjects, and are referenced throughout this book for further information. Appendices A, B, and C of this book identify information/database resources, investigation technologies, and cleanup technologies, respectively. Detailed research and use of these resources is also helpful for evaluating site-specific needs, new information and applications, and innovative technologies.

1.3.5 The Human Element

Chapter 7 of this book is focused on the human element behind environmental decision making, leadership, and teamwork. Do not underestimate the importance of these concepts (Figure 1-5). Embracing the fact that people accelerate cleanup through decision making, as much as technology does through science, is critical in order to maximize acceleration and streamlining success. Without agreement, teamwork, trust, and direction, progress is slowed down significantly. Chapter 7 describes the human element of the environmental industry, including topics on communication, teamwork, leadership, and trust, all of which are essential for scientists and engineers to efficiently and effectively accelerate response actions and decision making.

1.4 BOOK ORGANIZATION

This book is primarily organized by chapter based on government agency and industry information. Models, approaches, and methods used to accelerate investigation and cleanup of contaminated sites are presented from a government perspective in Chapters 2 through 5, and include information from EPA, the U.S. Department of Defense, the U.S. Department of Energy (DOE), and state programs. Private business and professional organization models and approaches are described in Chapter 6. The majority of the technical information/applications are in these first six chapters. In some cases, information from other government or private sources is described in order to support the principle described in each respective chapter.

Figure 1-5 Dr. Martin Luther King's *I Have a Dream* speech showcased his leadership and communication skills and helped bring unity, understanding, and direction to an important American movement. (Source: Archive Photos, with permission.)

Chapter 7 covers the human side of accelerating cleanup where behavior and attitude are entered into the acceleration equation. Chapter 8 includes the elements of strategic planning and lowering environmental costs. The information presented in Chapters 7 and 8 can "make or break" a proposed plan to accelerate cleanup and/or streamline decision making.

Appendix A contains numerous environmental resources and databases that can be used for the researching the latest policy information and innovative technologies. Appendices B and C contain listings of investigation and remediation technologies, including how to contact various equipment vendors.

CHAPTER 2

U.S. Environmental Protection Agency

The U.S. Environmental Protection Agency (EPA) is probably the most well-known environmental agency worldwide. EPA's great power extends not only to environmental issues in the U.S., but also to global environmental concerns such as the greenhouse effect and acid rain. EPA has been involved in one way or another in most of our lives, perhaps because of water quality concerns related to the Clean Drinking Water Act, or because we live by a hazardous waste site, or maybe because we breathe smog generated from the cars and industry of our cities. And as with many large agencies, problems can arise, including public dissatisfaction. Since the early 1990s, EPA has been under tremendous pressure from the public and Congress to improve its ability to work with liable parties, itself, and to efficiently and effectively investigate and clean up contaminated sites. EPA and other federal agencies have received substantial criticism for studying sites too long and failing to move ahead with response actions (EPA et al. 1994a). The concerns of the public and Congress have not fallen on deaf ears, and there have been many programs and approaches developed and used by the EPA to hasten protection of human health and the environment.

Yet the overall success of EPA's acceleration initiatives are not well known or obvious by standards most would hope. What is missing? Perhaps what is missing may simply be a more concerted effort, both internally and externally of the EPA, to embrace and apply their acceleration programs and approaches. In addition, the acceleration programs that have been successful may require more touting to the public and Congress in order to gain their attention. Most would agree that applying the existing programs and models at a majority of sites would help realize the full potential of accelerated cleanup in America. As of 1996, there are over 13,000 sites listed in the Comprehensive Environmental Response, Compensation and Liability Information System (CERCLIS) and over 1200 National Priorities List (NPL) sites. Thousands of more sites are in other programs such as RCRA and state-led abandoned mine reclamation programs. The sheer number of sites implies the need for accelerating investigation and cleanup in order to lower the number of sites required to meet the goal of the various federal and state environmental programs.

For EPA's acceleration initiates to work, project teams must be willing to step away from the old ways of conducting environmental work and begin using the acceleration methods and approaches that are available and demonstrated. Many successful and proven techniques are described in this chapter for fast-tracking cleanup. While EPA carries the greatest burden of implementing their own programs, and must enforce environmental regulations, win–win proposals can be generated to fast-track cleanup and reduce costs on the responsible party side if they choose to take a proactive approach for site restoration. Combining information in this chapter with information related to the human component of environmental decision making (presented later in this book) fosters successful implementation of EPA's acceleration methods and approaches. The following information represents EPA's most important approaches and methods used to accelerate investigation and cleanup. This chapter is not a complete compilation of EPA programs, initiatives, and approaches. Other EPA information is presented in other chapters of this book. Of particular importance are EPA's Brownfields Initiative and land use guidance, which are presented in Chapter 8 under environmental strategies. Additional information on EPA is also presented throughout this book.

2.1 EPA'S INFORMATION TRANSFER AND TECHNOLOGY DEMONSTRATION PROGRAMS

Accelerating cleanup begins in the investigation stage by streamlining the data collection efforts. Collecting real-time data is one example of a highly effective way of streamlining and focusing site characterization efforts. Supplementing site characterization data with data for remedial design is useful to propose cleanup alternatives for a remedial action early in the regulatory process. In the remedial investigation/feasibility study (RI/FS), remedial technologies are evaluated from a cost, effectiveness, and implementability standpoint. Selecting one remedial technology over another can involve consideration of technologies that will clean up a site faster or more effectively, which are important aspects of accelerating cleanup. Under Superfund regulations, for example, the National Oil and Hazardous Substances Pollution Contingency Plan (NCP) (40 Code of Federal Regulation (CFR) 300.430 [d][5]) requires that lead agencies consider using innovative technologies that "offer the potential for comparable or superior performance or implementability; fewer or lesser adverse impacts than other available approaches; or lower cost for similar level of performance than demonstrated technologies." The avenue is open for environmental professionals to propose new and better ways to clean up sites that accelerate the cleanup process. The avenue is also open for improved investigation technologies, which should never be overlooked. This section outlines EPA's most significant information transfer and demonstration programs that can help you make decisions that streamline the investigation and cleanup process. This section is divided into remedial technologies and accelerated investigation subsections. Appendix A contains environmental resource contacts, databases, bulletin boards, and other links from the EPA and many other government and private sources. The information covers not only technology subjects, but also many other environmental information subjects that can be used to find information or data, or answer environmental-related questions you have.

2.1.1 Remedial Technologies

EPA's Superfund Innovative Technology Evaluation (SITE) program is a state-of-the-art innovative treatment technology research and demonstration program (EPA 1994b). The program was formed to promote development and implementation of (1) innovative treatment technologies for hazardous waste site remediation and (2) monitoring and measurement technologies for evaluating the nature and extent of hazardous waste site contamination. The SITE program was established by EPA's Office of Solid Waste and Emergency Response (OSWER) and the Office of Research and Development (ORD). The remedial technologies program includes the following three components:

- Demonstration Program — Conducts and evaluates demonstrations of promising innovative technologies to provide reliable performance, cost, and applicability information for site cleanup decision making.
- Emerging Technology Program — Provides funding to developers to research efforts from the bench- and pilot-scale levels to promote the development of new innovative technologies.
- Technology Transfer Program — Disseminates technical information on innovative technologies to remove impediments for using alternative technologies.

A fourth SITE component is related to investigative technologies and is called the Monitoring and Measurement Technologies Program (MMTP). This program is presented in the accelerated investigation subsection of this chapter.

SITE Demonstration Program

The SITE demonstration process typically consists of five steps: (1) matching an innovative technology with an appropriate site; (2) preparing a demonstration plan and test plan, sampling and analysis plan, quality assurance project plan, and health and safety plan; (3) performing public activities; (4) conducting the demonstration (ranging in length from days to months); and (5) documenting results in an innovative technology evaluation report (ITER), a technology capsule, other demonstration documents, and a demonstration videotape.

Cooperative agreements between the EPA and the developer set forth responsibilities for conducting the demonstration and evaluating the technology. Developers are responsible for operating their innovative systems at a selected site, and are expected to pay the costs to transport equipment to the site, operate the equipment on site during the demonstration, and remove the equipment from the site. EPA is responsible for project planning, waste collection and pretreatment (if needed), sampling and analysis, quality assurance and quality control, preparing reports, and disseminating information. In essence, the program is a symbiotic relationship where both the technology vendor and EPA benefit.

Demonstration data are used to assess the technology's performance, the potential need for pre- and postprocessing of the waste, applicable types of wastes and media, potential operating problems, and the approximate capital and operating costs. Demonstration data can also provide insight into long-term operating and maintenance costs and long-term risks.

SITE Emerging Technology Program

Under the Emerging Technology Program, the EPA provides technical and financial support to developers for bench- and pilot-scale testing and evaluation of innovative technologies that are at a minimum proven on the conceptual and bench-scale levels. The program provides an opportunity for private developers to research and develop a technology for field application and possible evaluation under the Demonstration Program. A technology's performance is documented in a final report, journal article, project summary, or bulletin.

Technologies are solicited yearly for the Emerging Technologies Program through requests for pre-proposals. After a technical review of the pre-proposals, selected candidates are invited to submit a cooperative agreement application and detailed project proposal that undergoes another full technical review. The cooperative agreement between the EPA and the technology developer requires cost sharing.

SITE Technology Transfer Program

In the Technology Transfer Program, technical information on innovative technologies in the Demonstration Program, Emerging Technology Program, and MMTP is disseminated through various activities. These activities increase the awareness and promote the use of innovative technologies for assessment and remediation of contaminated sites, independent of the environmental program being followed. The goal of technology transfer activities is to promote communication among individuals requiring up-to-date technical information.

The Technology Transfer Program reaches the environmental community through many media, including:

- Program-specific regional, state, and industry brochures
- On-site Visitors' Days during SITE demonstrations
- Demonstration videotapes
- Project-specific fact sheets to comply with site community relations plans
- ITERs, demonstration bulletins, technology capsules, project summaries
- The SITE exhibit, displayed nationwide at conferences
- Networking through forums, associations, regions, and states
- Technical assistance to regions, states, and remediation cleanup contractors

SITE information is available through the following on-line information clearinghouses:

Alternative Treatment Technology Information Center (ATTIC)
System operator: 703-908-2137 or 908-321-6677

Vendor Information System for Innovative Treatment Technologies (VISITT)
Hotline: 800-245-4505 or 703-883-8448

Cleanup Information Bulletin Board System (CLU-IN)
Help Desk: 301-589-8368; Modem: 301-589-8366

Technical reports may be obtained by contacting the Center for Environmental Research Information (CERI) in Cincinnati, OH at the address or telephone number provided below. Additional SITE documents become available throughout the year.

CERI/ORD Publications
26 West Martin Luther King Drive (G72)
Cincinnati, OH 45268
Operator: 513-569-7562

The SITE Program is administered by EPA's ORD, specifically the Risk Reduction Engineering Laboratory (RREL) in Cincinnati, OH. For further information on the SITE Program or its component programs you can write the EPA, to the attention of the component program you are interested in at the above address. Administrative questions can be directed to Office of Research and Development Risk Reduction Engineering Laboratory, Operator at 513-569-7696.

VISITT

Another important source of remedial technology information is the Vendor Information System for Innovative Treatment Technologies (VISITT). VISITT is a free electronic yellow pages of innovative treatment technologies and vendors. The system was developed by EPA's Technology Innovation Office (TIO) to interface with environmental professionals who need to evaluate potential treatment technologies for site restoration. VISITT is a detailed database that contains over 300 treatment technologies from over 200 vendors. Users can easily screen through and assess remediation technologies with the database program using the available customized search capabilities. The database is updated regularly and can be downloaded from the Internet and some on-line services. Appendix C is a matrix prepared by the Department of Navy that contains primarily the VISITT release 4.0 innovative technologies, and other source information, organized alphabetically by treatment technology type. The matrix is useful to scan through the various treatment technologies and contact vendors who offer them. To obtain the most up-to-date information on VISITT, request the programmed version of VISITT, or the most recent release of VISITT, contact the EPA at:

U.S. EPA/NCEPI
P.O. Box 42419
Cincinnati, OH 45242-0419
Help Line: 800-245-4505 or 703-883-8448

2.1.2 Accelerated Investigation

EPA is one of the largest proving grounds of investigation and analytical tools used in the environmental industry. Under EPA's SITE program, the MMTP helps accelerate the development, demonstration, and use of new and emerging technologies that can be used to investigate contaminated sites. The technologies assessed under the MMTP emphasize faster, safer, and more cost-effective methods than conventional technologies for producing real-time or near-real-time data. MMTP works with technologies that detect, monitor, and measure hazardous and toxic substances in soil (saturated and unsaturated), air, biological tissues, wastes, surface water, as well as technologies that characterize the physical properties of sites. General technologies of interest include

- Chemical sensors for *in situ* measurement
- Groundwater sampling devices
- Soil and core sampling devices
- Soil gas sampling devices
- Fluid sampling devices for the vadose zone
- *In situ* and field portable analytical methods
- Expert systems that support field sampling or data acquisition and analysis

The table in Appendix B lists the MMTP projects. There are many MMTP projects that cover a wide berth of contaminants and media. For the purposes of this book, use Appendix B as one inventory of EPA's efforts in developing measuring and monitoring technologies, which may be useful to help design an investigation program at your sites. Another primary source of these types of tools is EPA's Vendor FACT, described later. EPA's work has been instrumental in helping the environmental industry develop and demonstrate new technologies. Specific application and use of the MMTP technologies is available from the EPA or the associated vendor contact listed in Appendix B. You can contact EPA's MMTP headquarters at the EPA Environmental Monitoring Systems Laboratory in Las Vegas, NV:

U.S. EPA
Environmental Monitoring Laboratory Systems—Las Vegas
P.O. Box 93478
Las Vegas, NV 89193
702-798-2432

The SITE MMTP is perhaps the largest EPA source of new technology information to accelerate investigation at environmental sites. Your EPA region may also have other technologies that are demonstrated to work in the physiographic or geologic setting you work in. Therefore, contacting your regional EPA headquarters may also be an important source of information. Also, Appendix A contains a resource guide you can use to gain access to other databases and agencies that will help you find the information you need.

TIO

As described in VISITT discussion previously, one particularly useful information source is EPA's TIO. In general, the TIO is a clearinghouse of environmental technology information offering databases to transfer information to customers, generally at no direct costs. The TIO's focus includes

- Developing and integrating state-of-the-art information on innovative remediation technologies for customers and delivering it through targeted, state-of-the-art means (printed publications, electronic, conferences, training, etc.).
- Developing and supporting creative approaches that streamline "getting to yes" for regulatory and engineering acceptance of innovative technologies.

In the accelerated investigation area, the TIO and the National Exposure Research Laboratory in Las Vegas, NV developed "Vendor FACTS" to provide environmental professionals information on applicability, performance, and current use of innovative site characterization technologies. Vendor FACTS is a Windows™-based system that allows the user to make one-to-one vendor comparisons of technology cost and performance data and assists the user in selecting the most appropriate technology. The database is user-friendly and allows the user to screen and search technologies by parameter such as contaminant type, data quality, media, risk assessment, cost, and development status. Version 1.0 of Vendor FACTS was completed in September 1995 and contains approximately 80 technologies provided by 71 vendors and Version 2.0 is scheduled for release in 1997. Technologies are grouped into four categories: analytical, geophysical, chemical extraction, and sampling. Participation in the database is free. For CLU-IN data access, call 301-589-8366. You can also access CLU-IN via the Internet at the cleanup up information page (www.CLU-IN.com) CLU-IN.EPA.GOV or telnet at 134.67.99.13. For voice help call 301-589-8368. After you have accessed CLU-IN and registered, you may begin the download process. You will need to fill out the Vendor FACTS registration questionnaire to get the password to complete the downloading process. Type REG from the "Main Board Command?" prompt and choose the option for Vendor FACTS registration. The user manual for Vendor FACTS is in a file called VFACMAN.ZIP, and is in WordPerfect™ format. Download the file called VFACTS1.ZIP to use the database. If you encounter problems using this method, call the Vendor FACTS Help Line at (800) 245-4505 or (703) 883-8448. Information is available for other ways to download Vendor FACTS, including America Online (AOL), Defense Environmental Network for Information eXchange (DENIX), EPA FTP Site, and other on-line access sites. Some costs may be associated with utilizing these other methods to download the database.

There are many techniques available to collect/analyze real-time or near-real-time data and foster quick decision making for contained and uncontaminated sites. Advantages and disadvantages of selected near-real-time innovative analytical methods are in Table 2-1. Knowing there are ways to collect/analyze data more quickly and efficiently (both in quantitative and qualitative level) and applying these technologies significantly accelerates the investigative stage of a project. The critical

Table 2-1 Advantages And Disadvantages Associated with Innovative Soil Sample Analytical Methods

Innovative Analytical Method	Contaminants Measured	Comparative Conventional Analytical Methods	Advantages	Disadvantages
Energy dispersive X-ray fluorescence	Trace metals	Hot-plate digestion with AA and ICP	• Field transportable • Fast (5 samples per hour) • Cost effective[a] • Does not require digestion of soil	• Sample preparation (e.g., drying and homogenization) required • Poor detection limits on some metals
Immunoassay techniques	CPs, PAHs, and PCBs	Soxhlet extraction and GC/MS or GC/ECD	• 4 to 20 samples per hour • Cost effective[b] • Good correlation with comparative techniques: PAHs (82%), PCBs (86%), PCP (89%) • Minimal sample preparation	• False negative results may be high (10% to 20%) • Confirmatory analysis required • Cross-reactivity with other analytes
Microwave-assisted digestion	Trace metals	Hot-plate digestion	• Field transportable • Fast (12 samples per hour) • Accuracy of results comparable to hot-plate technique	• High operating cost[c]
Rapid extraction followed by HPLC/fluorescence	PAHs	Soxhlet extraction and GC/MS detection	• Field transportable • Fast (2 samples per hour) • Cost effective[d]	• Does not quantify all of the individual PAHs

Method	Analyte	Technique	Advantages	Disadvantages
Solid-phase microextraction	BETX	Headspace with GC/FID	• Field transportable • Rapid (1 sample per hour) • Full resolution of BETX	
Supercritical fluid extraction	CPs and PAHs	Soxhlet extraction	• Field transportable • Faster than Soxhlet extraction[f] • Increased extraction efficiency • Minimal solvent usage	• Regulatory acceptance may be difficult • Cost[e] • Nonselectivity of the FID • Wet soil needs to be dried • Operating costs higher than Soxhlet extraction[g] • Low recoveries from some analytes

Notes: AA = atomic adsorption; BETX = benzene, toluene, ethylbenzene, and xylenes; CP = chlorinated phenolic; GC/ECD = gas chromatography/electron capture; GC/FID = gas chromatography/flame ionization detection; GC/MS = gas chromatography/mass spectrometry; HPLC = high pressure liquid chromatography; ICP = inductively coupled plasma; PAH = polycyclic aromatic hydrocarbon; PCB = polychlorinated biphenols; PCP = pentachlorophenol.

[a] Typical instruments are 30% to 40% of the cost of ICP detection, and since no sample digestion step is required, considerable savings should be realized.
[b] Cost per sample using immunoassay techniques will vary with the type of kit used and the per-diem cost of the field personnel, but on average, costs are 40% to 60% less than comparative methods.
[c] Capitol costs of the microwave and digestion vessels would make this method 5 to 10 times more costly than comparable hot-plate digestion.
[d] Typical cost savings should approach 50% or more over the conventional Soxhlet extraction and GC/MS detection.
[e] Cost per sample is comparable to BTEX analyses by headspace method.
[f] Supercritical fluid extraction is capable of 8 samples per hour, while the Soxhlet method requires overnight extraction.
[g] Cost effectiveness of supercritical fluid extraction is reflected in the increased number of samples analyzed rather than a reduction in operating costs.

Source: ASL Analytical Laboratories Ltd., Vancouver, B.C.; the B.C. Ministry of Environment, Lands and Parks; and Environment Canada.

aspects of selecting innovative technologies are (1) identifying the right technology that will yield the information you need, (2) properly implementing the technology in the field or in the laboratory, and (3) making sure the data you collect meets the data quality objectives for your project. Be sure you know the level of data quality you need and what the data objectives are to make remedial decisions at your site. EPA defines specific levels of data quality you should be familiar with before trying a new or innovative analytical technology which may or may not meet your site's data quality objectives. Under the strategy chapter in this book (Chapter 8) more information is available on data quality and making remedial decisions based on data quality objectives.

2.2 CERCLA EARLY ACTIONS AND ACCELERATED CLEANUP

In 1980, Congress enacted the Comprehensive Environmental Response, Compensation and Liability Act (CERCLA), commonly known as Superfund, to respond to threats poised by releases of hazardous substances in the environment. In 1986, Congress enacted the Superfund Amendments and Reauthorization Act (SARA) which did not change the basic structure of CERCLA, but did modify many of the existing requirements and added new ones. References made to CERCLA in this book should be interpreted as meaning "CERCLA as amended by SARA." Section 105 of CERCLA requires that the EPA establish criteria for determining priorities among releases or threatened releases of hazardous substances for the purpose of taking remedial action. To meet this requirement, EPA developed the Hazard Ranking System (HRS) to evaluate sites for the NPL. Sites listed on the NPL should represent the most serious potential threat to public health or the environment and are eligible for Superfund-financed remedial actions. Funding is of great importance under Superfund, because CERCLA efforts mostly focus on past or historic releases where there may be no identified liable parties to pay for the investigation and cleanup of contamination in soil or groundwater. The classic example of a Superfund site is when buried 55-gallon drums are uncovered and found to contain toxic waste, dumped and covered up by someone illegally. Often these sites can pose serious threats to the public and environment, because no one knew they were there and exposure to contaminants is likely through incidental contact, or consumption of drinking water. Recall Love Canal, which recently has been given a clean bill of health so that people are able to again live in the area; it is perhaps the most well known of all contaminated sites impacted by illegal dumping. In cases where there are liable parties, CERCLA contains broad provisions related to making liable parties conduct work and ways for EPA to seek reimbursement of costs associated with the administration of the program and damages to natural resources.

A greater portion of this chapter is related to the CERCLA process. The rationale for focusing on CERCLA is that there is a larger percentage of perceived ineffectiveness and inefficiency associated with the Superfund program and that several fast-tracking approaches have been developed by the EPA under CERCLA. The following subsections outline EPA's approaches and models that accelerate environmental action

U.S. ENVIRONMENTAL PROTECTION AGENCY

and decision making. Figure 2-1 is a simplified flow chart of the Superfund process one can take and the avenues available to impler close sites. At nonfederal facilities, EPA and/or state regulatory agencies mine the path of action for CERCLA and NPL sites. Both the regulatory commu... and responsible parties, if there are any, need to fully understand the opportunities available to achieve site closure. All involved parties need to identify the path of least resistance for protection of human health and the environment. If a longer, more complicated, and more expensive path is followed, it is often the public that is the loser.

In respect to federal facilities, CERCLA subsection 120 and Executive Order 12580 establish certain unique requirements, where the responsible federal facility takes on the lead agency role for their sites. The U.S. Department of Defense (DoD), for example, has developed its own program for CERCLA actions and ways to accelerate cleanup actions and decision making. These programs, although they are CERCLA-related, are described by individual government agencies in this book from a DoD, DOE, and to some degree a state-led program aspect.

2.2.1 Removal Actions

Removal actions are actions that are intended to reduce immediate risk to human health and the environment. In cases where there is obvious contamination or threat of contamination, removal actions can be proposed to eliminate or mitigate the threat of contamination. A removal action by itself is generally not necessarily considered the final remedy for site cleanup, whereas remedial design/remedial action (RD/RA) is considered a final remedy, if properly administered; however, removal actions can be used as the final remedy to close a site as long as risk is reduced sufficiently to protect human health and the environment and applicable or relevant and appropriate requirements (ARAR) are met. Since the late 1980s, EPA has advocated the use of removal actions as a significant opportunity to accelerate response actions at NPL sites (EPA 1989a; 1989b). More recently, removal actions were identified as one of the fundamental components of EPA's acceleration initiatives, and are grouped into a class of actions called early actions. To understand the appropriateness for implementing removal actions, meeting one or more of eight factors listed in the NCP (40 CFR 300.415 [2]i through [2]viii) must be documented to propose a removal action. The eight factors used to consider the appropriateness of a removal action are:

1. Actual or potential exposure to nearby human populations, animals, or the food chain from hazardous substances or pollutants or contaminants.
2. Actual or potential contamination of drinking water supplies or sensitive ecosystems.
3. Hazardous substances or pollutants or contaminants in drums, barrels, tanks, or other bulk storage containers that may pose a threat of release.
4. High levels of hazardous substances or pollutants or contaminants in soil largely at or near the surface, that may migrate.
5. Weather conditions that may cause hazardous substances or pollutants or contaminants to migrate or be released.
6. Threat of fire or explosion.

STRATEGIES FOR ACCLERATING CLEANUP AT TOXIC WASTE SITES

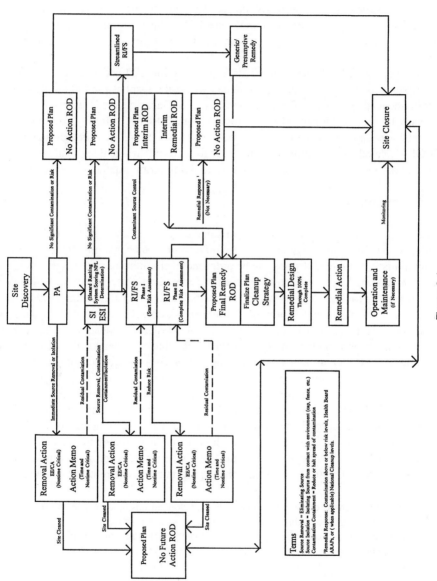

Figure 2-1

7. The availability of other appropriate federal or state response mechanisms to respond to the release.
8. Other situations or factors that may pose threats to public health or welfare of the environment.

The eight criteria are extremely broad and flexible. In general, if you suspect there is a threat to public health, workers, or the environment, whether it is based on analytical data or visual observations, you can, and perhaps should, propose a removal action. Upon determination by the EPA or lead agency that the proposed action is appropriate, the removal action process can be initiated to clean up a site, or portion of a site, and reduce risk.

There are three types of removal actions, including emergency, time-critical, and non-time-critical. Emergency removal actions are generally not proposed unless there is a bona fide reason to initiate cleanup immediately, such as an imminent threat to the public. Time-critical removal actions and non-time-critical removal actions can be proposed during all investigation and design stages of the CERCLA process (but usually none in the beginning of projects), where the emergency action is reserved for the very beginning of a project or if there is a significant finding during work at the site (such as uncovering buried drums). Table 2-2 lists the document preparation, public participation, and the time table associated with CERCLA nonemergency response actions. For all removal actions, an action memorandum is drafted and serves as the primary decision document substantiating the need for a removal response, identifying the proposed action, and explaining the rationale for the removal action, and serves also to reserve funds for a removal action in cases where there is no responsible party to cover the costs (EPA 1990a). Public comment and the timing for each type of removal action varies. In general, the emergency removal action is the easiest action to initiate; however, a clear need for such an immediate action should be obvious, such as newly discovered metal-laden soil migrating off site through wind dispersion.

Time-Critical Removal Actions

The time-critical action is generally reserved for actions that present an immediate threat requiring a quick response. In addition, they are usually relatively simple actions requiring only a limited planning period of up to 6 months, after which the construction phase of the action should begin. In cases where more than 6 months is required to plan the action, increased public involvement is required and possibly an engineering evaluation/cost estimate (EE/CA) should be prepared. In most cases, a detailed EE/CA is not necessary to conduct a time-critical removal action; however, the cost, implementability, and effectiveness of the removal alternatives must be

---

Figure 2-1   Avenues and milestones under the CERCLA process including the traditional approach (center) and opportunities to conduct early actions, streamlined investigation. The figure outlines four general avenues for no action, no further action, and remedial action needed to close site. Other avenues that can streamline decision making and accelerate actions, such as institutional controls, are inherent in the process based on site-specific conditions. (Source: PRC Environmental Management, Inc.)

Table 2-2 Cercla Response Action Matrix

	Type of Action		Triggers for Action	Documentation	CR Requirements	Example
	Early	Long-Term				
Time Critical Removal Action	X		• Meets one or more removal criteria • Action *must begin within 6 months to protect human health and the environment*	• Removal Site Evaluation • Action Memorandum • Close-Out Report • OSC Report, if requested	• Designate spokesperson • Notice of availability of AR within 60 days of starting action • CRP if on-site activities greater than 120 days • Public comment if lead agency determines appropriate	• Removal of corroded drums of waste • Removal of plating shop waste • Removal of free product from groundwater • Capping contaminated surface soil
Non-Time Critical Removal Action	X		• Meets one or more removal criteria • Planning period of *6 months or more is available without further threats* to human health and the environment	• Removal Site Evaluation • EE/CA Approval Memorandum • Action Memorandum • EE/CA • Close-Out Report	• Designate spokesperson • Notice of availability of AR by the time EE/CA Approval Memorandum is signed • CRP — before EE/CA completion • Public comment of EE/CA 30-45 days • Responsiveness Summary — part of Action Memo	• Removal and on-site treatment of contaminated sediments • On-site treatment and disposal of contaminated surface soil

Action Type		Criteria	Documents	Community Relations	Examples
Interim Remedial Action	X	• Qualitative or quantitative risk assessment indicates action is necessary • Exceedance of health-based ARAR • Environmental damages	• Site Assessment Data • Focused Feasibility Study or Proposed Plan that evaluates alternatives • Risk Assessment • Proposed Plan • ROD	• CRP • Notice of availability of AR prior to public comment • Public comment — 30–60 days • Responsiveness Summary — part of ROD	• Alternative Water Supply • Groundwater Plume Control • Temporary Protective Covers
Final Remedial Action	X X	• Baseline Risk Assessment indicates unacceptable risk • Exceedance of health-based ARAR • Environmental damages	• RI • Baseline Risk Assessment • FS • Proposed Plan • ROD	• CRP • AR established and available when RI starts • Public comment — 30–60 days • Responsiveness Summary — part of ROD • Fact Sheets — throughout the project	• Capping landfill and leachate and gas control • Groundwater extraction, on-site treatment, discharge to river • Lagoon sludge and contaminated soil treatment with on-site disposal of residuals

Notes: CRP = Community Relations Plan; CR = Community Relations; EE/CA = Engineering Evaluation/Cost Analysis; CERCLA = Comprehensive Environmental Response, Compensation, and Liability Act; RI = Remedial Investigation; FS = Feasibility Study; AR = Administrative Record; ROD = Record of Decision.

Source: PRC Environmental Management, Inc.

considered before selecting the preferred alternative. An action memorandum (the decision document for removal actions) is required for all time-critical removal actions, and must be finalized before the construction phase of the action begins. In comparison to an emergency removal action, the action memorandum is generally drafted after the action is completed, since a release characterized as "emergency" requires an immediate response to protect human health and the environment. Table 2-2 shows the document trail and public involvement under the removal authority for time-critical removal actions and other nonemergency response actions.

In cases where there is a serious potential threat to humans or the environment, but an immediate response is not necessary because current conditions, for example, limit exposure potential, a time-critical removal action can be considered for making a cleanup decision, as long as the action is relatively simple. Relatively simple response actions do not warrant an EE/CA, and the time-critical removal action process can be used to initiate cleanup efforts. For example, polychlorinated biphenyl (PCB) soil contamination falls into this category in many cases because it can be a serious threat to human health at elevated concentrations, often it is isolated from public exposure within fenced compounds, and since there are only a limited number of remediation alternatives for PCB soil contamination. While the PCB soil contamination may not have a direct pathway to humans, the fact that some day it will have to be remediated or conditions mitigated may warrant conducting a removal action to excavate and remove the PCBs vs. a remedial action. Based on only limited site characterization information, it is possible to propose and plan a removal action under a time-critical removal action and fast-track the cleanup process. Another example would be a closing federal facility that will be transferred to the public on a prescribed date. To meet the transfer schedule, you could conduct a time-critical removal action to fast-track transfer of the property, assuming that an EE/CA or more detailed engineering evaluation is unnecessary.

Non-Time-Critical Removal Actions

Non-time-critical removal actions are reserved for actions that require more forethought before a preferred cleanup alternative can be selected and where a more detailed consideration of cost, effectiveness, and implementability are necessary. In addition, in most non-time-critical removal action scenarios, there are no immediate concerns related to threats associated with site contaminants, but still an early response is warranted. An EE/CA and action memorandum both need to be prepared for a non-time-critical removal action, where the EE/CA is considered a focused feasibility study (FS) under the removal authority. The non-time-critical removal action process has a 6 month or longer planning period, which generally is used to evaluate cleanup alternatives and prepare the associated construction design packages and removal action work plan documents. Often, non-EPA federal facilities mix and match parts of the removal authority to develop a program that promotes cleanup, yet still meet the requirements of the NCP. One example would be to increase public awareness for a time-critical removal action by conducting an open house allowing the community to visit the site. While promoting public visitation of the site is not required for a removal action, it may be beneficial to involve the public and gain community support. Use of a non-toxic critical removal action can fast-track the cleanup process similar to the time-critical removal action alternative.

Application of Removal Actions

A misconception involving removal actions is that these actions are restricted by their name: "removal" in its purest sense. Too often, the removal action alternative is passed up to conduct the long-winded RD/RA process because the final action requires more than isolation of wastes or excavation of contaminants. Removal actions can be simple or comprehensive in design for conducting risk reduction actions. For example, removal actions can be as simple as putting a fence around a site to keep people out, excavating contaminated soil and containing it in bins for future treatment, designing and installing an *in situ* soil vapor extraction (SVE) system, or implementation of a complicated process for soil or groundwater treatment that reduces risk. While there may be varying opinions on just how far the removal authority can be used, the regulations governing removal actions do not, in an explicit sense, limit their use in the NCP.

Since the NCP is flexible in its application of removal actions, and there are a wide range of opportunities available to implement removal actions. Does this mean environmental professionals should use them for all actions? Absolutely not. Under the best conditions, removal actions should be used when there is a high potential of success. In addition, there is a $2 million cap on costs and a 1-year duration time limit for removal actions. EPA must justify removal projects that exceed the cost and time duration limits to obtain an exemption for projects, which can make the remedial authority more effective in long-term cleanup decisions compared to the removal authority.

An analogy helps make the point of when removal actions make sense. In some cases, apples are relatively easy to pick, for a portion of the fruit grows at a height within arms' reach, and these apples can be gathered without much risk. To pick higher apples in larger trees, most people get a ladder and gain elevation safely. Climbing up the tree without a ladder can be a risky business. Most often, there are some easy "pickings" when it comes to removal actions and you should be implementing them if the action reduces risk, moves the project toward closure, or can reduce the overall cost of the project. In some cases, a site can be closed out through a removal action because the cleanup alternatives are relatively simple and straightforward. At the remaining sites, where cleaning up contamination is less understood or more complicated, you must revisit the orchard with your ladder — taking a closer look at site conditions, contaminants, risk, and public concerns to develop a site-wide remedy for all contaminants. Do not try to use the removal authority for all cleanup actions unless you are relatively certain it will be successful and cost-effective. In cases where a risk assessment has not been prepared and actions levels are not known, performing a removal action can be somewhat risky because the action has not been defined in relation to the final closure goals. If not properly administered, a removal action may not be consistent with the final site remedy, making long-term planning and final remedy considerations important aspects of planning a removal action. Target action levels can be developed without a risk assessment and can be based on standards, federal or state remediation goals, or other criteria. In turn, the target action levels can be used to approximate where the removal should begin and end, which should interface with the long-term strategy for site closure. More information on removal action authority is discussed in the Superfund Accelerated Cleanup Model (SACM) subsection of this chapter. Developing short-term and long-term strategy for accelerating cleanup is described in Chapter 8.

Removal Action Case Study

Metal-laden sandblasting grit was thought to be present at Lot 645 at Fleet and Industrial Supply Center Oakland (FISCO), California. The site was still in the discovery phase, along with 24 other sites. Based on the green color of the sandblasting grit and experience with other nearby military operations where sandblasting grit was used for fill, the conditions at Lot 645 strongly suggested that it contained elevated chromium concentrations. The sandblasting grit covered the top 2–3 in. of soil across the site and was easily discernable from the underlying red-colored soil. To fast-track the decision-making process, a flexible field work plan was prepared to characterize soil conditions quickly. The plan called for either a relatively comprehensive sampling approach to characterize the site for long-term remedial decisions or, after collecting five samples and having them quickly analyzed, changing the focus of the field effort to initiating the removal action process to remove the sandblasting grit, if it was shown to be contaminated.

As suspected, after analyzing the initial five samples, the green sandblasting grit was shown to be contaminated with high levels of chromium. Upon validating the data, the site conditions were determined to meet four of the eight appropriateness criteria in the NCP for conducting a removal action, which provided the necessary justification to move forward with the removal action. A time-critical removal action was proposed because of the relatively simple natural of the action, lack of contact by humans at the site, and potential for off-site migration. An emergency action was not warranted, nor was a non-time-critical removal action based on site conditions and threat to humans and the environment. An action memorandum was prepared, which included a brief comparison of alternatives. In the end, the action included scraping about 500 yd^3 of sandblasting grit off the surface and recycling the materials at an off-site facility (Figure 2-2). As the removal was completed, confirmation samples were collected and the site was further characterized.

Although the removal action was successful in removing the sandblasting grit, the action did not close the site. The underlying shallow aquifer system was contaminated with low levels of dry cleaning solvents and other contaminants. The removal action goal was to remove the most significant pathway of contaminants to humans and the environment by ridding the surface soil that surpassed EPA Region IX's industrial Preliminary Remediation Goals (PRG) for chromium. Pending additional investigation, and a risk assessment, additional cleanup actions may be considered at the site based on an industrial risk setting.

Figure 2-2 A front end loader takes a load of metal-contaminated soil as part of construction work for a removal action at Lot 645 Fleet and Industrial Supply Center Oakland, California. (Photograph by Dan Shaffer.)

2.2.2 Interim Remedial Actions

Interim remedial actions are actions that include contaminant and source control objectives along with risk reduction measures. Similar to removal actions, interim remedial actions are not necessarily considered to be final remedies for site cleanup, whereas RD/RA, and in some cases "early remedial actions," are considered the final site cleanup action (EPA 1991). However, if properly administered, an interim remedial action can be used to close out a site if risk is reduced sufficiently to protect human health and the environment, ARARs are met, and source control is successful. Interim remedial actions differ from removal actions in that they fall under the remedial authority of CERCLA. Generally, a streamlined or focused RI/FS is completed prior to conducting an interim remedial action and a proposed plan and interim remedial action record of decision (ROD) are prepared. In general, the interim remedial action process requires more public involvement compared to removal actions and often include complicated construction designs. The ROD for an interim remedial action is more focused and less comprehensive than a final remedy ROD for a remedial action, and cannot be used alone to close out an entire site. More specific information on interim remedial actions, including the need for a focused RI/FS and closing site under an interim remedial action ROD are discussed under the SACM and ROD subsections of this chapter (Sections 2.2.3 and 2.2.6).

2.2.3 Superfund Accelerated Cleanup Model

On July 7, 1992, EPA's Office of Solid Waste and Emergency Response (OSWER) issued OSWER Directive Number 9203.1-03: *Guidance on Implementation of the*

Superfund Accelerated Cleanup Model (SACM) under CERCLA and the NCP to accelerate cleanup of Superfund sites. Separate EPA guidance (1994a) was issued for federal facilities. The vision for SACM is to "build public confidence through prompt and appropriate hazardous waste cleanup that protects the health of people and the environment" (EPA 1992a). EPA concluded that their vision was possible by accelerating and streamlining the Superfund process to provide risk-based cleanup at the greatest number of both NPL and non-NPL sites. The name SACM lost popularity by the mid 1990s and was less publicized than previously. However, the concepts and approaches were and remain state-of-the-art for acceleration actions under CERCLA. To ensure that SACM concepts and strategies kept monmentum, OSWER created Acclerated Response Centers for EPA regions. Through the Accelerated Response Centers, SACM approaches are followed and information and strategy for acceleration measures are used and transferred through integrated branches of the EPA using teamwork and partnership models for improved response times. Overall, SACM is the foundation of EPA's acceleration efforts under CERCLA. For this reason, the SACM concepts are summarized in this book. However, while this summary contains the concepts and overall focus of SACM, a thorough review of EPA's guidance referenced in this subsection is recommended to completely understand SACM.

There are four types of actions or assessment used with SACM that accelerate and streamline the Superfund process:

- Removal actions
- Interim remedial actions
- Integrated assessments
- Presumptive remedies

The actions are applied through five key elements of the SACM process:

- One-step screening and risk assessment
- Regional decision teams
- Early actions
- Long-term cleanup
- Enforcement and community involvement

Under SACM, the Superfund authorities (remedial and removal) are grouped together (EPA 1993a; 1993b). Rather than viewing removal and remedial actions as parts of separate programs, they are viewed as separate legal authorities with different, but complementary, application at Superfund sites (EPA 1992a; 1992b). The intent of SACM is to take better advantage of the flexibility conferred through CERCLA and the NCP in implementing these two authorities. Traditionally, project teams entered the Superfund program through one of two doors marked "remedial" or "removal." Under SACM, project teams enter one door called "Superfund."

Site Screening under SACM

All site assessments take place in one program, combining as appropriate elements of removal actions, preliminary assessments (PA), site inspections (SI), remedial investigations (RI), and risk assessments (EPA 1992c). This step is sometimes referred to as the SI/RI stage. The one-step assessment approach, referred to most often as integrated assessments, can (1) substantially streamline the Superfund process, (2) combine action and investigative tasks, (3) initiate public involvement, and (4) allow EPA to immediately initiate enforcement search and notification activities. The integrated assessment subsection of this chapter (Section 2.2.4) provides more insight into the single assessment program under SACM. Regional decision/management teams are used to (1) unite management experience, (2) prioritize efforts for a shared goal, (3) help identify and direct sites for implementing cleanup actions as soon as possible and, if necessary, (4) score sites where long-term restoration activities are expected to be necessary (EPA 1992d). EPA has used the phrase "traffic cop all sites" (EPA 1992h) when referring to the regional decision/management teams, where the approach is to keep projects moving forward without delay.

Early Actions under SACM

At any point during or after the assessment process, the regional decision/management team may consider short-term actions to address threats to human health and the environment. Short-term actions include cleanup activities that will generally take no more than 3 or, at the most, 5 years to complete. EPA felt that for short-term actions, a 5-year time frame was reasonable based on the program's ability to identify and address immediate risks to human health and the environment. These actions are coined "early actions" and are assumed to be actions that are quickly implemented at Superfund sites. In some cases, for example, early actions can close sites through elimination of the majority of human health or ecological risk by treating or disposing of exposed hazardous waste on the ground surface.

In general, early actions are responses performed under the removal or remedial authority to eliminate or reduce threats to human health or the environment, or threat of release of hazardous substances, pollutants, or contaminants. Early actions can be completed as emergency, time-critical, or non-time-critical removal actions, or interim remedial actions. EPA, however, refers to "early remedial actions" differently than early actions, where early remedial actions can apply not only to short-term cleanup objectives, but also to long-term objectives consistent with final remedies. Early actions include such activities as waste and soil removal, preventing access, relocating people, and providing alternate drinking water (Table 2-3). Actions that have the effect of immediately reducing threats to public health and safety are addressed under this part of the SACM process. While standardized or traditional cleanups for similar sites expedite cleanup, innovative technologies should be considered whenever it is faster, more efficient, more acceptable to the public, less expensive, or more effective (EPA 1992e). Figure 2-3 illustrates the SACM process including site screening, early actions, long-term actions, and site closure stages. Under long-term actions, the EPA informs the community that these actions require years to complete but do not necessarily pose an immediate threat to human health or the environment.

In terms of cost factors under SACM, because either removal or remedial authorities may be used, there is no maximum dollar cap on the cost of an early action. EPA regions, however, must follow the current rules for justifying and obtaining an

Table 2-3 Example of Early Actions and Long-Term Actions under SACM.

NCP Terminology	Early Actions				Long-Term Actions	
	Emergencies	Time Critical	Non-Time Critical			
Funding Source and Authority	Removal	Removal	Removal	Remedial	Remedial	Remedial
Types of Actions Generally Performed Under Each Authority	Classic emergencies	Site access Direct threats Water supply Visible soil contamination Remove surface Structure and debris	Source control (soil and waste) Treatment (incineration and other technologies) Containment Capping DNAPL source extract GW plume containment Post-removal site control (by PRP or state)		Restore Groundwater Surface Water Sediments Wetlands/estuaries Large mining sites Actions with extended O&M Property Acquisition Permanent relocation Institutional controls	
Enforcement Vehicles Available	↕	↕	Administrative Order of Consent Consent Decree Unilateral Order	↕	↕	
State Role	←——— Notification, consultation, and optimal participation ———→				(Required participation)	
					Required	Required
Cost Share	Not required	Not required	Optional		Required (or get waiver)	
Requirements ARARS	←——— Extent varies based on urgency and duration ———→ ←——— Required to extent practicable (or get waiver) ———→			Required		
Community	No requirements	Comment after removal begins	Required on EE/CA		Required on RI/FS	
Public Relations		Risk documented in action memo	Informal under EPA policy		Generally Required	
Baseline Risk Assessment		As time allows (but not required)	Preferred under EPA policy		Required	
Preference for Treatment						
NCP Terminology	Action Memo	Action Memo			Quick ROD	Full-Scale Rod
	Emergency Waiver Administrative Record (after the fact)	Emergency Waiver Consistency Waiver Administrative record	Consistency Waiver		HRS scoring NPL listing RI/FS CA/SSC Administrative Record	

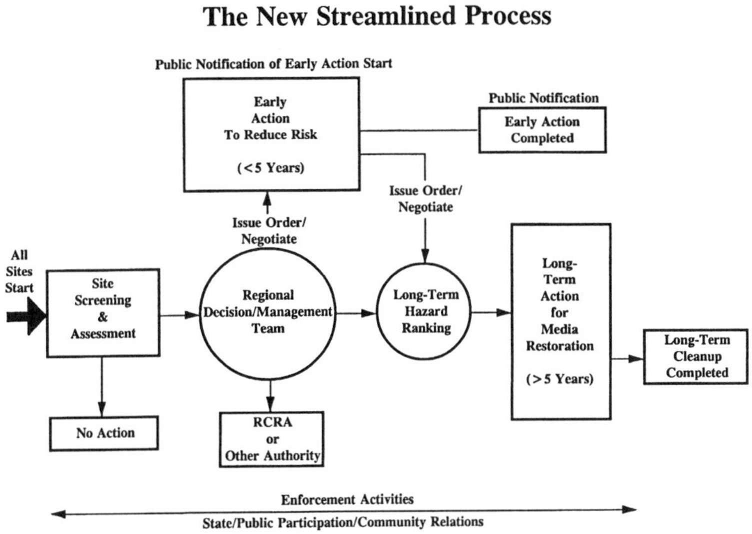

The New Streamlined Process

Figure 2-3 The Superfund Accelerated Cleanup Model process overview. (Source: presentation overheads developed by EPA OSWER/OERR, March 1992.)

exemption for removal actions estimates over $2 million or schedules that exceed more than 1 year in duration (for nonfederal facilities only). In general, the EPA will support projects up to $5 million under the current policy until it is updated, even if there is no state participation (EPA 1992e). Those projects over $5 million would need a compelling case for consideration in the absence of state participation.

EPA's enforcement policy does not change under SACM. Like that for typical Superfund sites, EPA has a comprehensive enforcement policy that they aggressively pursue for the removal and remedial authorities (EPA 1992f). Under SACM, the EPA maintains their aggressive policy for conducting potentially responsible party (PRP) searches, issuing notice letters, maintaining an administrative record, *de minimis* settlements, cost recovery, and negotiating with PRPs to conduct an action through unilateral or consent orders or consent decrees (Figure 2-3). The lead time available for non-time-critical actions should allow for comprehensive PRP searches and subsequent negotiations. Overall, SACM should fast-track EPA's enforcement policy.

Risk Management under SACM

Risk management is critical under SACM, since removal and remedial action levels and cleanup may differ. If an early action will be conducted and long-term response actions will be considered, the early action should include (if possible) the long-term cleanup goals. For example, when performing a source removal to mitigate a direct contact threat at a site that also has a groundwater threat, it may be prudent to consider removal of additional soil contaminants consistent with project groundwater cleanup goals to eliminate leachate generation. In turn, this could eliminate the need for additional source control actions during future response actions and reduce the ongoing release of contaminants to groundwater.

Risk management under SACM is of particular importance for sites that may become an NPL site. The attempt to evaluate sites more quickly, and to initiate response actions earlier, may have some impact on a site's scoring and possibly listing on the NPL (EPA 1992g). Lowering the HRS score early in the SACM process could determine whether a site is of NPL caliber. Only sites listed on the NPL are eligible for fund-financed remedial actions (40 CFR 300.425(b)(1)), and therefore the HRS score plays into the funding issues that affect long-term remedial actions. (However, removal and remedial actions carried out by private parties pursuant to EPA's authority may be conducted at NPL or non-NPL sites, since private funding may be available). Under the HRS, the physical removal of a hazardous substance from a site may reduce the site's HRS score, but only if the action occurs prior to the remedial SI phase of the site assessment. HRS scoring would not consider the absence of the hazardous substance if the response action occurred after the SI phase (even though the waste is gone and no further action may be warranted). However, the timely removal of all hazardous substances (before scoring) would always result in the HRS score of 0, such as a voluntary cleanup by a responsible party. Mitigating hazards, such as providing alternative drinking water, would not affect the HRS score. Careful consideration and planning of short-term and long-term actions is needed for complicated and highly contaminated sites of NPL caliber to ensure fast-tracked cleanup is not impeded by funding blocks.

Advantages of SACM

The advantages of SACM are that (1) the approach reduces the majority of risk from Superfund sites up front, (2) more money is dedicated to cleanup vs. study, (3) the program is more efficient and effective since it is geared for results, (4) substantial cost savings and time savings can be realized, and (5) it is an achievable program (EPA 1992i). SACM is a substantial improvement over the traditional Superfund process since it can be applied to all Superfund sites. Importantly, the current focus on Superfund sites is to move away from the old way of studying and cleaning up sites and into the SACM approach. A concerted effort by private firms, federal agencies, and state regulatory agencies to implement SACM at Superfund sites would greatly streamline the process and accelerate closure of sites. Technical descriptions of approaches the EPA developed to implement accelerated actions under SACM are contained in two subsections of this chapter. Like removal actions and interim remedial actions, in many cases these approaches can be used within or outside the SACM approach, and to some degree, other environmental programs such as RCRA.

2.2.4 Integrated Investigations

The integrated assessment was developed by the EPA to accelerate investigation and cleanup of Superfund sites. EPA refers to integrated assessments as the integration of site assessment activities for the removal and remedial programs, also described in the SACM subsection in this chapter (Section 2.2.3). The continuous and integrated assessment of potential and known hazardous waste sites is a critical element of SACM; however, it can be used in some cases outside of the SACM approach under *other* environmental programs. One example is investigations conducted at military facilities, where an integrated approach for site characterization has been used at several facilities before SACM was developed. The concept of integrating removal and remedial assessment activities was introduced by the EPA in *Assessing Sites Under SACM* (EPA 1992g). EPA has developed additional guidance related to integrated assessments and its relation to SACM (EPA 1993a; 1993b). Guidance of particular interest is *Integrating Removal and Remedial Site Assessment Investigations* (EPA 1993a), which establishes guidance on integrating removal and remedial assessment programs while maintaining separate legal authorities for the two programs, and examines the duplication and key differences between the two programs.

The basic goals of integrated assessment activities include

- Eliminate duplication of effort
- Expedite the process, and at a minimum avoid delays for time-critical removal actions
- Minimize the number of site visits and other steps in the process
- Collect only the data needed to assess the site appropriately

As part of these SACM goals, objectives of the integrated assessment process are (1) accelerated assessment and response times, (2) improved efficiency of project resources, (3) improved communication both internally and in the community, and (4) improved assessment and response decisions (EPA 1992g). Under the integrated assessment approach, EPA focuses on the following five tasks:

- Determination of CERCLA eligibility
- Documentation of the presence and type, or absence of uncontrolled hazardous substances on-site, and/or those substances that have a potential to be released to the environment
- Collection of source, receptor (target), and site characterization information
- Determination of a site's preliminary HRS score
- Determination of the potential need for removal action

Seven Steps for Applying the Integrated Assessment Process

The five tasks of the integrated assessment process are implemented following a seven-step approach. The following is list of the seven steps, followed by a narrative describing the implementation of each step. The seven steps, in this approximate order, include the following:

1. Site reconnaissance and background review.
2. Comprehensive plan and strategy development.
3. Field activities (and emergency/time-critical cleanup action).
4. Report generation and preliminary scoring.

5. Full HRS scoring (if necessary).
6. Early action analysis, planning and implementation.
7. Community relations (throughout the process as required).

Step 1. Site reconnaissance and background review. Before developing site-specific documents for an integrated assessment, EPA first reviews and evaluates information that is already available. In most cases, a preliminary assessment may have been performed or limited sampling data may have been collected by a state agency. In particular, previous analytical data for the site is useful for developing planning a sampling strategy for the project, including strategy for data gaps and HRS scoring.

Step 2. Comprehensive plan and strategy development. After reviewing available site background information, EPA determines the scope of the integrated assessment. After the scope of the integrated assessment has been established, four site-specific planning documents are used in conducting the investigation. These include a work plan, a field operations plan (including quality assurance/quality control (QA/QC) requirements), a health and safety plan, and an investigation derived waste (IDW) plan. Clear, concise planning documents are prerequisites for obtaining quality analytical data and making reliable conclusions under the integrated assessment, since it is a comprehensive approach to collecting site data. As part of these plans, a site-specific sampling and analysis plan is included in the work plan to describe the site hazardous waste contamination, identify pathway(s) of concern, identify data gaps and the proposed sampling strategy, and describe site-specific activities and team responsibilities. The integrated assessment strategy must include evaluation of preliminary HRS information, identification of pathways of concern (i.e., groundwater, surface water, soil and/or on-site exposure, and air), and associated data gaps of concern. Appropriate sampling strategies are evaluated and proposed so that the activities unique to each site and program are served. In addition, appropriate QA/QC procedures should be developed and used to ensure reliability, accuracy, and defensibility of the data collected during the integrated assessment. As part of the QA/QC procedures, EPA determines the type of analysis required and determines if Contract Laboratory Program (CLP), non-CLP, field screening, mobile laboratory, or fixed laboratory capabilities are necessary, and the level of data validation to be conducted.

As part of the integrated assessment, the EPA generally prepares a Preliminary Draft Package consisting of a standard integrated (removal/remedial) assessment report, HRS evaluations (standard PA Score/SI work sheets), and a scoping strategy/sampling plan. Note, that a removal action that physically removes (not mitigates) hazardous waste from a site can be implemented before the HRS scoring is completed, and reduce the overall score and NPL listing potential. As part of the Preliminary Draft Package, RCRA information on Areas of Concern (AOC) are evaluated for use in completing the NPL Characteristics Data Collection Form with regard to the occurrence or potential for releases of hazardous substances from AOCs. EPA will generally include a brief description of the AOCs, the start-up and closure dates of the AOCs, the types and volumes of waste managed at the AOCs, containment systems, release controls, and operational practices at the AOCs, and a brief release history for each AOC. At this point in the integrated assessment process, if the HRS score is less than 28.5 and site sampling is not warranted, a no action decision document can be proposed.

Step 3. Field activities. If there is a potential for a human or ecological threat, the field effort begins, which may include an emergency removal action and/or investigation activities. After integrated assessment field work is completed, EPA evaluates the data and integrates it with other site data obtained in earlier phases of work, such as preliminary assessments, SIs, state investigations, or emergency removal action data. All data are evaluated for validity and applicability with regard to EPA protocols and guidance, similar to the way data has always been evaluated for preparing the HRS score. Simultaneously, the EPA evaluates potential human health and environmental risks associated with the site based on the data obtained during the integrated assessment. The remedial portion of the integrated assessment investigates possible threats to the four migration and exposure pathways (groundwater, surface water, soil, and air). Based on the results of this part of the integrated assessment, the EPA may consider a time-critical removal assessment or action and use an abbreviated plan to initiate quick actions at any time and to ensure data quality objectives are met and that document generation does not hinder the removal action process.

Before proceeding with the integrated assessment report preparation and HRS scoring, the EPA compiles all data (i.e., PA, earlier SI, state investigations, emergency response/removal actions, owner/operator investigations). All data used in the integrated assessment process are evaluated for its validity and applicability with regard to EPA protocols and guidelines. Much of this evaluation process is generally completed in the earlier phases of the project in order for the investigator to determine data gaps and sampling requirements needed to fulfill HRS data needs. The evaluation process continues through the project as additional data become available (i.e., current sampling results) prior to preparation of the HRS evaluation package for the site.

Steps 4 and 5. Report generation, preliminary HRS scoring, and full HRS scoring. The integrated assessment report includes all analytical data and highlights background information and significant findings, including a site description, history and nature of waste handling at the site, known hazardous substances, pathways of concern for these substances, the impact to human and environmental targets, and the potential need for a removal action. The final integrated assessment package include the integrated assessment report, preliminary HRS scoring information, and the NPL Characteristics Data Collection Form, if appropriate. If the preliminary HRS score is greater than 28.5, the EPA may consider recommending the site for a full HRS evaluation as a potential NPL caliber site. If the preliminary HRS score is less than 28.5, the EPA may classify the site a no-action site.

Step 6. Early action analysis, planning, and implementation. Identification of early actions is a critical component for integrated assessments. Removal actions (see the removal action subsection in this chapter) or presumptive remedies (see also the presumptive remedy subsection in this chapter) should be maximized at sites in order to accelerate and streamline project cleanup. Often, conditions at sites vary from obvious surface contamination posing the greatest threat to human health and the environment to residual concentrations of contaminants that pose a minimal threat and must undergo a comprehensive evaluation to make that determination. While the responsible party(s) may identify potential early actions under the integrated assessment approach, the EPA will determine what, if any, cleanup action should be completed, what regulatory process should be followed, and whether innovative technologies are available to streamline efforts and lower costs.

Step 7. Community relations. Building a good relationship and maintaining lines of communication with persons living near, affected by, or concerned with a hazardous waste site is an important part of the Superfund program. The public must be kept informed of the activities at the site and given an opportunity to participate in key decisions regarding the site. Although not required for emergency or in most cases time-critical removal actions, public participation may be considered to foster a proactive working relationship between the regulatory agencies, responsible parties, and the community. A community relations plan that describes plans for allowing concerned persons an opportunity to be involved in activities at the site may be appropriate or required by the EPA, depending upon site-specific conditions and the relative public awareness surrounding the site.

2.2.5 Presumptive Remedies and Plug-In Remedies

EPA developed the presumptive remedy to accelerate cleanup actions and decision making as part of SACM (EPA 1993c). The presumptive remedy, or the "plug-in remedy" when there are multiple sites with similar contaminants and site conditions, is a "standardized solution" project teams can use for site restoration (EPA 1993c; 1994a). The presumptive remedy offers fast-track cleanup by limiting the cleanup alternatives evaluated at a contaminated site and is a default or point-of-departure remediation strategy for potential application to more commonly occurring types of sites (Artrip 1996). Certain types of sites and contaminants, such as older inactive municipal landfills or soil with volatile organic compound (VOC) contamination, often have nearly identical remedy selections in the record of decision (ROD) stage, since there are proven solutions to these types of contaminated sites. In addition, it is EPA's position that the presumptive remedy is entirely within the regulatory scope of the NCP (40 CFR 300.420(b)(iv)) as required by law. A compilation of presumptive remedies is in Table 2-4.

Traditionally, the Superfund remedy selection process is site-specific. Each site is considered a unique problem that is independently investigated after evaluating a number of potential solutions to propose a site-specific remedy. Usually, the EPA characterizes the nature and extent of contamination under an RI, evaluates and compares several remedial alternatives in a feasibility study (FS), proposes one of those alternatives to the public in a proposed plan, addresses public comment on the proposed plan, then selects a preferred alternative for a final remedy ROD (Figure 2-4). After finalizing the ROD, the exact technical specifications and construction detail of the remedy are designed during the remedial design stage, and finally, the cleanup takes place in the remedial action phase. Under the traditional remedy selection, several alternatives are matched, or evaluated, for a single site. Site characterization is usually almost entirely completed before a final decision is made for a remedy selection. This approach is used in the traditional remedy

Table 2-4 Summary Of Presumptive Remedies And EPA Contacts

Site Type/ Information	Presumptive Remedy	Selected Resources	EPA Contact
General policy and procedures	Not applicable	Presumptive Remedies: Policy and Procedures, OSWER Dir. 9355.0-47FS, 1993, EPA/540/F-93/035	Shahid Mahmud Headquarters (703) 603-8789
Volatile organic compounds in soils	Soil vapor extraction, thermal desorption, incineration	Presumptive Remedies: Site Characterization and Technology Selection for CERCLA Sites with Volatile Organic Compounds in Soils, OSWER Dir. 9355.0-48FS, 1993, EPA/540/F-93/048	Shahid Mahmud Headquarters (703) 603-8789
Wood treaters	Organics: incineration, bioremediation, dechlorination	Presumptive Remedies for Soil, Sediment, Sludge at Wood Treating Sites, Dir. 9200.5-162, 1995, EPA/540/R-95/128	Frank Avvisato Headquarters and ERD (703) 603-9052
	Inorganics: immobilization	Technology Selection Guide for Wood Treater Sites, EPA 1993, OERR Pub. No. 9360.0-46FS	Harry Allen ERD (908) 321-6747
Municipal landfills	Containment (includes capping, leachate collection and treatment, institutional controls, etc.)	Presumptive Remedy for CERCLA Municipal Landfill Sites, OSWER Dir. 9355.049FS, 1993, EPA/540/F-93/035 Application of the CERCLA Municipal Landfill Presumptive Remedy to Military landfills, OSWER Dir. 9355.0-62FS, 1996, EPA/540/F-96/007	Andrea McLaughlin Headquarters (703) 603-8793
Contaminated groundwater	Pump and treat (preferred treatment technologies and overall approach)	Presumptive Response Strategy and Ex-Situ Treatment Technologies for Contaminated Groundwater at CERCLA Sites, OSWER Dir. 9283.1-12, 1996, EPA/540/R-96/023	Ken Lovelace Headquarters (703) 603-8787
Region VII pilots: PCB sites, coal gas sites, grain storage sites, metals in soil (DOE cooperative effort)	Final remedies pending	Presumptive remedy guidance is scheduled to be released in 1997 for PCB sites, coal gas sites, and grain storage sites. Metals in soil is a new pilot and a final date of guidance release is undetermined.	Diana Engerman Region VII (913) 551-7746

Notes: PCB = polychlorinated biphenols; ERD = Emergency Response Division; CERCLA = Comprehensive Environmental Response, Compensation, and Liability Act. See also EPA's home page (http://www.epa.gov/superfund/index.htm) under products and Superfund reform.
Adapted from EPA 1993c.

selection, because a decision that is based on inadequate data, or unknown characteristics of the site may render a selected remedy ineffective. In this sense, there is a lower risk of failure when making a remedy selection, but a lot of time and money is expended to clean up a site. Under some conditions, however, it is possible to

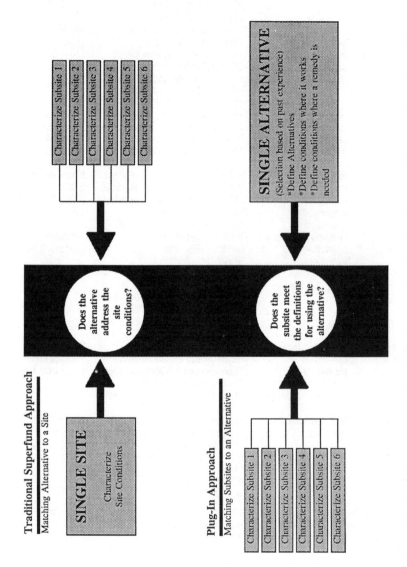

Figure 2-4 Comparison of traditional and plug-in approaches for evaluating remedial alternatives. This model was used at the Indian Bend Wash Superfund Site, South Area, Tempe, AZ. (Source: Indian Bend Wash Superfund Site, South Area ROD.)

maintain a low risk of failure for remedy selection and accelerate the cleanup and decision making process.

One of the objectives of the presumptive remedy is to shorten the FS process by eliminating the need for the traditional, exhaustive remedial technology identification and screening of alternatives. In many cases the RI process can also be streamlined since the remedy is assumed closer to the beginning of the project, making data collection efforts focused on the proposed remedy, and not solely focused on site characterization (Artrip 1996). However, until the streamlined RI is completed, the project team cannot always be certain a cleanup is actually warranted. Prior to proposing a presumptive remedy, justification for cleanup in the form of regulatory limits or risk drivers must therefore be identified. Once the streamlined RI is completed, a focused FS or EE/CA is usually sufficient to propose the remedy when employing the presumptive remedy approach (EPA 1994a).

Volatile Organic Compound Presumptive Remedy

In the case of VOCs in sandy soil, the problem is generally not in finding a remedial technology to treat the VOCs, since soil vapor extraction (SVE) (under a wide range of soil conditions) has been shown to be highly effective in reducing levels of contamination if properly designed and installed. Rather, the difficulty lies in moving away from the traditional approach of investigating sites and using the VOC presumptive remedy. In addition, difficulty can arise from administering many similar yet distinct subsites. In cases where soil is amenable to SVE, geologic conditions across the contained site(s) are relatively similar, and VOCs in soil present an unacceptable level of risk to humans or the environment, the VOC presumptive remedy approach should be used to fast-track site characterization and decision making. However, a word of caution is important. If more contaminants other than VOCs are present in soil (such as metals), future actions and associated remedial costs and risk of all contaminants should be considered in the decision-making process. Evaluating the risk or potential risk associated with all contaminants is important to ensure that, while VOCs can be remediated under a presumptive remedy, the remaining contaminants do not have to be cleaned up under a separate action where costs of the associated actions are prohibitive or where the results of the VOC presumptive remedy are better managed under a final remedy approach where all contaminants are cleaned up under one action. In this sense, a site-specific action may be more prudent if multiple types of contaminants are present and the VOC presumptive remedy would be redundant in the long term.

Planning a streamlined or focused RI for the VOC presumptive remedy should include field testing the SVE potential. As part of the RI work plan, plan on installing several SVE pilot wells and temporarily installing a mobile blower on site to test air permeability and area of influence characteristics. These data are critical components of the streamlined RI, since they provide important engineering data needed to propose a full-scale operation. Geotechnical analyses such as permeability testing and sieve analysis data also should be collected, for example, during the streamlined RI. Under a traditional approach, field testing the remedy has been reserved for the RD/RA stage and efforts were strictly focused on characterizing the contaminant and hydrologic setting. The presumptive remedy approach expands the scope of the RI into the RD/RA stage by incorporating data collection for major design components and processes too. Careful thought is needed to ensure engineering data needs are part of the streamlined RI to accelerate the overall process and eliminate the need to revisit the site to collect engineering data.

Landfill Presumptive Remedy

In the case of CERCLA (inactive) municipal landfill sites, the presumptive remedy involves contaminant source control and containment (Artrip 1996). EPA determined that because of the heterogeneity and volume of waste involved at these types of sites, treatment is impracticable. EPA lists the following source control and contaminant remedies that should be considered at landfills:

- Landfill cap
- Source area groundwater control for plume containment
- Leachate collection and treatment
- Landfill gas collection and treatment
- Institutional control to supplement engineering controls

Under the presumptive remedy approach for landfills, the presumptive remedy is selected early in the project (such as the Agreed Order of Consent stage). A streamlined or focused RI is implemented to not only characterize the site, but also to collect engineering design data to provide technical justification to support elimination of one or more of the five remedies listed above. A landfill cap is considered a likely presumptive remedy for most landfills, since it is a fundamental regulatory component of RCRA Subtitle D landfill closure requirements, which would be considered an ARAR. Subtitle D contains specific requirements for landfill cap design, which are often the least stringent in most states for landfill closures. If the scope of the RI can be used to eliminate one or more of the above source control or contaminate remedies, a substantial time and cost savings can be realized, both in the investigation stage, design stage, and in the long-term operation and maintenance stage. Similar to the VOC presumptive remedy, careful planning is needed to ensure engineering data needs are met in the RI stage. Figures 2-5 and 2-6 show common remedial actions and a selection guide for covers at municipal landfills, respectively.

Benefits of Presumptive Remedies

Overall, the benefits of the presumptive remedy are increased efficiency and effectiveness of the RI/FS process, expedited design, and accelerated cleanup action. Substantial cost savings are likely too. Table 2-5 lists the effect of the presumptive remedy on the CERCLA process. When there are multiple sites and a plug-in presumptive remedy approach is being considered, the plug-in approach allows the remedial action to begin without entering into a redundant remedy selection process. Also, the plug-in approach can be used to initiate action at sites that are in different

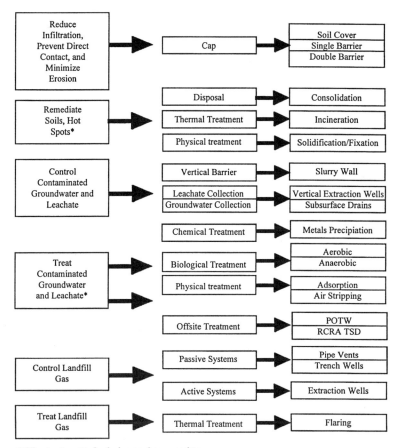

Figure 2-5 Technologies frequently used for remedial actions at CERCLA municipal landfills. (Source: EPA. 1990e. Streamlining the RI/FS for CERCLA Municipal Landfill Sites. OSWER Dir. 9355.3-11FS.)

stages of completeness. As the streamlined RI is completed at the subsite, the presumptive remedy can be initiated immediately without slowing action down at other sites not as far along in the investigation process. Lastly, the plug-in process for multiple sites is a way of sharing information from subsite to subsite, eliminating duplicate data collection, which saves money and time on projects.

EPA has a number of presumptive remedies in addition to VOCs and municipal landfills (Table 2-4). In general, these presumptive remedies do not always save money or are not necessarily innovative, and they all do not fall under the CERCLA process (such as coal gasification). Incineration, for example, is a presumptive remedy that can be considered for treatment of solids at CERCLA sites. Incineration is highly effective; however, the cost of incineration is often prohibitive, making it less desirable. In some cases, incineration is the only practicable alternative for a particular contaminant or volume of wastes and is the preferred alternative. As new technologies are tested, other alternatives may become available.

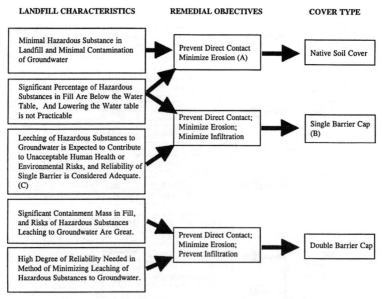

Figure 2-6 Landfill cover selection guide. (Source: EPA 1990e. Streamlining the RI/FS for CERCLA Municipal Landfill Sites. OSWER Dir. 9355.3-11FS.)

Professional risk is greater when you pursue a presumptive remedy at contaminated sites. In some cases a presumptive remedy is not the correct path to follow, and the traditional approach must be followed in order to propose a comprehensive cleanup plan at complicated sites, such as sites with multiple contaminants. Presumptive remedies are oriented toward contaminants that have not migrated off-site, and therefore a presumptive remedy may not be protective or sufficiently comprehensive. Another potential danger is not considering available innovative technologies for site remediation, when there may be a more efficient or effective remedy available (Artrip 1996). As more RODs are finalized and more success stories are compiled, it may become clear that other remedies are successful and should be considered as a presumptive remedy or as an alternative to a presumptive remedy under certain conditions. Soil from underground storage tank (UST) sites, for example, is often treated by landfarming and *ex situ* bioremediation, which are exceptionally successful in reducing levels of petroleum hydrocarbons. More presumptive remedies will likely be approved as data become available to support them. Currently, EPA and other federal agencies are working on other presumptive remedies, such as remedies for jet fuel spills (EPA 1994a).

2.2.6 The Record of Decision

This section outlines the type of RODs that can be drafted and the basic components needed to finalize a ROD. However, before going through the ROD process,

Table 2-5 Common Effect Of Presumptive Remedies On Cleanup Processes

Phases of the Cleanup Process	Effect on Cleanup Processes
Preliminary assessment/site inspection or removal site evaluation	Focused
Scoping	
• Collect and analyze existing data	Not impacted
• Identify initial project operable units and remedial action objectives	Streamlined
• Identify range of likely alternatives	Streamlined
• Identify potential ARARs	Not impacted
• Identify initial data quality objectives	Not impacted
• Prepare project plans	Streamlined
Remedial Investigation	
• Conduct field investigation	Focused; streamlined for municipal landfills
• Define nature and extent of contamination	Not impacted; streamlined for municipal landfills
• Identify ARARs	Not impacted
• Conduct baseline risk assessment	Not impacted; streamlined for municipal landfills
Remedy Selection	
• Identify potential treatment technologies and containment/disposal requirements	Eliminated
• Screen technologies	Eliminated
• Assemble technologies into alternatives	Eliminated
• Screen alternatives as necessary to reduce number subject to detailed analysis	Eliminated
• Further refine alternatives as necessary	Streamlined
• Analyze alternatives against the nine criteria and each other	Streamlined
Proposed Plan	Streamlined
Record of Decision	Streamlined
Remedial Design	Streamlined

Source: Presumptive Remedies Policy and Procedures OSWER Dir 9355.0-47FS

its important to understand where the CERCLA process has been in relation to decision making — and more importantly where it needs to go in relation to fast-tracking cleanup. The court system is often an avenue used by regulatory personnel and/or responsible parties to resolve different points of view. This process is important in some cases; however, it lengthens the time it takes to get to the decision-making stage, delays actions necessary to protect human health and the environment, and delays closing sites. Averting this type of approach is critical for accelerating cleanup and decision making. Managers must look toward teamwork and cooperation in order to protect human heath and the environment, make practical environmental decisions, and minimize project costs. Actions are negotiable and need to be treated not only from a protective standpoint (depending on the level of actual risk), but from a practicable, scientific, and common sense standpoint. The usual approach of spending vast sums of money on perpetual investigation and court battles does not protect human health and the environment or reduce costs. Project dollars that are focused on cleanup in the beginning of the project, however, are generally more protective and cost-effective in the long-term. The ROD process is an avenue to make up-front decisions and initiate a limited or comprehensive response at CERCLA sites.

Open dialogue related to site data, public concerns, and available funding is important to develop a ROD that offers a win–win solution to all parties. In some cases, the respective parties may call this approach seeking a compromise, yet with so many sites in need of response actions or closure, postponing short-term or long-term decision making is unacceptable from a community standpoint or from the current focus on accelerating actions. Environmental attorneys may disagree with me, since they are in their element when they are in the court room, rather than in the field. And, I am certain there are some sites where it may be better (cost-wise) to collect more and more data, and argue until project decisions are determined in our court system. However, for the majority of sites the opposite is true, and it is more cost-effective and better for public relations to close sites through early actions, for example, which can also help more complicated sites reach the final-remedy ROD stage more quickly. In essence, project teams need to make decisions that resolve the most serious health risks to the public or environment in the beginning of a project, thereby reducing the regulatory pressure EPA must apply and limiting the possibility of a prolonged court battle.

The Record of Decision Process

When a CERCLA site reaches the ROD stage, it is a significant milestone for a project — no matter what type of ROD it is. Using the traditional Superfund approach, it can take years and often more than a decade to reach the ROD stage. Half the battle of accelerating environmental actions is often accelerating the time it takes to get to the cleanup decision-making stage. Negotiating a decision for site restoration is absolutely critical for selecting environmental actions that will be efficient and effective for cleaning up a site. Throughout a project, there should always be some level of discussion of where the project is going and where it will end. For example, is the EPA or the state going make a responsible party move 100,000,000 yd^3 of contaminated soil from one spot to another? In rare cases, perhaps yes. But often the answer is no, since moving this amount of waste is usually cost-prohibitive and can be shown to be prohibitive with relatively simple calculations and conservative assumptions. Many bits of information contribute to determining the answer to a question like this. Do not wait until the bitter end to find out that 5 years ago the decision was relatively clear and close at hand.

A clear plan for site actions and closure in the beginning of the project is paramount. As time goes by you need to revisit your plan and make adjustments based on technical data, site conditions, future land use, public concerns, available resources, regulatory concerns, risk considerations, and available remedial technologies to update your plan. You must ask yourself, what is the final decision going to be based on what I know today? When your plan becomes clear, even if it is early in the CERCLA process, you may want to decide whether an additional investigation is really warranted or if it is better to accept a calculated level of potential failure in the interest of accelerating the decision-making process. Often the 80/20 rule applies to sites where 20% of the data leads to 80% of the cleanup. An operable units (OU) strategy, for example, to break out work and identify relatively easy cleanup decisions can be used to fast-track the ROD process for one portion of a site. This approach will show progress, accelerate the decision-making process, and help ease public concerns. More information on OU strategy is provided in Chapter 8.

EPA issues the ROD as the final remedial action plan for a site or OU. The ROD is both a technical document and a legal document (EPA 1989c). The ROD summarizes the problems posed by the conditions at a site, the alternatives considered for addressing those problems, and the comparative analysis of those alternatives against nine evaluation criteria (EPA 1990b). The ROD then presents the selected remedy and provides the rationale for that selection, specifically explaining how the remedy satisfies the requirements of Section 121 of CERCLA of 1980, as amended by SARA of 1986. In order for the EPA to prepare a final-remedy ROD, there is generally a relatively comprehensive RI/FS completed, a baseline human health risk assessment completed showing the presence of unacceptable risk, and a proposed plan prepared which is reviewed by the public (Table 2-2). The proposed plan is prepared after the RI/FS is completed and provides a brief analysis of remedial alternatives under consideration, identifies the preferred alterative, and provides members of the public with information on how they can participate in the remedy selection process (EPA 1990c). In addition, fact sheets are generally prepared throughout the project to keep the public informed. Upon completion of these documents and findings, the EPA can draft the remedy ROD. Figure 2-7 shows a 180-day suggested time frame for the remedy ROD process after the draft FS is completed. A 180-day time frame is necessary because a significant amount of time is needed for the community relations component of the ROD process. The nine evaluation criteria for the remedy ROD alternatives as stated in the NCP (subsection 300.430) are

- Overall protection of human health and the environment
- Compliance with ARARs
- Long-term effectiveness and permanence
- Reduction of toxicity, mobility, or volume through treatment
- Short-term effectiveness
- Implementability
- Cost
- State acceptance
- Community acceptance

One important consideration for making a practicable remedy selection is the anticipated land use. When the nine evaluation criteria are used to evaluate alternatives, the future land-use scenario must be realistic and accurately identified. Using a conservative approach, and assuming a residential use scenario when more than

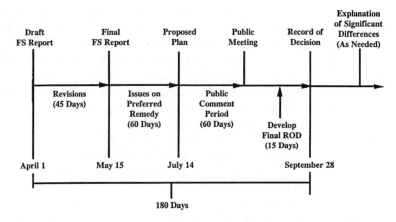

Figure 2-7 Optimal time frame for the record of decision process. (Source: PRC Environmental Management, Inc.)

likely the site will be used for recreation, can significantly limit the number of potential remedial options, and increase the amount of resources and time it will take to close out a site, because the effectiveness of many remedial technologies in the 10^{-6} human health risk range is often poor. Depending on the site conditions, meeting more stringent residential risk criteria will drive the cost up and impact the implementability factors for most cleanup alternatives. In some cases, deed restrictions and land use planning are important considerations that the EPA and responsible parties use to develop final decisions. Limiting future activities at a site can preclude impacts to human health and the environment and foster finalizing a ROD. However, waivers for some ARARs may be necessary if relatively high action levels are proposed or a deed restriction is pursued for a ROD. For those who pursue deed restrictions to foster a less stringent cleanup decision, site-specific legal language is required to properly outline the deed restriction in the ROD. Identifying cleanup levels that are one or more orders of magnitude less than residential risk criteria can open up many opportunities for early cleanup and site closure. However, migration of contaminants to adjacent residential settings must be considered before making such remedy decisions. Controlling off-site migration of contaminants may be a critical component of the decision-making process to ensure no off-site hazards or potential for a hazard. EPA has several types of RODs in addition to the final-remedy ROD. The following text outlines the no-action ROD, interim remedial action ROD, and contingency remedy ROD. Other types of CERCLA RODs and fast-tracked decision making are discussed in Chapters 3 and 5.

No-Action RODs

No-action decisions are one way to close a site and show progress. In cases where there has been a relatively large amount of effort or funds focused on investigation, or early actions have mitigated site risk and met ARARs, closing a site through the no-action ROD process is important. However, there are specific requirements that must be met to prepare a no-action or no-further-action ROD. EPA (or the lead agency) may determine that a no-action decision is warranted under the following criteria (EPA 1991):

- When a site or a specific problem or area of a site (such as an OU) poses no current or potential threat to human health or the environment.
- When CERCLA does not provide the authority to take remedial action.
- When a previous response eliminated the need for further remedial response.

Situations where no action or no further action may be appropriate could include

- Where the baseline risk assessment concluded that conditions at the site pose no unacceptable risk to human health or the environment.
- Where a release involved only petroleum product that is exempt from remedial actions under CERCLA section 101.
- Where a previous removal action or interim remedial action eliminated existing and potential risks to human health and the environment such that no further action is necessary.

In cases where there has been a cleanup, the term "no further action" ROD is used instead of no action to imply a cleanup of some type has been completed and that no additional action beyond the early action, for example, is warranted. Also, under a no-action decision, monitoring water or air quality is considered "no action," even though additional data are necessary to continue supporting a no-action decision. The benefits of completing no-action RODs are that it officially closes a site or OU, through a legally binding document, that otherwise could be a future point of concern even though the current parties agree there is no significant risk associated with the site or OU. In addition, closing sites through the no-action ROD process shows progress and effective use of resources, since it reduces the overall number of CERCLA regulated sites and need for remedial actions.

No-action decisions do not include proposing natural attenuation of groundwater contaminants, even if no man-made processes are used to enhance site restoration. Natural attenuation is considered a normal remedial decision for site restoration, because contaminant levels are reduced over time and the level of risk to human health or the environment is reduced, similar to engineered alternatives. Further, supporting data are necessary to show natural attenuation is active, ARARs are met, and that natural attenuation is a viable alternative. Generally, the case for natural attenuation is reserved for a final or interim remedy ROD, rather than the no-action ROD process. More information in Chapter 3 is presented on the factors required to support the feasibility of natural attenuation for groundwater contamination.

Interim Remedial Action Decisions

During the initial investigation efforts, or at other points in the RI/FS, the EPA (or lead agency) may determine that an interim remedial action is appropriate to initiate risk reduction actions and/or control contaminant migration. Alternatively, a removal action can or should be considered instead of an interim remedial action to address short-term risks at sites, and an action memorandum is prepared as the decision document instead of an interim remedial action ROD. Both the interim remedial action and removal action processes are excellent avenues for accelerating cleanup, and depending on the complexity of the action and the need for community involvement, can be used to initiate an early action. In comparison to a final remedy ROD, an interim remedial action is limited in scope and is generally proposed to address areas, media, or OUs that will potentially require a final remedy ROD and cleanup action (EPA 1991). Reasons for taking an interim remedial action could include to

- Take quick action to protect human health and the environment from an imminent threat in the short-term (while a final remedial solution is developed)
- Institute temporary measures to stabilize the site or OU and/or prevent further migration of contaminants or additional environmental impacts

While interim remedial actions can be implemented to address individual OUs, or are a component of a final remedy ROD, they can be comprehensive enough to close out an entire site or a portion of a site. Depending on the success of the interim action, they can be the final remedy for a site or OU as long as they meet the final ROD requirements of (1) providing long-term protection of human health and the environment, (2) fully addressing the principal threats posed by the site or OU, and (3) addressing the statutory preference for treatment that reduces the toxicity, mobility, or volume of wastes. In a situation where the interim ROD does meet these criteria, a final remedy ROD would be required to document missing information and the decision to use the interim remedial plan as the final remedy, even though no additional remedial actions would be necessary. Examples of possible interim actions include

- Installation of groundwater extraction wells to stress an aquifer, thereby mitigating migration of a contaminant plume (with the possibility of installing more wells or conducting other actions to address the final remedy selection).
- Providing a temporary alternate source of drinking water with the possibility of later remediating the contaminant source and/or contaminated groundwater.
- Constructing a temporary cap to control or reduce exposures, until a final remedy can be selected.
- Relocating contaminated material from one area of a site to another area of the site for temporary storage until a decision can be made on how to best manage site wastes.

In a case where an interim remedial action will result in closure of a portion of a site, such as excavating all wastes from one area, the interim ROD should contain the final action components of the area to be closed. This is important, since it helps fast-track the final closure decision and focuses efforts on the remaining contaminated areas. In addition, an interim remedial action may be taken early in the CERCLA process when there may not be sufficient time to prepare a comprehensive RI/FS report. Preparation of a comprehensive RI/FS is not required for interim action; however, for the purpose of fulfilling the NCP's administrative record requirements, there must be documentation that supports the rationale for the action. Available site data should be used to summarize the threat the site presents, and remedial alternatives should be evaluated in a fashion similar to an EE/CA to select a preferred remedy.

EPA (1991) clarified the differences between interim remedial actions vs. "early remedial actions." EPA points out that interim remedial actions should not be confused with early remedial actions, which may be either interim or final actions. For example, an interim action might include providing a temporary alternate water supply and sealing wells that are pumping from a contaminated aquifer. An early remedial action might involve a final action where there is complete removal of drums and contaminated soil that could result in migration of contaminants.

Contingency Remedy Record of Decision

EPA (or the lead agency) may decide to incorporate a contingency remedy in a ROD to help ensure successful site cleanup and closure. Use of a contingency ROD may be appropriate when there is significant uncertainty about the ability of the remedial options to achieve remediation levels (EPA 1991). For example, a contingency ROD may be appropriate when the performance of an innovative treatment technology appears to be the most promising option, but additional testing is needed during the remedial design stage to verify the technology's performance capabilities. In this example, a more "proven approach" could be identified as a contingency remedy. The use of contingency remedies should be carefully considered. Site managers should perform the necessary steps of treatability studies, field investigations, and even pilot testing to evaluate the technology's performance capabilities during the RI/FS. The contingency ROD must specify under what circumstances the contingency remedy would be implemented. EPA (or the lead agency) will use these criteria to decide whether to implement the contingency option as opposed to the selected remedy in cases when there is a failure to achieve the desired performance goals. Under ideal conditions the contingency ROD is useful to implement an innovative technology to more quickly remediate a site, while still having a full back plan using more traditional remedies if the innovative technology fails to yield the desired results or cost savings.

2.2.7 Superfund Case Study for Accelerated Cleanup

The following case study was reported in an EPA memorandum from Myron Knudson, Director of the Superfund Division, to Elliott Laws, Assistant Administrator for Solid and Hazardous Waste Response.

> In an August 7, 1995, ceremony in Bartlesville, OK the Oklahoma Department of Environmental Quality (ODEQ) signed an agreement for the remedial action at the National Zinc Company Superfund site. The city of Bartlesville and one of the responsible parties for the site, Cypress-Amax, are also parties to the agreement. Execution of this agreement marks the successful completion of a pilot project in which the responsible parties completed all site investigations, remedy selection, and remedial design work with oversight provided by the State of Oklahoma. These activities were completed with the same speed and quality achievable had the Federal government conducted the work under the Superfund law.

The National Zinc project was a tremendous demonstration for the concept of community-based environmental protection. The Bartlesville Coalition, a group of area residents, civic leaders, and local elected officials, took a leadership role in developing and implementing a solution to the problem. The Coalition was actively involved in keeping the Environmental Protection Agency (EPA), ODEQ, and the responsible parties focused on developing an acceptable solution to the problem from the smelter. The Coalition also provided a focal point for communicating project activities with the public and, in particular, the residents living near the smelter in west Bartlesville.

There were two main themes characterizing the success of the pilot project expressed during the ceremony. The first theme was an appreciation for the Environmental Protection Agency's innovative application of the Superfund law, postponing promulgation of the site on the National Priorities List as long as the responsible parties conducted all necessary work in a timely manner. This sentiment was highlighted in a letter from Congressman Ernest J. Istook presented during the ceremony.

The second theme, noted by several speakers, was the phenomenal coordination and communication among all parties involved in the project, including federal, state, and city government officials, the responsible parties, and the residents of the city of Bartlesville. Without the involvement and contributions of all of the stakeholders, the Bartlesville project could not have been the success celebrated on August 7. This level of communication and coordination transformed the community from one that did not want Superfund involvement to a community proud of its involvement in the process and of the cleanup underway.

Background

The city of Bartlesville is located in northeast Oklahoma and has a population of approximately 40,000 people. For nearly 70 years, the National Zinc Company smelter operated without air pollution controls. Very high levels of lead, cadmium, and arsenic from smelter air emissions were deposited on surface soils primarily on the west side of the city. Initial interest in potential problems from the site was the result of a series of articles in a Tulsa, OK, newspaper in August 1991. The State of Oklahoma, the Agency for Toxic Substances and Disease Registry, and EPA began initial health and environmental evaluations in fall 1991.

Based on the results of the these initial evaluations, Region 6 of the EPA began a SACM pilot project in July 1992. This project was undertaken to integrate EPA and state investigation efforts and to mitigate the exposure of children to the high concentrations of lead near the smelter more expeditiously. In the first year of this effort, the following activities were accomplished:

- An initial removal action addressed contaminated soils in 29 "high access" areas, including schools, city parks, and day care centers, and 22 residences of children with elevated blood lead levels.
- Field sampling was completed to provide 90% of the data needed for the long-term remedial investigation and risk assessment. Approximately 6000 soil samples were collected over 36 mi^2 surrounding the smelter to define the extent of contamination.

- Blood-lead studies were completed by the Oklahoma State Department of Health through a grant from Agency for Toxic Substances and Disease Registry (ATSDR). The state found increased blood-lead concentrations in children living in areas of high soil lead concentrations near the smelter. Children living away from the smelter did not exhibit elevated blood lead concentrations.

Despite more than a dozen public meetings during the initial removal/site assessment activities, the reception by the community to the proposal addition of the National Zinc Site to the National Priorities List has been mixed.

In response to constituents' fears of negative economic impacts to the city associated with being a Superfund site, members of the Oklahoma congressional delegation asked EPA to consider allowing studies and site cleanup to proceed under state oversight of the potentially responsible parties. On September 28, 1993, representative of EPA's Regional Office and Headquarters programs met with Congressman Ernest Istook, Senator David Boren, and Senator Don Nickles to discuss the site as a possible candidate for a state and potential responsible party pilot project under EPA's Administrative Improvements initiative. The congressional delegation also asked that EPA postpone inclusion of the site on the NPL while the responsible parties conducted the work.

In response to the request of the congressional delegates, EPA developed parallel enforcement processes for responsible party involvement in the site.

- The potentially responsible parties (PRP) were given the opportunity to conduct the upcoming removal action and complete the long-term RI/FS with federal oversight (the traditional Superfund process).
- The state and potentially responsible parties were also given the opportunity to develop an innovative state/responsible party project, whereby the state would oversee the activities of the responsible parties at the site.

EPA acceptance of the state oversight pilot project was based on the ability of the state to address certain issues, including

- A demonstration that the state had adequate legal authority to oversee the potentially responsible parties.
- A demonstration that both the state and potentially responsible parties were willing and able to commit the resources and qualified, capable, staff to the site project.
- A demonstration that the same quality and speed of study and remediation would be achieved as if it were conducted through EPA's application of the Superfund law.
- Initial acceptance by, and continued involvement of, the Bartlesville community was necessary.

On March 1, 1994, the Oklahoma Depatment of Environmental Quality (ODEQ) submitted a proposal, which the EPA regional office subsequently approved, for the pilot project. In the proposal, ODEQ demonstrated its ability to meet each of the listed requirements. Concurrent with the state/PRP pilot project, EPA negotiated an agreement with the PRPs to remove soil from the most highly contaminated residential properties. This agreement was structured to allow the PRPs to continue to remove contaminated soil from yards until a permanent remedy was selected and a formal agreement for implementing the remedy could be executed between the state and the PRPs. Such an agreement was executed at the August 7 ceremony.

With the execution of the Consent Agreement and Final Order, EPA oversight of PRP activities in Bartlesville ceased. The regional office's future role in the project will be limited to reviewing periodic progress reports to ensure that the cleanup goals set for the site are met. As these goals are met, the region will be responsible for "unproposing" the site for the NPL through the Federal Register.

2.3 RCRA AND OTHER EPA PROGRAMS

Thus far, Chapter 2 has focused on EPA's information transfer programs and CERCLA related acceleration efforts. EPA is responsible for numerous other environmental programs that include many different types of contaminants than the wastes listed under CERCLA (40 CFR Table 302.4). EPA also oversees many types of contaminated sites and facilities, processes within operating facilities, and other media such as air. In most cases, these EPA programs are not within the scope of this book. RCRA treatment, storage, and disposal (TSD) facilities, however, are closely related to the theme of accelerated cleanup. In addition, the number of TSD facilities requiring corrective actions rival the number of CERCLA sites at over 6000 facilities and 81,000 Solid Waste Management Units (SWMU) (these counts are 1993 data, the most recent information reported on EPA's RCRA hotline). Projected cleanup costs vary for all TSD facilities, depending on the source of information, but all reported costs are indeed high, generally in the hundreds of billions of dollars for full restoration to be completed. Timing of restoration at TSD facilities is also a concern. Estimates for the required time necessary to complete response actions at TSDs extend well into the 21st century. For this reason, the RCRA program is included in this chapter and opportunities for accelerated actions under RCRA are described.

In general, Section 2.3 of this chapter is devoted mostly to RCRA and a limited number of other EPA programs, where the EPA has developed acceleration and cost-saving measures. While not all encompassing of EPA's efforts, the remainder of this chapter should provide a sample of efforts EPA has initiated to address cost reduction and acceleration efforts, while still protecting human health and the environment. In some cases, the approaches described in this chapter (and other chapters of this book) are, from a distance, not fundamentally different from those methods described for CERCLA actions and decision making. Yet, the regulatory process and terms between programs differ substantially. Making this observation and becoming familiar with many EPA programs is important, because it allows one to apply a common strategy for fast-tracking actions at your sites. Combining acceleration approaches across programs is one way to develop a site strategy that maximizes cleanup effort, reduces costs, and fast-tracks closure.

2.3.1 Resource Conservation and Recovery Act

RCRA was developed primarily as a prevention-oriented program, with the primary objective of preventing new releases that contaminate sites and facilities. Following this objective, a strict set of standards were developed to ensure protection of human health and the environment from active waste management. The Subtitle C regulations of RCRA are specified as uniform, national standards that must be complied with by all RCRA-regulated facilities. These standards are generally considered very stringent. They are designed to ensure adequate level of protection nationally and to prevent or minimize environmental releases over a wide range of hazardous waste types, environmental conditions, operational contingencies, and other factors.

The Hazardous and Solid Waste Amendments of 1984 (HSWA) strengthened the RCRA pollution prevention program by adding the land disposal restrictions (LDR) and minimum technology requirements (MTR), which have become central features of the RCRA prevention program. These features added incentives to generators to minimize the amount of waste being generated by providing technology-based standards. LDRs set technology-based standards for the treatment of hazardous waste before disposal in a land-based unit, and the MTRs established technology-based standards for the design of land-based disposal units such as landfills, surface impoundments, and waste piles. However, in 1996, the EPA proposed new rules to loosen some of the LDR and MTR regulatory requirements, meaning additional changes under RCRA are likely.

In addition to the prevention-oriented provisions of RCRA, the HSWA corrective action program created a very different and new mandate for the RCRA program: cleaning up releases from SWMUs at TSD facilities. A lay person may call the RCRA corrective actions program the "active" facility remediation program and the CERCLA remedial action program as the "inactive" facility remediation program. As described in the beginning of Section 2.3, the number of TSD facilities requiring corrective actions is quite high, outnumbering NPL sites and rivaling the number of CERCLA sites. The concept of parity between RCRA and CERCLA should be realized in the relationship between remedial solutions, because both programs deal with similar problems. While both programs differ from a regulatory aspect, there should be parity in the remedial solutions that are applied in both programs, including fast-tracking actions and decision making.

2.3.2 RCRA Corrective Actions

The RCRA corrective actions program focuses primarily on TSD facilities with hazardous waste, and the HSWA provisions are written in general terms and allow the EPA considerable latitude to develop a comprehensive regulatory program to oversee the corrective actions program (Ovenden and Nixon 1993). Owner/operators of RCRA permitted facilities must carry out, as conditions of their final RCRA permits, whatever corrective actions are necessary on SWMUs where releases have occurred. Nonpermitted facilities may also be subject to the corrective action provisions under a voluntary cleanup or may have to undertake cleanup under other state or federal regulations. The corrective actions program has many similar investigation and design components of the Superfund program, and in fact, Superfund guidance is often used as the framework for similar activities for investigation activities. The terms differ between the two programs where CERCLA has a preliminary assessment vs. an RCRA facility assessment or CERCLA feasibility study vs. an RCRA corrective measures study (CMS), for example. The science and approaches used in RCRA are similar to Superfund, but the regulatory approach to administering the program and sites is very different. From an accelerating cleanup perspective there are several important courses that should be considered for corrective actions:

- Innovative investigation and remedial technologies
- Presumptive remedies
- Corrective action management units (CAMU) and temporary units (TU) (note: the CAMU rule may become obsolete in the future under proposed rule changes)
- Interim measures
- Comprehensive contingency plans
- Site stabilization, point of compliance, and mixing zones

In general, compared to other programs, the RCRA corrective actions program is wide open for opportunity to fast-track cleanup. The regulatory framework, while stringent in some respects, is flexible when it comes to proposing and initiating cleanup. The following courses of action illustrate this point.

Innovative Investigation and Remediation Technologies

Innovative investigation and remedial technologies are discussed in Section 2.1 of this chapter. In relation to RCRA, innovative investigation approaches, such as horizontal monitoring well installation, collecting real-time data, or improved analytical methods, can be used to fast-track investigation of sites and reduce project costs. Similarly, innovative remedial technologies can be used to accelerate the remedial process, reduce the restoration time of a SWMU, and provide greater efficiency to achieve stringent action levels. Project teams should evaluate the opportunities innovative investigation and remedial technologies offer during all aspects of the project, regardless of the environmental program a site is regulated under.

Presumptive Remedies

Presumptive remedies, discussed in Section 2.2 of this chapter, is a cross-program adaptable approach for accelerating investigation activities, decision making, and cleanup. Project teams need to evaluate presumptive remedies under the RCRA corrective actions program as an avenue to focus project funds on actions and determining whether a full-scale RCRA facility investigation (RFI) and CMS are necessary, for example, at an RCRA landfill or at SWMUs were there is a VOC release.

RCRA Corrective Action Management Units (CAMU)

In 1993 the EPA relaxed some of the requirements that apply to remediation under RCRA. Also in 1993, the CAMU rule was challenged. However, the challenge has been stayed pending publication of the final Hazardous Waste Identification Rule for Contaminated Media (HWIR-media). EPA expects that the HWIR-media rule will eliminate the need for the CAMU rule, and the EPA may withdrawal the CAMU regulations as part of the HWIR-media proposal. In the meantime, CAMUs may be used to support efficient and protective cleanup. Including this discussion is important, because the strategy behind establishing CAMUs for accelerating cleanup will likely apply to future RCRA rule changes and other environmental programs to fast-track cleanup — even though CAMUs themselves may not be a viable regulatory approach in the future. When finalized, the HWIR-media proposal will likely offer project teams an opportunity to use a simple and perhaps even more flexible CAMU strategy to accelerate actions.

EPA introduced the CAMU and TU provisions for RCRA corrective actions (EPA 1993d). Three main differences between "as-generated waste" and "waste managed" at remediation sites were cited by the EPA in developing a new regulatory framework for waste generated at these sites:

- The level of regulatory oversight for remedial actions is much greater than for as-generated wastes, thus there is far less need for strict, uniform national standards that address many types of wastes being generated at a wide range of facilities.
- Subtitle C requirements often impose counterproductive constraints and disincentives on the remedy selection process at remediation sites. These requirements were developed as technology-based standards to provide incentives to generators to reduce the amount of waste being generated, whereas cleanup activities are driven by health-based standards.
- Remediation often involves management of large volumes of contaminated media, such as soil and groundwater, and the physical characteristics of contaminated media can be different from those of as-generated wastes.

The CAMU and TU rules add substantial flexibility for managing contaminated media at facilities and also provide another avenue to fast-track RCRA corrective actions at TSD facilities. Importantly, the rule serves as an ARAR at sites managed under CERCLA, and therefore can assist, in some cases, in management of contaminated media at these sites. An example of how CERCLA and RCRA are integrated under the CAMU rule is described in the integration of environmental programs subsection of this chapter (Section 2.3.5). Please note, however, that states must be authorized by the EPA to implement the RCRA program, and they may or may not have adopted the CAMU and TU rules, because these are less stringent compared to the normal waste management requirements under RCRA. Check first to find out if the state you are working for is authorized by the EPA to implement the RCRA program and whether or not the rule has been adopted by the state before proposing a CAMU or TU.

EPA defines CAMUs as "... an area within a facility that is designated by the Regional Administrator under 40 CFR Part 264 Subpart S, for the purpose of

Before Remedial Activities

The remedial goal at this facility is to treat the wastes in each of the solid waste management units and consolidate the wastes from the SWMUs in the flood plain to a more protective location.

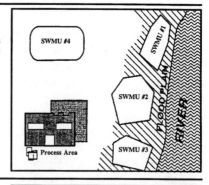

The Resolution Using CAMUs

(A) The regional administrator or state director designates SWMU #4 as a CAMU.

(B) The remediation wastes from the four SWMUs are then removed and treated in a temporary on-site treatment unit.

(C) SWMU #4 is retrofitted with a liner.

(D) The treatment residuals can be placed in the CAMU without meeting the land disposal restrictions. Specific treatment standards and other design, operation, closure, and postclosure requirements for the CAMU would be specified according to the criteria in the CAMU regulation

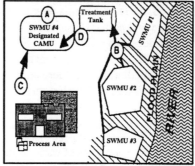

Figure 2-8 How CAMUs and TUs may be used. (Source: EPA. 1993f. EPA Issues Final Rules for Corrective Action Management Units and Temporary Units. EPA/530/R-93/0800 OSWER Publication 9234.2-25.)

implementing corrective action requirements under SubSection 264.101 and RCRA Section 3008(h). A CAMU shall only be used for the management of remediation wastes pursuant to implementing such corrective action requirements at the facility." The definition does not specify a CAMU as being contiguous areas of contamination, which has a significant impact on the applicability of LDRs on waste management in a CAMU. This allows the noncontiguous areas of contamination with similar characteristics to be managed as one unit, and waste can be moved from one contamination area to another without triggering the LDRs (Figure 2-8). Waste management in a CAMU also does not trigger other unit-specific requirements that apply to hazardous waste land disposal units. The primary benefits of CAMUs are that

- Placement of remediation wastes into or within a CAMU does not constitute land disposal of hazardous wastes.
- Consolidation or placement of remediation wastes into or within a CAMU does not constitute creation of a unit subject to minimum technology requirements (MTR).

LDRs and MTRs will not be triggered under these conditions:

- Movement or consolidation of remediation wastes within a designated CAMU.
- Placement of remediation wastes into a designated CAMU that were generated at the facility but outside a designated CAMU.
- Movement and subsequent placement of remediation wastes from one designated CAMU to another CAMU.
- Excavation of remediation wastes from a CAMU, treatment on site in another unit, and redeposition of those wastes or residuals into the CAMU.

Non-land-based units, such as tanks, that are physically located within the boundaries of the CAMU are not part of the CAMU and must meet the applicable Subtitle C tank requirements. TUs can also be located either inside or outside the physical boundaries of a CAMU, as long as they meet the applicable TU requirements (see the TU provisions below for more information).

In addition to the waste management activities outlined above, other land-based activities may be covered by the CAMU rule. For example, wastes are often excavated and staged in piles before being transported to a treatment unit. Under a CAMU, the area where the wastes are piled would not be considered a separate waste pile unit for RCRA, although technical standards, such as liners, wind dispersion controls, closure, would still apply. Similarly, areas of a CAMU could also be used for land-based treatment processes, such as bioremediation systems that involve structures or equipment to maintain optimal treatment conditions.

The following requirements must be met for CAMUs (Ovenden and Nixon 1993):

- The CAMU shall facilitate the implementation of reliable, effective, protective, and cost-effective remedies (note: evaluation of this criterion does not require a cost/benefit or other quantitative analysis).
- Waste management activities associated with the CAMU shall not create unacceptable risks to humans or to the environment resulting from exposure to hazardous wastes or hazardous constituents (note: a quantitative risk assessment is generally not required to satisfy this criterion).
- The CAMU may include uncontaminated areas of the facility only if including such areas for the purpose of managing remediation waste is more protective than management of such wastes at contaminated areas of the facility.
- Where wastes remain in place within the CAMU after closure of the CAMU, those areas shall be managed and contained so as to minimize future releases, to the extent practicable (i.e., the long-term reliability and effectiveness).
- Timeliness of remedy implementation (i.e., does it expedite the remedy selection?).
- The CAMU shall enable the use, when appropriate, of treatment technologies (including innovative technologies) to enhance the long-term effectiveness of remedial actions by reducing the toxicity, mobility, or volume of wastes that will remain in place after closure of the CAMU.
- The CAMU shall, to the extent practicable, minimize the land area of the facility upon which wastes will remain in place after closure of the CAMU.

Along with these requirements, there are specific design criteria that must be specified for a CAMU. There are also closure and post-closure requirements for CAMUs. Existing SWMUs, regulated units, and other contaminated areas of a facility may be designated by the EPA regional administrator as CAMUs or as parts of CAMUs (Ovenden and Nixon 1993). Regulated units include landfills, surface impoundments, waste piles, and land treatment units that receive hazardous waste after July 26, 1982. Designation of a regulated unit as a CAMU has two limitations:

- Only regulated or closing units may be designated. Operating regulated units, including regulated units continuing to operate under delay of closure provisions, are not eligible for designation as CAMUs.
- Regulated units may be designated as CAMUs only if doing so will enhance implementation of an effective, protective, and reliable remedy for the facility.

Regulated units designated as CAMUs must continue to comply with applicable 40 CFR part 264 or part 265 groundwater monitoring, closure and postclosure, and financial responsibility requirements. When incorporation of a CAMU is initiated by the owner/operator, it will generally be approved by the Class III permit modification process. It may be possible to include a CAMU provision "up front" in the Corrective Actions Plan during the original permit generation process as part of the interim contingency corrective action plan, bypassing the need for a full Class III permit modification process. This somewhat maverick approach, to my knowledge, has not been attempted. Designation of a CAMU as an interim status facility may be done through a section 3008(h) order or possibly through a section 7003 order.

Provisions for Temporary Units

TUs are for short-term operation of tanks and containers storage units for the treatment or storage of remediation wastes. These units may be used only for remediation wastes, and they must be located at the facility where the remediation is taking place (Figure 2-8). TUs do not include incinerators, non-tank thermal treatment devices, or units regulated under 40 CFR part 264 subpart X (miscellaneous units). The corrective action regulations provide that for temporary units, an alternative design, operating, or closure standard may be allowed rather than the standards that normally apply to permitted facilities. The following factors are considered by the EPA regional administrator for establishing standards for TUs:

- Length of time the unit will be in operation
- Type of unit
- Volume of wastes to be managed
- Physical and chemical characteristics of the wastes to be managed in the unit
- Potential for releases from the unit
- Hydrological and other relevant environmental conditions at the facility that may influence the migration of any potential release
- Potential for exposure of humans and environment receptors if releases were to occur from the unit

Community input is generally required for a TU, unless the provision for a TU were included up front in the corrective action plan or contingency plan of the facility permit. Again, this is a maverick approach and likely has not been attempted under the RCRA permitting process. In most cases, therefore, a Class II permit modification and the associated public involvement are required to designate a TU, unless the TU designation is initiated by the EPA (such as a section 3008(h) order). Wastes can be stored in a TU for up to 1 year; however, extensions are available on a case-by-case basis where

- Continued operation of the unit will not pose a threat to human health and the environment.
- Continued operation of the unit is necessary to ensure timely and efficient implementation of remedial actions at the facility.

One example of an effective use of a TU would be a central container storage area for investigation-derived waste from all remediation efforts at a facility. The storage area would have to meet several basic design and operational standards, but it would not have to meet all the design and contaminant standards applicable to typical storage areas permitted under RCRA. If the storage TU were located within a designated CAMU, investigation-derived waste could be stored for up to 1 year, or possibly longer, and then removed from storage and treated or disposed of back in the CAMU without triggering LDRs. This would allow the investigation-derived waste to be managed with other soils or groundwater being treated and disposed of on site.

Interim Measures

Interim measures are an important avenue to accelerate cleanup at TSD facilities. The corrective actions process can take years to complete and in many ways mirrors the time it takes to complete the CERCLA process for site restoration. However, RCRA interim measures can be proposed to initiate cleanup actions early in the corrective actions process, propose plume containment, supply alternate drinking water, initiate removal of significant threats at TSD facilities, and propose other interim actions. In many respects, RCRA interim measures are similar to the CERCLA early actions under SACM. Using the SACM strategy, it is important to effectively identify and propose interim measures at a TSD facility early in the corrective action process, and the scope of the RCRA facility assessments and RCRA facility investigations should include identification of potential interim measure opportunities. Once an interim measure is identified, the scope of the facility investigation should be focused on not only SWMU characterization, but also on key success factors associated with implementing the interim measure, such as evaluation of soil properties for SVE or extent of contamination for excavation activities. In addition, interim measure alternatives should be evaluated under an interim CMS or similar study to consider the effectiveness, implementability, and cost of alternatives. This would be similar to a focused RI/FS under CERCLA. After implementing the interim measures, facility compliance would generally be conditional on completing the entire RCRA corrective actions process, since interim measures are generally not the final facility remedy. However, similar to the CERCLA removal action or early action process, an interim measure could be adopted as the final remedy based on performance, regulatory compliance, and protection of human health and the environment components.

In relation to timing of interim measures, a release that is not of an emergency nature could be addressed under an interim measure assuming it is not already included in a planned corrective action. The decision to conduct an interim measure would be approved through an EPA order or follow the permit modification process. In cases where no corrective action plan is in place for the permit, the Class III permit modification process would be followed if the owner/operator requested the change. In cases where there is a corrective action plan, and an interim measure is desired, the interim measure would be added to the plan following the Class II permit modification process. However, upon request of the owner/operator, the regional administrator may, without prior public notice and comment, grant the owner/operators temporary authorization for changes in the permit. Temporary authorization must have a term not more than 180 days, although an additional 180-day term may be granted on a case-by-case basis by the regional administrator. In this case, an interim measure may be able to be proposed by the owner/operator and implemented without up-front permit changes as long as the action can be completed in 180 days or up to a total of 360 days. In cases where there must be a Class II or III permit modification, and an interim measure would have a substantial benefit to human health and the environment, the regional administrator may be inclined to approve a temporary authorization to fast-track the decision process. In cases where the EPA initiates the change, it would be done through, for example, a section 3008(h) order. Interim measure offer owner/operators an avenue to more quickly initiate cleanup actions, however, a clean understanding of the permitting an authorization process is essential.

Developing Comprehensive Contingency Plans

The permitting process under RCRA is an opportunity for owner/operators to contemplate a proactive corrective actions program as a means of reducing the overall cost of restoration. EPA can require more up-front planning and inclusion of acceleration measures for corrective actions plans under facility permits. Ideally, both the owner/operator and the EPA would seek this type of TSD facility permit, where prudent actions to protect human health and the environment can be undertaken at the time of discovery that are not an emergency-class action. Often, however, permits for facilities contain the minimum requirements under RCRA and are a "boiler plate" document based on other facility permits. In addition, owner/operators are not always in agreement with the EPA on the number of SWMUs or AOCs, making it difficult to work through the permitting process and consider acceleration measures. Also, most of the EPA regions have their own permit format they follow, resulting in differences in each region's RCRA permits, and in some cases offering less opportunity to include acceleration measures. Bringing EPA regions into alignment on acceleration measures may be an important future consideration to fully realize fast-tracking approaches under RCRA.

Expanding permits for TSD facilities, specifically the contingency plan, to include acceleration measures in both the investigation stage and the corrective actions stage can be a substantial cost-saving approach for the TSD facilities and would allow cleanup to occur more quickly. Importantly, the scope of the contingency plan must be agreed upon by both the EPA and the owner/operator ahead of time for it to be successful in reducing cost for the owner/operator and expediting risk-reducing measures or site stabilization for the EPA. Embracing the fact that cleanup will need to be addressed at some point in the future is important, and including measures to deal with releases as soon as possible reduces the time needed to complete actions, thus saving money in most cases. A consensus in the number of SWMUs, and the relative threat or potential threat they present, must be gained to develop a comprehensive, facility-wide acceleration program. In this sense, an "early action" approach could be adopted in the contingency plan of the permit, outside of the emergency procedures, to allow the owner/operator to move forward with studies and actions. The contingency plan could include provisions for non-emergency cleanup under interim measures when data, site conditions, and agreement between parties for conducting the action is achieved to support such actions. These provisions could include establishing CAMUs as necessary and avoid the Class III permit modification process; initiating plume containment actions to quickly stabilize the site and mitigate off-site migration of contaminated groundwater; evaluating mixing zones; implementing the presumptive remedy process; and other actions. In most cases, for these example acceleration measures to be included in an RCRA permit, there would need to be a "trusting" relationship between the EPA, the owner/operator, and the community. Meeting the needs of all parties and establishing this type of partnership is possible, but requires all parties to allow flexibility, (within regulatory limits) to initiate studies and actions that are in line with protection of human health and the environment, to help ensure compliance and cost-effectiveness.

Importantly, the RCRA permitting process generally involves up-front public participation. Acceleration measures or an acceleration program can be proposed to the public up front to gain their approval, thereby limiting the number of future permit modifications that would be necessary for implementing the corrective actions program. Allowing the owner/operator to act proactively may be viewed as risky by some in the environmental field, yet there are examples where it works. EPA Region X headquarters, for example, performance standards are generally included in facility RCRA permits. The performance standards establish criteria that facilities must meet under their RCRA corrective actions program. How facilities in Region X meet standards varies, yet the common goal of meeting the facility standards is usually met, and accordingly, actions are protective of human health and the environment, comply with the permit requirements under RCRA, are often cost-effective.

Balancing the needs of compliance and protecting human health and the environment with practical solutions for TSD facility corrective actions must be considered to establish a trusting relationship between all parties, including the public. Using this approach and establishing a comprehensive contingency plan for RCRA facility permits can accelerate decision making, and cleanup actions, and reduce costs.

Site Stabilization Initiative, Points of Compliance, and Mixing Zones

As the EPA and state environmental agencies have gained more experience at RCRA facilities, it has become clear that at many sites final cleanups were difficult and time-consuming to achieve and that an emphasis on final remedies at a few sites could divert limited resources from addressing ongoing releases and environmental threats at many other sites (Federal Register 1996). As a result, the EPA established the site stabilization initiative as one of the primary implementation objectives for the corrective action program. The goal of the site stabilization initiative is to increase the rate of corrective actions by focusing on near-term activities to control or abate threats to human health and the environment and prevent or minimize the further spread of contamination. With this approach, the EPA will achieve an increased overall level of environmental protection by implementing a greater number of actions across many facilities, rather than following the traditional process of pursuing final, comprehensive remedies at a few facilities. In essence, site stabilization focuses on controlling or managing contamination within a facility, rather than focusing on final site cleanup.

Controlling exposures to humans or the environment or the migration of a release can stabilize a facility, yet the facility may not be completely cleaned up. At some facilities that are stabilized, contamination is still present and additional investigation and remediation or mitigation may be necessary. However, as long as the stabilization measures are maintained, stabilized facilities should not present unacceptable near-term risks to human health or the environment. EPA together with the owner/operator of the facility now have the opportunity to shift their resources in an effort to conform to the stabilization initiative. Stabilization actions should be a component of, or at least consistent with, a final remedy approach. Once a facility is stabilized, the EPA or state regulatory agency may reduce the level of oversight to focus effort on facilities that present a more imminent threat to human health or the environment. Examples of site stabilization include

- Installation of grout curtains to control migration of contaminants in groundwater.
- Capping surficial contaminants that pose a threat to workers or a threat of groundwater contamination.
- Controlling dust migration through excavation or cover construction.
- Installation of recovery trenches that intercept floating product on the water table.
- Stressing the aquifer to control groundwater flow direction or intercepting groundwater contamination before it leaves the facility point of compliance.
- Implementing an interim measure to mitigate the level of risk to human health or the environment.

Important aspects of site stabilization include not only risk elements associated with protecting human health and the environment, but also the point of compliance, concentration limits, and establishment of a mixing zone for facilities with groundwater contamination. Decisions for site stabilization can vary greatly, depending on where the point of compliance is physically located and the respective concentration limits assigned to the facility and point of compliance. At facilities where there is a downgradient area that can be identified as a mixing zone for groundwater contamination, it may be possible to propose a site stabilization plan because dilution of groundwater contaminants may substantially reduce these concentrations within the facility. In instances were there is little or no room for mixing of ambient and contaminated groundwater to lessen the off-site impacts of groundwater migration, a final remedy is likely may be more appropriate. Stabilizing off site of the facility is in most cases not in agreement with EPA's stabilization initiative, since its purpose is to keep contaminants managed on the facility. In cases where impacts outside the point of compliance are, or will be, an issue, interim measures or corrective actions need to be considered to replace water supplies (for example), and prevent public exposure to contaminants.

In some cases, the owner/operators of facilities have purchased adjoining land for the purpose of establishing larger mixing zones. In these zones, the owner/operator mitigates the impact of contaminated groundwater migration by increasing the area of dilution, or mixing zone. In respect to the stabilization initiative, this approach has been a point of contention, both from an EPA and community standpoint, since it expands the area of contamination, rather then keeping wastes managed on the facility proper. However, in most cases, stabilization is an approach that can bring the EPA personnel and the owner/operator together to more quickly protect human health and the environment. In addition, the stabilization initiative can accelerate cleanup and in some cases substantially reduce costs of cleanup. The critical component of site stabilization is identifying approaches to efficiently and effectively manage the contamination within the facility to ensure off-site impacts are mitigated.

2.3.3 USTs and RCRA

Cleanup actions under UST programs are often prime examples for how to accelerate cleanup. There are numerous examples of UST sites where corrective actions have been initiated without delay and the respective actions were successful in substantially reducing impacts in soil and groundwater. In addition, there is sometimes "freedom" associated with UST sites allowing environmental professional experience to act quickly in comparison to, for example, an NPL site. The ability to act quickly stems from how UST programs are managed from state to state, state funding reimbursement for addressing leaking USTs, and from the aspect that most releases into soil and groundwater are similar from site to site, and are comprised of mostly petroleum hydrocarbons. In essence, a presumptive remedy approach is applied by some states for investigating and sometimes cleaning up UST sites. For example, when filling my car up with gasoline in my home state, often the hum of an SVE system is obvious in the background, which is removing VOCs from the vadose zone. SVE is exceptionally successful in removing gasoline contaminants from porous soil and is widely used to reduce contaminant levels. Importantly, common petroleum hydrocarbon contaminants such as benzene can be a serious threat to human health and the environment, and there are practical approaches used at UST sites to fast-track investigation and cleanup efforts. In addition, USTs by themselves are a major contributor to groundwater contamination, and appropriate response actions should be initiated as soon as practicable upon discovery.

In relation to RCRA and the EPA, UST programs are generally overseen by state or local regulatory agencies, who control funding and technical approach for UST restoration and compliance. However, RCRA contains specific UST regulatory requirements that need to be considered for distinguishing between new and old USTs. Requirements for new USTs (40 CFR 280) relate to

- Corrosion protection and leak detection
- Spill and overflow performance standards
- Installation and record-keeping requirements
- Reporting releases and taking corrective actions
- Insurance requirements
- Closure requirements

New USTs must meet these requirements, while older USTs must be removed or closed in place if they are not in compliance. Investigation of USTs generally follows protocol developed by each state UST program for all aspects of UST installation, operation, removal, and closure. An example of a state protocol is a requirement for a minimum number of soil samples needed for a UST removal, based on the tank size, that must be collected at predetermined locations to document whether there has been release. Opportunities for accelerated actions can be considered prior to, and during, UST removal to fast-track protection of human health and the environment and reduce cost. Information on accelerated cleanup and decision-making opportunities for USTs are discussed in Chapter 5, under state environmental programs, since the majority of UST work is through state regulatory agencies. Other examples of UST strategy for accelerating actions are in Chapter 6.

2.3.4 RCRA Waste Minimization

One of the easiest ways to accelerate cleanup action is not to be in a situation where corrective actions are necessary in the first place. The term "good housekeeping" is used by many professionals working under the RCRA regulatory umbrella to imply maintaining a well-organized facility where spills or releases are less likely because there are strict rules in place and the facility staff work hard at keeping the facility free of unnecessary wastes. Good housekeeping reduces the potential environmental liability at a facility, reduces the likelihood of an unsatisfactory EPA audit, and increases likelihood that the public will have a positive perception. Along the same lines, waste minimization should be considered at operating facilities. Incentives for waste minimization include to

- Save money by reducing waste treatment and disposal costs, raw material purchases, and other operating costs
- Meet state and national waste minimization policy goals
- Reduce potential environmental liabilities
- Protect public health and worker health and safety
- Protect the environment
- Improve public image

Of particular importance is the saving money incentive for waste minimization. Facilities undergoing RCRA corrective actions often have enormous environmental cleanup costs, and identifying ways to reduce other facility costs can be used to offset costs associated with cleanup. In this example, waste minimization is a potential avenue for cost reduction. Evaluating a facility from all aspects is helpful to accelerate cleanup, including aspects that on the surface may not be directly connected to the cleanup actions at hand. Chapter 8 information on strategy and lowering cleanup costs provides other examples of cost-saving approaches you can use to offset expensive investigation or remedial efforts.

Also of particular importance is public and employee perception. The RCRA corrective actions process generally involves public participation. In some cases, poor public perception related to a facility can result in increased cleanup costs. In these cases there is a lack of trust between the public and the facility, and a more stringent remedial plan may be opted for by the regulatory agency to address their concerns, even if the added cost is substantial and may not provide a substantial improvement in protecting human health or the environment. Unfortunately, science and practicality do not always win out. Public perception must be and is considered when making remedial decisions. Along the same lines, facility employees who enjoy their jobs and have a positive perception of the facility, fully understanding the commitment the owner/operator has made to protect the environment, are more likely to work safer and decrease the likelihood of a spill. Indirectly, waste minimization and good housekeeping can positively affect other environmental decision making, such as RCRA corrective actions. Involving the employees with waste minimization planning because they understand the day-to day-process will help instill a positive perception with employees.

The working definition used by the EPA for waste minimization consists of source reduction and recycling (EPA 1988). Of the two approaches, source reduction is usually preferable to recycling from an environmental perspective, since it eliminates wastes rather than reusing them. Source reduction and recycling are comprised of a number of practices and approaches (Figure 2-9), and a complete explanation of these is not within the scope of this book. When developing a waste minimization program, realize that all pollutant emissions into air, water, and land must be considered as part of your waste minimization program. Also, the transfer of pollutants from one medium to another is not waste minimization. For example, the removal of organics from wastewater using activated carbon, in and of itself, is not waste minimization, since the pollutants are merely transferred from one medium (wastewater) to another (carbon, as solid waste). EPA developed guidance (EPA 1988) for those who seek an opportunity in waste minimization as a way to leverage minimization efforts to positively impact operations and corrective actions at facilities. As is described in Chapter 3 for the DoD, other federal agencies also have developed ways to reduce waste and costs through waste mimimization techniques.

2.3.5 Integration of Environmental Programs

Minimizing or maximizing the number of environmental programs that affect site cleanup and decision making can play an important role in how fast a site will

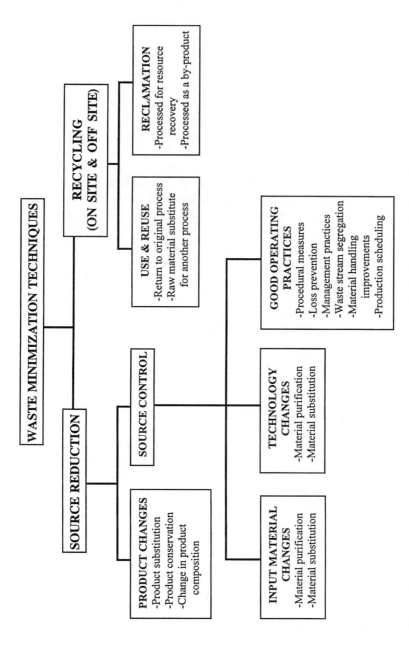

Figure 2-9 Waste minimization techniques. (Source: EPA. 1988. Waste Minimization Opportunity Assessment Manuals EPA/625/7-88/003.)

move from "cradle to grave." I have often been told that there are duplicate environmental programs and when they cross paths, it slows down progress at sites. In some cases this is true, particularly when, for example, cultural resources are considered during remedial decisions where historic railroad tracks overlie PCB contamination and there are conflicting views on whether the railroad track should be removed. However, in some cases it can be to your advantage to use one aspect of one program to investigate a site and follow up with cleanup through another program. The difficult part of cross-program integration is increasing the number of decision makers for one site. Agreement between the regulatory programs, the facility, and community is surprisingly important for actions to proceed smoothly. The following are two examples of integrating programs.

Example A — Integrating Programs

Using RCRA CAMU for a CERCLA remedial effort can be a significant cost- and time-saving approach. Examining the difference between these two programs leads to the integration of the programs. Under RCRA, "facility" generally refers to all contiguous property under control of the owner/operator of the facility that is permitted or has interim status under Subtitle C of RCRA. Under CERCLA section 101(9), however, the definition of facility is any building, structure, installation, equipment, pipe or pipeline (including any pipe into a sewer or publicly owned treatment works), well, pit, pond, lagoon, impoundment, ditch, landfill, storage container, motor vehicle, rolling stock, or aircraft, or any site or area where a hazardous substance has been deposited, stored, disposed of, placed, or otherwise come to be located, but does not include any consumer product in consumer use or any vessel. Likewise, the RCRA definition of "on site" differs from that of CERCLA. Under RCRA, "on site" refers to the larger scope of the facility, whereas "on site" under CERCLA is limited to the aerial extent of contamination and all suitable areas in close proximity to the contamination necessary for implementation of the response action. The CERCLA facility is often referred to as an area of concern (AOC). It is important to remember the differences in these terminologies when applying RCRA as an ARAR for remediation at a site regulated under CERCLA, because the RCRA requirements are applicable to the larger facility as a whole, not just the AOC. In this case a CAMU could be designated by the EPA for the purpose of treating or storing CERCLA wastes from AOCs. In this scenario, the transport of the wastes to the CAMU will not trigger land disposal restriction (LDR) requirements you would normally have to consider if the waste is moved "off site" of the CERCLA facility. Keeping the waste within the RCRA portion of the facility has many cost- and time-saving advantages. In this integration example, CERCLA wastes can be managed permanently at a CAMU without triggering LDRs or MTRs. More information is provided on CAMUs in the RCRA subsection of this chapter (see CAMUS in Section 2.3.1).

Example B — Integrating Programs

Asbestos is a CERCLA-listed hazardous substance (40 CFR Table 302.4) and could, in some cases be present in friable form with other types of CERCLA-listed hazardous substances nearby, such as tetrachloroethylene (PCE). Theoretically, in a situation where a remedial or removal action is proposed because the PCE presents a substantial threat to the environment, the action could include asbestos abatement as part of the CERCLA action. Based on the level of documentation and requirements under CERCLA, the normal asbestos abatement program should be used, since it is likely more efficient and cost-effective to complete compared to CERCLA. In this approach, separating activities under both programs is critical to propose cost-effective asbestos abatement.

Guidance for integrating programs is available from the EPA, state agencies, and sometimes local regulatory agencies. In cases where there is no guidance, you will need to review regulatory issues within the programs you are working in to determine if there is an opportunity to find common ground and whether the common ground offers an advantage to your project. An example of available guidance is EPA's Toxic Substances Control Act (TSCA) related to conducting remedial action under CERCLA with PCB contamination. While TSCA is not focused around CERCLA, it does have important regulations related to Superfund (40 CFR section 761.60 to 761.79, Subpart D: Storage and Disposal), and is an ARAR. EPA (1990d) prepared the guidance document, *"Remedial Action for Superfund Sites with PCB Contamination"* to explain how you consider TSCA as an ARAR for CERCLA sites and integrate the two programs to conduct a remedial action. Depending on your specific site conditions, programs related to air quality, storm water management, underground storage tanks (UST), radiological contamination, or other programs may offer acceleration and cost-saving avenues. Poor planning, or unanticipated integration of programs, can result in a substantial slowdown of investigation and cleanup activities.

2.3.6 Evaluating the Impracticability of Groundwater Cleanup

EPA developed guidance (1993f) for evaluating the impracticability of groundwater cleanup and identifying other alternatives to cleaning up or mitigating groundwater contamination in bona fide cases. Experience gained since the inception of groundwater restoration activities under CERCLA and RCRA shows that complete restoration of contaminated groundwater is not always achievable due to the limitation of available remedial technologies. Furthermore, complex and heterogeneous hydrogeologic settings and the presence of dense non-aqueous phase liquids (DNAPL) substantially affect our ability to clean up groundwater. Feldman and Campbell (1994) referred to these limitations as "technical impracticability from an engineering perspective," meaning we sometimes cannot design solutions for groundwater cleanup that will restore an aquifer to pristine conditions, whether it is an extraction or *in situ* treatment application.

EPA lists two major limitations for groundwater cleanup as (1) hydrogeological factors and (2) contaminant-related factors, both of which are considered when evaluating the impracticability of groundwater cleanup. The presence of free-phase DNAPL is perhaps one of the most important considerations when evaluating the impracticability of groundwater restoration, since it sinks and pools on layers and lenses of low permeability materials within hydrostratigraphic units (Figure 2-10). Those who are

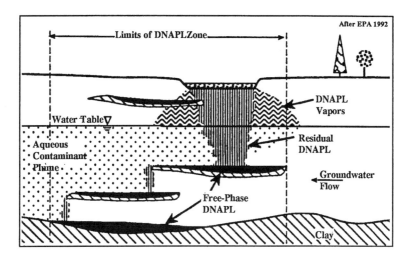

Figure 2-10 Types of contamination and contaminant zones at DNAPL sites. Remediation of DNAPL can be exceedingly difficult and expensive under many common conditions. (Source: EPA. 1992k. General Methods for Remedial Performance Evaluations. ORD, EPA/600R-92/002.)

unfamiliar with the dynamics associated with DNAPL and recovery of DNAPL should first understand the difficulties associated with characterization and remediation of DNAPLs before proposing, for example, a product recovery system. Few success stories are available for recovery of free-phase DNAPL contamination in groundwater.

To justify the technical impracticability of groundwater cleanup, there are several criteria related to site characterization and remedy performance that must be addressed in order to present a complete presentation and argument for limited action (Feldman and Campbell 1994). EPA (1993f) identifies these components as

- Identification of ARARs that must be waived in support of technical impracticability decisions.
- Identification of the aerial extent over which the technical impracticability decision will cover.
- Presentation of a comprehensive conceptual model showing the site physical and chemical conditions.
- Demonstration that contaminant sources have, or will be, identified and removed to the extent practicable.
- Presentation of remedial action performance describing the suitability and performance of any completed or ongoing groundwater remedial actions for adequacy, effectiveness, enhancement performance, and trends in concentrations.
- Presentation of the estimated time frame required to achieve groundwater restoration where longer time frames may be considered in technical impracticability decisions.
- Demonstration that no other remedial technologies or strategies would be capable of achieving groundwater restoration at the site, which may include treatability studies or pilot testing to help determine the effectiveness of remedial technologies.
- Presentation of the existing or proposed remedy options including present worth of construction, operation, and maintenance costs.

Based on these components necessary for making a technical impracticability presentation, most sites will need to go through interim or full-scale groundwater remediation efforts to determine the actual effectiveness of a remedial system, which can be extremely expensive and time consuming. In general, sites with less site characterization data would likely fall into this category. However, in some cases, a technical impracticability decision can be made prior to the remedy implementation (Feldman and Campbell 1994). These "front-end" decisions must be supported adequately by detailed site characterization and data analysis which focus on the most critical limitations for groundwater restoration. In these cases, the decisions are based on predictive performance, rather than remedy performance, and therefore the evaluation must be thorough. In cases where complete groundwater restoration is not technically practicable, the EPA will select an alternative remedial strategy that is protective of human health and the environment, and satisfies the regulatory requirements of the respective environmental programs (Feldman and Campbell 1994). Alternate remedial actions typically address two or more of the following options:

- Prevent exposure to contaminated groundwater.
- Remediate/mitigate contaminant sources.
- Remediate and/or contain aqueous plume.

Consideration of institutional controls, timing to complete actions, less stringent cleanup levels, natural attenuation, and natural flushing within the point of compliance are important aspects for alternative remedy selection the EPA considers. In all cases, future land use and migration of contaminants within drinking water supplies are critical components for the alternative remedy selection. Developing a comprehensive plan to investigate a site, documenting technical impracticability of groundwater restoration, and considering alternative remedies are vital to meet all state and federal requirements. EPA guidance (1993f) provides more detail related to evaluating the technical impracticability of groundwater restoration and should be understood before making an argument for limited groundwater restoration.

2.3.7 Additional EPA Information

Appendix A contains a list of contacts that are available through computer links and direct telephone calls. Use Appendix A as a resource guide to help identify the most innovative and cost-effective approaches in the environmental industry for investigation and cleanup activities. The environmental industry is extremely large and dynamic, and is constantly evolving as new information is reported and success stories are promoted. Information presented in this book should be augmented with the most up-to-date information you can acquire, thus ensuring you are making informed decisions. Appendix A contains many sources of information for new cleanup and investigation technologies and identifies information transfer centers available to help identify the newest regulatory changes, cleanup models and strategy, and approaches for accelerated decision making. Often, the sources in Appendix A can offer valuable insight one can use on projects.

2.4 REFERENCES

Artrip G. 1996. Superfund's Presumptive Remedy. Pollution Engineering. February. Vol 28, No. 2, pp. 48-49.

Feldman, P.R. and Campbell, D.J. 1994. Evaluating the Technical Impracticability of Groundwater Cleanup. Proceedings of the 8th National Outdoor Action Conference and Exposition, Minneapolis, MN, May 23-25, pp. 595-608.

Ovenden, T. and Nixon, J. 1993. CAMUs: EPA's New Tools for Cleanup Managers. Journal of Environmental Regulation (currently know as Environmental Regulation and Permitting), Summer, pp. 407-412.

Federal Register. 1996. Proposed Rules. Vol. 61. No. 85. May 1.

U.S. Environmental Protection Agency (EPA). 1988. Waste Minimization Opportunity Assessment Manual. EPA/625/7-88/003.

EPA. 1989a. Use of Removal Approaches to Speed Up Remedial Action Projects. July 6 memorandum from Jonathan Z. Camron (OSWER Dir. 9355.0-25A) to selected EPA environmental programs.

EPA. 1989b. Accelerated Response at NPL Sites Guidance. PB 90-258302, OSWER Dir. 9200.2-02.

EPA. 1989c. Guidance on Preparing Superfund Decision Documents. Interim Final. OSWER, EPA/540/G-89/007.

EPA. 1990a. Superfund Removal Procedures Action Memorandum Guidance. OSWER, EPA/540/P-90/004.

EPA. 1990b. A Guide to Developing Superfund Record or Decisions. OSWER Dir. 9335.3-02FS-1, May.

EPA. 1990c. A Guide to Developing Superfund Proposed Plans. OSWER Dir. 9335.3-02FS-2.

EPA. 1990d. Guidance on Remedial Action for Superfund Sites with PCB Contamination. EPA/540/G-90/007, PB 91-921206.

EPA. 1990e. Streamlining the RI/FS for CERCLA Municipal Landfill Sites. OSWER Dir. 9355.3-11FS.

EPA. 1991. Guide to Developing Superfund No Action, Interim Action, and Contingency Remedy RODs. OSWER Dir. 9355.3-02FS-3.

EPA. 1992a. Superfund Accelerated Cleanup Model — Vision for the Future. October 19 memorandum from Henry L. Longhest II OERR to all OERR personnel.

EPA. 1992b. Superfund Accelerated Cleanup Model (SACM) The New Superfund Paradigm. OSWER concept paper dated August 5.

EPA. 1992c. The Superfund Accelerated Cleanup Model (SACM). PB 9203.1-021. Vol. 1. No. 4.

EPA. 1992d. Status of Key SACM Program Management Issues — Interim Guidance. OSWER PB 9203.1051. Vol. 1. No. 1.

EPA. 1992e. Early Action and Long-Term Action under SACM — Interim Guidance. OSWER PB 9203.1-051. Vol. 1. No. 2.

EPA. 1992f. Enforcement Under SACM — Interim Guidance. OSWER PB 9203.1-051. Vol. 1. No. 3.

EPA. 1992g. Assessing Sites Under SACM — Interim Guidance. OSWER PB 9203.1-051. Vol. 1. No. 4.

EPA. 1992h. SACM Regional Decision Team — Interim Guidance. OSWER PB 9203.1-051. Vol. 1. No. 5.

EPA. 1992i. Guidance on Implementation of the Superfund Accelerated Cleanup Model (SACM) under CERCLA and the NCP. July 7 memorandum from Don R. Clay and Lisa K. Friedman of OSWER to selected EPA environmental programs.

EPA. 1992j. Evaluation of Ground-Water Extraction Remedies, Phase II. PSWER PB 9355.4-05, Vol. 1-2.

EPA. 1992k. General Methods for Remedial Operations Performance Evaluations. EPA/600/R-92/002.

EPA. 1993a. Integrating Removal and Remedial Site Assessment Investigations. EPA/540/F-93/038. OSWER Dir. 9345.1-16FS.

EPA. 1993b. Integrating Removal and Remedial Sites Assessment Investigations. October 2 memorandum from Henry Longhest II OERR to selected EPA environmental programs.

EPA. 1993c. Presumptive Remedies: Policy and Procedures. EPA/540/F-93/047. PB 93-963345.

EPA. 1993d. Corrective Action Management Unit and Temporary Units; Corrective Actions Provisions Under Subtitle C; Final Rule. OSWCAB. Federal Register. Volume 58.

EPA. 1993e. EPA Issues Final Rules for Corrective Action Management Units and Temporary Units. EPA/530/F-93/001.

EPA. 1993f. Guidance for Evaluation the Technical Impracticability of Ground-Water Restoration. EPA/540/R-93/080. OSWER Publication 9234.2-25.

EPA. Department of Energy (DOE) and Department of Defense (DoD). 1994a. Guidance on Accelerating CERCLA Environmental Restoration at Federal Facilities. August 22 memorandum from Steven A. Herman, Elliot P. Laws (EPA), Thomas P. Grumbly (DOE), and Sherri W. Goodman (DoD) to selected personnel in EPA, DOE, DoD environmental programs.

EPA. 1994b. Superfund Innovative Technology Evaluation Program Technology Profiles, Seventh Edition. EPA/540/R-94/526.

CHAPTER **3**

U.S. Department of Defense and Military Branches

As directed by the President and Congress, the U.S. Department of Defense (DoD) is the agency responsible for establishing the U.S. as the most powerful military country in the world, and protecting the U.S. from aggressors. Through DoD's efforts to protect the U.S., DoD is also responsible for a legacy of environmental contamination in both the U.S. and overseas (DoD 1993a). Substantial environmental contamination has resulted from military operations and day-to-day base operations. During war times, environmental impacts were not usually considered when making decisions related to military operations, and in many cases extenuating circumstances justified efforts that protected or saved lives but were not sensitive to long-term environmental concerns. In addition, environmental contamination is commonplace at military facilities and bases, and in some cases has been detrimental to the environment. During peace times, releases were commonplace as a result of "poor housekeeping" practices where hazardous materials and wastes were not properly stored, releases were not properly cleaned up and disposed of, or contamination was not discovered until environmental awareness became an issue. The military has become a significant agency responsible for environmental restoration in the U.S., and has undertaken the large task of characterizing past releases, cleaning up contamination, and revamping its entire environmental program to limit future releases and conserve natural resources. As a direct consequence, DoD has improved its environmental practices and is currently engaged in cleanup at approximately 4000 sites in the U.S. and overseas, of which about 100 sites are on the National Priorities List (NPL) in the U.S. Through these efforts, DoD and its military branches have been able to accelerate cleanup and decision making.

3.1 DEPARTMENT OF DEFENSE

The effort of restoring or stabilizing contaminated sites at military bases covers the entire gamut of the environmental industry related to site remediation. In relation to cleaning up contaminated bases, there are a wide variety of contaminants present in soil, sediment, and water, some of which are unique to DoD. Common contaminants at DoD facilities include

- **Fuel and solvents:** gasoline, diesel, jet fuel, degreasers, and cleaning compounds
- **Toxic and hazardous waste:** metals such as lead and mercury, caustic cleaners, dyes, paints, and strippers
- **Unexploded ordnance (UXO):** unexploded bombs and artillery shells containing high explosives (HE)
- **Low-level radioactive wastes:** radium-treated equipment such as dials and gauges

3.1.1 The Environmental Security Program

Building on the commitments of former Presidents Reagan and Bush and former Secretary of Defense Cheney, President Clinton, Vice President Gore, and Secretary Perry created the Office of the Deputy Under Secretary of Defense for Environmental Security in 1993. Their goal was to focus and energize the environmental efforts of the defense agencies and military services to fully incorporate environmental security into the U.S. defense mission. In short, environmental security is now an essential part of the U.S. defense mission and a high priority for DoD. DoD's environmental security program fulfills four overriding and interrelated goals: (1) comply with the law, (2) support the military readiness of the U.S. armed forces by ensuring continued access to the air, land, and water needed for training and testing, (3) improve the quality of life for military personnel and their families by protecting them from environmental, safety, and health hazards, and maintaining quality military facilities, and (4) contribute to weapon systems that have improved performance, lower cost, and better environmental characteristics (Perry 1995).

DoD's overall strategy for environmental security contains four components:

- Cleaning up contaminated bases
- Complying with environmental law in day-to-day operations
- Conserving natural resources
- Preventing pollution

Of the above, cleaning up contaminated bases is the primary strategic component discussed in this chapter for accelerating cleanup; however, the remaining three strategic components are, in many cases, related and integrated into the overall DoD approach for fast-tracking environmental decision making and accelerating cleanup. Through DoD's environmental security program and the Environmental Restoration Account, which Congress created specifically for defense cleanup in the U.S., there has been significant progress in the program now transitioning from initial investigations in the early 1990s to actual cleanup in the mid- to late-1990s. DoD's progress in substantially increasing the number of cleanups at their installations has been possible through DoD acceleration programs. Overall, the environmental security program has made strides in the following four general areas: (1) accelerating cleanup, (2) fostering new environmental technology, (3) establishing partnerships, and (4) involving community (Perry 1995).

Accelerating Cleanup

DoD has shortened the traditional lengthy cleanup schedules at numerous installations through the use of innovative contracting methods and close coordination with regulatory agencies. This approach has paid dividends at Marine Corps Station Tustin, CA; Naval Air Weapons Station Point Mugu, CA; and Defense Distribution Region West, Defense Distribution Depot San Joaquin (Sharpe Site), Lathrop, CA.

Fostering New Environmental Technology

DoD is reducing cleanup costs and accelerating cleanup through innovative technologies for both site characterization and treatment. Example environmental technological advances have been applied at Hickam Air Force Base, HI, and Presidio of San Francisco, CA.

Partnerships

Successful partnering efforts with other federal, state, and local agencies have streamlined the cleanup process and reduced costs by fostering early involvement, technology sharing, and jointly determined cleanup goals. Example military bases with successful partnering programs include Air Force Plant #6, Marietta, GA; Keesler Air Force Base, MI; and Naval Air Station, North Island, CA (see also Chapter 7).

Community Involvement

Current DoD policy related to community involvement calls for providing comprehensive cleanup information to local communities early in the process, and seeking community input through Restoration Advisory Boards (RAB). Community involvement has been a significant factor at Fort Wingate Depot Activity, New Mexico; Massachusetts Military Reservation, Massachusetts; and Charleston Naval Base, South Carolina.

As a result of DoD's efforts to improve their environmental programs and address cleanup at military installations, DoD has contributed significantly to the environmental industry as a whole, not only accelerating cleanup and developing innovative technologies, but providing environmental leadership, reducing costs, and fast-tracking environmental decision making. The remainder of this chapter presents example programs and approaches in which DoD and its military branches have accomplished these achievements. Not all DoD approaches, methods, technology transfer programs, and strategies are presented in this chapter. Rather, the programs that are innovative and unique are included in this chapter to augment the information contained in other chapters of this book.

3.1.2 Lead Agency Responsibility

Executive Order 12580 gives federal facilities specific authority and responsibilities under the Comprehensive Environmental Response, Compensation and Liability Act (CERCLA) as amended by the Superfund Amendments and Reauthorization Act (SARA) of 1986. The President delegated authority primarily to the U.S. Environmental Protection Agency (EPA), but has also specifically delegated authority to other federal facilities, such as DoD, with regard to releases on and from, for example, DoD's facilities and vessels. The Executive Order requires DoD to follow CERCLA in general, and CERCLA section 120 in particular, which requires that federal facilities comply with CERCLA regulations, the National Oil and Hazardous Substances Contingency Plan (NCP) 40 Code of Federal Regulation (CFR) Section 300, and guidance documents when undertaking cleanup actions.

Legally, DoD is the lead agency for cleanup on their installations and property for CERCLA-regulated spills and releases. Under CERCLA-related compliance issues, DoD acts as EPA from a lead agency decision-making aspect. Further, DoD is the responsible party for releases on military installations and must fund its own environmental actions through budgets established by Congress. Congress created the Defense Environmental Restoration Account (DERA) to fund CERCLA and other related environmental cleanup actions. DoD, in turn, established the Installation Restoration Program (IRP) under each military branch to investigate and cleanup CERCLA-related contamination in the U.S. and overseas. Under CERCLA and the IRP, DoD retains great power to move projects forward and act proactively to clean up contamination, since it controls the budgets and has the lead agency responsibility. IRP actions are also funded in some cases through the Base Realignment and Closure (BRAC) account if the military installation has been slated for closure or realignment, which is discussed later in this chapter.

Many environmental programs and laws, in addition to CERCLA, govern environmental work at military installations, such as the Resource Conservation and Recovery Act (RCRA) and the Toxic Substances Control Act (TSCA). Since DoD is a federal agency, there are other advantages (in addition to Executive Order 12580) that allow the military additional flexibility under environmental programs for fast-tracking cleanup. Perhaps one of the most significant advantages the military has over other federal or state agencies is its chain of command. Innovative technologies, accelerated cleanup, and cost-reduction strategies can, and are, mandated by commanding military personnel as part of implementation of environmental programs. Required actions or "orders" are generally acted upon immediately in the military without question. This type of discipline is unique to DoD, since enlisted and civilian personnel are often responsible for direct orders related to environmental cleanup at their installations. Poor performance, in essence, is generally not tolerated and environmental problems are acted upon in many cases, from a "must-complete" perspective, similar to combat. Another example advantage DoD has is that most of the property it uses is federally owned property, where the government retains certain rights for the activities it carries out on its property, giving them more flexibility to operate.

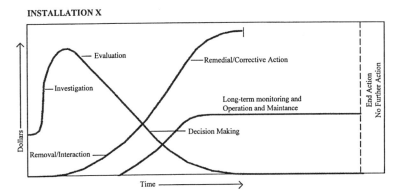

Figure 3-1 Conceptual illustration of funding requirements over time that are required for investigation, early response actions/long-term response actions, and operation and maintenance. The curves apply to a single source discovery and response action. Multiple sites and contaminants can skew such curves, because of the uncertainty encountered that requires more time and money to address. The curves illustrate the importance of up-front funding needed to move past investigation and into early and long-term response actions.

3.1.3 Budgeting as a Means of Fast-Tracking Cleanup

One avenue Congress has taken to ensure DoD conducts more cleanups and fast-track site closure is to mandate funding percentages for cleanup and investigation activities. For example, remedial or removal construction activities in 1996 were mandated as 80% of the total fiscal year funding as part of the Defense Appropriations Act. A mandate of 80% (or higher) construction activities ensures cleanup is prioritized above investigation and the majority of available dollars are spent on cleanup. The percentage of mandated cleanup dollars vs. investigation dollars varies from year to year, and a relationship of more cleanup dollars and less investigation dollars over time can be developed, for example, as an installation's IRP or underground storage tank (UST) program matures (Figure 3-1). In the beginning of a project or environmental program, most of the available funding is focused on characterizing site conditions, evaluating cleanup alternatives, and later making decisions related to short- and long-term cleanup. Dollars needed to move a project to the decision-making point wane and costs of conducting cleanup actions begin to wax; finally, investigation activities eventually cease and long-term monitoring or operation and maintenance costs bring the project or site to closure or no further action.

Mandating cleanup activities as part of a fiscal year budget forces the military remedial project managers (RPM), military program management, regulatory personnel, and supporting military contractors to be innovative and proactive. The efforts associated with identifying accelerated cleanup opportunities to meet mandated funding at sites help maintain ongoing work at sites and meet regulatory compliance requirements. More importantly, under an accelerated program, more up-front funding is needed to fund early actions, etc., and achieve cleanup more quickly. From this perspective, a constant flow of funding is critical to help meet mandated cleanup budgets vs. investigation budgets. Overall, DoD budgeting priorities as a means of fast-tracking cleanup have resulted in more sites cleaned up or closed at military installations.

One drawback to budgeting cleanup actions with priority over investigation tasks is that the approach can substantially slow cleanup at military sites where there are extenuating circumstances related to contamination or site conditions, and a more exhaustive investigation process is necessary to determine future cleanup alternatives and address threats to human health and the environment. In these cases, the respective military personnel are competing for fewer and fewer available funds for investigation activities, which could potentially result in unacceptable levels of risk to human health or the environment at some military sites. The military circumvents this potential concern by evaluating the relative risk of each site at each installation, and funding the investigation task (at non-BRAC installations) posing the greatest relative risk to human health or the environment (see also "Relative Risk Evaluation" below). Currently, the relative risk approach, in theory, should prioritize spending cleanup and investigation dollars; however, it is only partially effective, since most DoD sites are known to have a high relative risk to human health and environment because of the similarity of many of the operations. Risk should be evaluated, not only from an analytical standpoint, but also from a future land use perspective, and a level of confidence standpoint and other considerations, to accurately define the relative risk between sites and better determine funding priorities. Determining the actual risk and future land use is in itself an investigative task, which may or may not be funded at the time the information is needed. This illustrates the dilemma related to funding priorities for cleanup vs. investigation. Overall, however, earmarking the majority of available dollars for cleanup actions has the effect of forcing project teams to identify early actions and accelerate the decision-making process.

Relative Risk Evaluation

In 1993, DoD developed a risk management concept to help establish funding priorities for IRP projects, whereby the relative risk of sites is evaluated to categorize IRP sites as high-, medium-, or low-ranking sites (DoD 1993b). In turn, funding is based on the risk ranking, where high relative risk sites receive funding over medium and low relative risk sites. The ranking is based on the contaminant hazard, migration pathway, and human and ecological receptors. Overall, the concept of relative risk holds true for identifying funding priorities based on these aspects and assuming future land use is residential and the ambient or background concentrations of inorganic compounds are known. In cases where background metal concentrations are not known, there is potential for false high or medium relative risk determinations, because residential and industrial risk levels are relatively low for some metals. Natural concentrations of some metals can drive risk higher if not characterized as ambient concentrations. For example, in the San Francisco Bay area, arsenic is often a primary risk driver when evaluating all organic and inorganic analytical data at Navy installations next to the bay. Yet, arsenic is generally detected at similar concentrations on and off the military installations in the Bay area. Using arsenic data in the relative risk ranking process could therefore yield a false high-risk determination, if the determination of whether arsenic is "naturally" occurring at the detected concentrations has not been finalized.

Also, community concerns and sensitivity of sites are not considered in the risk generation equations. In some cases there may be arbitrarily high relative risk assigned using limited assumptions or unknowns (such as land use). Currently, DoD is in the process of reviewing the relative risk approach to potentially improve the process and better quantify the relative risk of sites. At closing bases, the relative risk approach is not applied, since community reuse is a fundamental factor for considering funding needs. However, relative risk can be used to determine which sites or installations require priority funding to fast-track protection of human health and the environment. Relative risk, the relative complexity of sites, community sensitivity, future land use, background chemical data, and current environmental status of sites must be considered as important components, in addition to contaminant hazard, migration pathway, and human and ecological receptors, for determining funding priority.

3.1.4 DoD's Aggressive Approach for Site Cleanup

The approaches and methods developed by DoD for fast-tracking cleanup are based on the strategy that cleanup opportunities exist during all phases of environmental projects. In order to maximize cleanup efforts, cleanup actions are initiated as soon as it is appropriate, such as CERCLA early actions or RCRA interim corrective actions developed by EPA. Currently, this type of approach is considered a nontraditional, or a proactive approach, veering away from the traditional linear methods developed in the 1970s and 1980s. Using this strategy, DoD evaluates contaminated sites by identifying the risk drivers associated with each site, such as "hot spots," and the regulatory avenues that exist for proposing cleanup actions. Identifying obvious cleanup actions that should be taken to protect human health and the environment expedites the overall cleanup schedule. The conventional approach and DoD's proactive, or "nonlinear," approach are illustrated in Figure 3-2. The conventional approach is the industry standard, where cleanup is generally initiated after the design stage is completed. Under the conventional approach, a relatively long period of time is spent on discovery, assessment, and the evaluation stages before the design stage is initiated (Payne and Ocampo 1994). Often, potential early actions are postponed until site-wide remedial decisions can be made. Typically, at least 3 years, and more commonly 5 to 10, pass before a remedial design is completed and cleanup is initiated, regardless of what regulatory framework is followed.

Utilizing DoD's aggressive or nonlinear approach, it is possible to identify early actions and implement them quickly in order to reduce risk at sites from either a comprehensive standpoint, where site closure is possible, or when addressing only part of the contamination problem, such as surface soil contamination. In general, the nonlinear strategy is consistent with how early actions are selected under the Superfund Accelerated Cleanup Model (SACM) described in Chapter 2 on EPA, which contains several other aspects associated with accelerating cleanup, such as

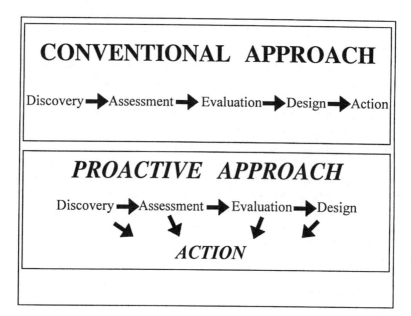

Figure 3-2 Comparison of the conventional approach and the military's proactive environmental approach. Both are scientifically based approaches; however, the proactive approach maximizes the use of regulatory avenues that fast-track response actions. (Source: Adapted from an overhead developed by the Department of the Navy, Naval Engineering Field Activity West, San Bruno, CA.)

integrated assessments and enforcement avenues. Actions such as removal actions can result in early closure of sites if the proposed actions are satisfactory in reducing risk and meet all regulatory requirements. Although simplistic in appearance, DoD's nonlinear strategy is a fundamental element of many acceleration programs described in this book. In order to work effectively, the strategy must be coupled with a strong regulatory understanding in order to identify acceleration avenues. Fundamentally similar approaches, such as SACM, also present the nonlinear approach from an implicit standpoint. It is DoD, and specifically the Navy, that has taken the initiative to graphically illustrate the fundamental concept of the nonlinear approach.

3.1.5 Base Realignment and Closure

Since the Cold War ended, the level of military need has diminished, and over time, many military bases have been slated for closure or realignment. As the number of bases slated for realignment and closure has increased since the early 1990s, the BRAC environmental process was developed to address significant potential economic impacts facing local communities with military base closings or plans for significant military downsizing. In an effort to consider not only the environmental needs of property transfer associated with closing bases, economic redevelopment and public participation were integrated into the BRAC process. From an environmental cleanup standpoint, the BRAC process is a comprehensive, cross-program approach that is designed to expedite and improve environmental response actions in order to facilitate the disposal and reuse of closing or realigning installations, while protecting human health and the environment. From a reuse standpoint, the BRAC process encourages public and local government participation to shape future reuse, participate in making cleanup decisions, identify ways to mitigate economic impacts, and identify opportunities for reuse.

Challenges under BRAC for the public, participating regulatory agencies and DoD are

- Finding ways to accelerate environmental cleanup
- Making on-site and real-time decisions at military installations to meet program goals
- Identifying impediments to decision making
- Integrating ongoing activities at bases being conducted under multiple environmental programs into a comprehensive, expedited and centralized response effort
- Ensuring that all environmental concerns at installations are addressed in an optimal manner

There must be agreement on:

- A reuse scenario that should be used to focus cleanup efforts
- Reuse priorities
- Cleanup levels
- Environmental risk
- Project schedules

In many aspects, the BRAC environmental program is similar in approach to fast-tracking cleanup under other environmental programs. However, with respect to focus, BRAC is much more comprehensive than other environmental programs, since it literally encompasses all environmental programs under one umbrella for closing installations. For this reason, the concepts and approaches DoD developed under BRAC are not fully detailed. A detailed presentation of BRAC could only be summarized in a much larger text, which is available from DoD (see DoD 1995a and 1995b). Highlights related to BRAC are provided, but a thorough review of the BRAC program and associated guidance is recommended in order to achieve a complete understanding of the program.

BRAC differs from other environmental programs in that the primary goal is property transfer, often referred to as "disposal." Under this premise, property transfer can be met after a remedial decision is shown to be effective, but not necessarily complete, where the cleanup action may continue for some time before it is completed. In addition, the future user can acquire the real property by accepting property that is impacted, without any cleanup, if they choose. The acceleration strategy developed for BRAC is therefore designed to reach the remedial decision-making stage quickly, initiating the cleanup as soon as possible, and demonstrating the action is successful. In general, the timing of completing cleanup actions is considered when remedial decisions are made under BRAC, since the military IRP (for example) it is a component of the environmental programs covered under BRAC. In cases where there are no significant reasons to fast-track site cleanup, a slower, less expensive cleanup decision may be made, as long as it is protective of human health and the environment. However, in some cases, the time it takes to clean up a site may be considered more strongly if it affects the long-term reuse of the closing base.

The Main Components of BRAC

Under BRAC, the key regulatory citations are under CERCLA 42 U.S.C. §§9601–9675 and the Community Environmental Response Facilitation Action (CERFA) of 1992 amended §120(h)(4) and (h)(5) and is codified in 42 U.S.C. §9620(h)(3)–(5). Regulatory information related to applicability and definitions for the above citations are in 40 CFR part 373 of the NCP. To address the regulatory issues associated with BRAC installation, DoD developed a process that included specific teams and documents (DoD 1995d). The process integrates the concept of expedited action as well as meeting the regulatory requirements. The primary BRAC-related teams and individuals established to help make BRAC decisions include

- **BRAC Environmental Coordinator (BEC)**: A single point of contact for environmental cleanup issues at closing bases who is generally employed by or enlisted in the military, and is involved in all, or most, aspects of environmental decision making at the installation.
- **BRAC Cleanup Team (BCT)**: A team of three environmental cleanup managers, including the BEC, as the military representative, and one state and EPA BRAC representative. The BCT is the main decision-making group for BRAC-related issues and is responsible for informing the community and other BRAC groups about issues and opportunities under BRAC. A BRAC project team of more regulatory personnel, project managers, contractors, and other as-needed personnel support the BCT and interface with other BRAC groups and individuals.
- **Base Transition Coordinator (BTC)**: A local DoD liaison that helps coordinate closure, cleanup, and reuse actions (such as property inventory, environmental cleanup, and community needs) to expedite the transition of the base for civilian or other use.
- **Office of Economic Adjustment (OEA) Project Manager (PM)**: A point of contact that provides assistance and supports the local community for redevelopment planning.
- **Local Redevelopment Authority (LRA)**: The group that represents the community in developing a plan for base reuse. The LRA is the entity recognized by DoD that is responsible for developing the community's redevelopment plan. The LRA works closely with the BCT, exchanging information on cleanup and redevelopment plans, priorities, and decisions.
- **Restoration Advisory Board (RAB)**: A local board comprised of interested community members, associations, or coalitions (DoD and EPA 1994) that comment on the proposed cleanup plan, reuse, and priorities for a BRAC installation. A member of the LRA is generally an active participant on the RAB.
- **General public**: Anyone in communities located in the vicinity of the closing installation.

To support and plan decisions made by the above teams and personnel, DoD developed several primary documents that must be generated for BRAC installations that help establish a consistent technical approach for closing bases and meeting the regulatory requirements. These documents include

- **BRAC Cleanup Plan (BCP)**: The BCP describes the status of the base environmental program and current reuse plan, identifies strategies and schedules that accelerate cleanup, summarizes funding needs to complete environmental programs necessary to transfer property, and integrates environmental cleanup actions with community reuse. The BCP is usually updated periodically to outline the most recent plan, schedules, and strategy. In terms of acceleration measures and strategy, the BCP is the most important document, since it outlines the strategy for fast-tracking cleanup for each environmental program. Examples of programs included in the BCP are the IRP, USTs, asbestos, lead paint, radon, polychlorinated biphenols (PCB), UXO, hazardous materials management and RCRA facilities, air emissions, small arms ranges, National Pollution Discharge Elimination System (NPDES) permits, lead in drinking water, radiological, and any other environmental programs that may affect base closure.
- **Local Redevelopment or Reuse Plan**: A comprehensive plan developed by the LRA that provides redevelopment alternatives, and the preferred redevelopment alternative, of real and personal property at a closing base and provides a schedule for such redevelopment.
- **Land Use Plan**: An element of the redevelopment plan that proposes land use, including zoning, and supports the community redevelopment goals. Often the land use plan is drafted prior to the redevelopment plan to help establish the potential long-term reuse alternatives.
- **Environmental Baseline Survey (EBS)**: A comprehensive review of all buildings, parcels, areas, etc., on a closing installation for the presence or likely presence of a release or presence of a hazardous substance. The environmental condition of property is categorized into seven potential listings (DoD 1995c) (Table 3-1), and recommendations are generated to further characterize existing or potential presence of hazardous substances.
- **Environmental Impact Statement (EIS)**: A comprehensive analysis of potential impacts from reuse alternatives related to the environment, and cultural and community resources. The EIS is developed under the National Environmental Policy Act (NEPA).
- **Finding of Suitability to Lease or Transfer (FOSL/FOST)**: A decision document supporting the transfer or lease of DoD property (see also Table 3-1).

The BRAC documents are designed to help plan and support base closure and reuse decisions and meet regulatory requirements. Figure 3-3 shows the relationship between generation of the BCP and other BRAC documents and BRAC decisions. To meet environmental challenges associated with this goal and to facilitate the implementation of cleanup and restoration actions, DoD developed a comprehensive guidance document (DoD 1995a) to create a consistent BCP development for all installations where integration of many environmental programs is key.

The overall BRAC disposal process is shown in Figure 3-4. With regard to accelerating cleanup and base transfer under BRAC, DoD outlines several different strategies. These include establishing partnerships and consensus between the BCT members, fostering a team approach with stakeholders, encouraging public involvement, conducting a bottom-up review of environmental programs, integrating environmental

Table 3-1 Finding of Suitability to Transfer Requirements for Notification, Convenants, and Access

Environmental Condition of Property	Notification Requirements	Covenant and Access Requirements	Relevant DoD FOST Guidance
Category 1: Areas where no release, disposal, or storage of hazardous substance or petroleum products has occurred	No notification required; may be identified under CERCLA §120(h)(4) as "CERFA-uncontaminated"	Covenant and access clauses as prescribed in CERCLA §120(h)	DoD Guidance on the Environmental Review Process to Reach a FOST for Property Where No Release or Disposal Has Occurred
Category 1 (cont'd.): Areas where only storage (less than 1 year) of hazardous substances or petroleum products has occurred, but no release or disposal has occurred	No notification of storage required	Covenant and access clauses as prescribed in CERCLA §120(h)	
Category 2: Release of petroleum but no removal or remedial action required; Release of petroleum products but required actions have not yet been implemented; Release of petroleum products and all remedial actions have been taken; or Release of petroleum products and remedial actions are underway but have not been taken.	Notification of storage, release, or disposal as prescribed in CERCLA §120(h)(1) for contracts for sale and (3) for deeds		
Category 3: Areas where storage or release of hazardous substances has occurred, but at concentrations that do not require a removal or remedial response			DoD Guidance on the Environmental Review Process to Reach a FOST for Property Where No Release or Disposal Has Occurred
Category 4: Areas where storage or release of hazardous substances has occurred, and all removal or remedial actions to protect human health and the environment have been taken			
Category 5: Areas where storage or release of hazardous substances has occurred, and removal or remedial actions are underway, but all required remedial actions have not yet been taken			
Category 6: Areas where storage or release of hazardous substances has occurred, but required actions have not yet been implemented		Not eligible for transfer by deed	
Category 7: Areas where a release of hazardous substances or petroleum products is suspended and requires further evaluation			

Note: By using categories based on environmental criteria, the military can more quickly identify the property/parcels that can be transferred at closing bases under the base closure process. Clean parcels can be transferred for redevelopment or reuse by the local reuse authorities. Progress would be much slower transferring the entire base because of potential or documented environmental problems that may affect relatively small areas.

Source: DoD. 1995. Fast Track to FOST: A Guide to Determining if Property is Environmentally Suitable for Transfer. Office of the Deputy Under Secretary of Defense Environmental Security, February.

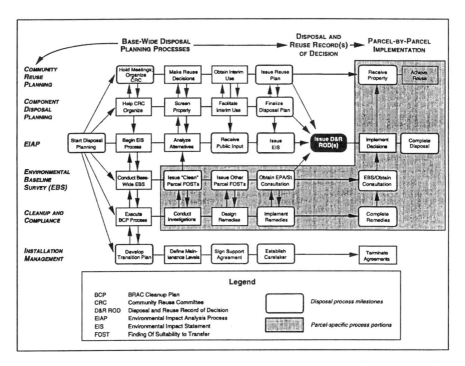

Figure 3-3 Conceptual flow chart for the base disposal process. Integration and interaction of base planning elements is essential in order to link the steps necessary to fast-track reuse of closing military bases and expedite environmental compliance and cleanup. (Source: DoD. 1995a. BRAC Cleanup Plan Guidebook. Revised BCP Guidebook. Fall.)

programs and reuse to expedite actions, and evaluating cleanup opportunities based on a need to accelerate cleanup and transfer. Some strategies are explicit in their application under BRAC; for example, public involvement in the process is explicit in that RABs are required under BRAC, and public outreach programs are encouraged by DoD. Integrating environmental programs and reuse under BRAC, on the other hand, is implicit, because each installation is in some way unique from other installations from a contaminant, site condition, and regulatory standpoint. Acceleration strategies are outlined in BRAC guidance, which help project teams integrate them into their BRAC program and, to a degree, develop a custom closure strategy. Since there is freedom to develop a custom strategy for each BRAC installation, it is the desire and ability of the BCT and project team that determines their ability to fast-track cleanup and transfer the base. A nonlinear approach or SACM strategy, for example, works well under the BRAC process, since these are focused on identifying early actions and early site closures. The conventional approach to site characterization and cleanup often impedes the closure process, unless it is the most appropriate avenue to follow due to site conditions and contaminants.

In terms of consensus and partnerships, the BCT (with help from support staff, regulatory personnel, and other individuals) is responsible for evaluating site conditions and developing decisions for cleanup and reuse. All three BCT members

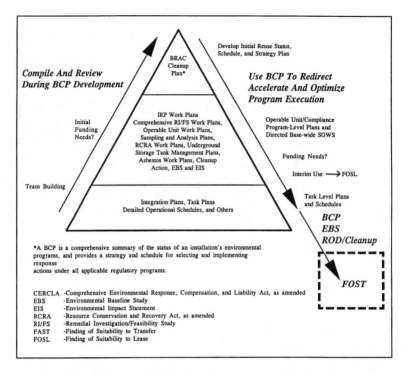

Figure 3-4 Relationship of the Base Closure Plan to other environmental plans. The Base Close Plan is the large-scale strategic plan that outlines the environmental status, integrates environmental programs, and outlines the overall environmental strategy and schedule necessary for addressing environmental issues affecting base transfer and closure. Base Closure Plans are essential because of the complexity of dealing with multiple environmental programs and community issues at closing military installations. Chapter 8 includes more information on importance of strategic planning for large projects. (Source: DoD. 1995a. BRAC Cleanup Plan Guidebook. Revised BCP Guidebook. Fall.)

must sign a consensus statement included in the BCP, which serves as an agreement to move projects forward. In reality, DoD retains all responsibilities for site restoration and is not indemnified from future cleanup at transferred sites.

The BRAC guidance for BCPs (DoD 1995a) suggests several acceleration mechanisms including source control alternatives and resultant land reuse restrictions (Figure 3-5); response mechanisms to expedite environmental restoration (Figure 3-6); methods to implement early actions consistent with SACM; establishment of zones to develop investigation strategy (Figure 3-7), and utilization of operable units (OU) to develop remedial strategies (Figures 3-8 and 3-9). DoD also recommends use of accelerated schedules to help project teams expedite review, development, and finalization of decision documents under environmental programs. In addition to cleanup under BRAC, DoD proposes using actions and approaches consistent with those presented in Chapter 2 of this book, such as early actions, interim actions, corrective action management units (CAMU), interim measures, and presumptive remedies are summarized.

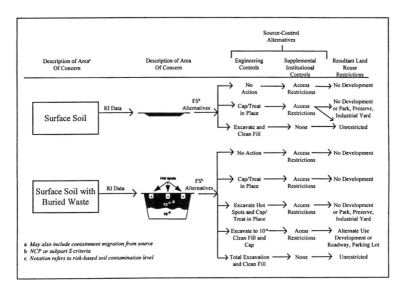

Figure 3-5 In cases where presumptive remedies are not easily implemented or applicable for sites, the project team may opt to limit or focus the range of alternatives during the feasibility stage to those that satisfy cleanup objectives. A hypothetical example of this recommendation is shown for source-control alternatives for typical source target areas and resultant land use restrictions. Depending on the engineering controls and supplemental institutional controls selected, the resultant land reuse, time needed to complete actions, and cost of the response action(s) greatly vary. In some cases, chemical, location, or action-specific standards may need to be waived for alternatives that have restrictive land use designation, yet are still protective. For this example, CAMUs and TUs, or similar future designations, help accelerate response actions and manage remediation wastes. (Source: DoD. 1995a. BRAC Cleanup Plan Guidebook. Revised BCP Guidebook. Fall.)

Summary of Accelerating Cleanup under the BRAC Process

With respect to other environmental programs, BRAC is more complicated because it is a combination of essentially all environmental issues and programs affecting military bases. For this reason, the BRAC process and associated guidance are exceedingly comprehensive and far-reaching. On the surface, DoD developed the BRAC process and guidance to cover the necessary strategy for accelerating cleanup, meeting regulatory requirements under numerous environmental programs, expediting transfer of closing bases, providing a consistent approach to close bases, and potentially stimulating economic growth. Yet, because the BRAC process is so complex, in many cases, the BRAC process is not the reason for successfully accelerating cleanup and transferring property. In fact, there are some installations where the project teams have reported that the BRAC requirements are responsible for delaying cleanup.

BRAC success is not limited to simply following the process; rather, it is a combination of (1) quickly identifying a planned or proposed reuse, (2) contracting

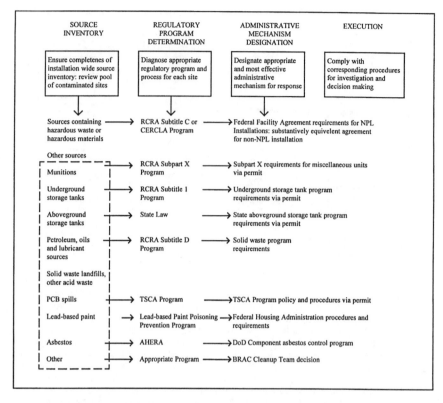

Figure 3-6 Conceptual flow chart for determining administrative response mechanisms that expedite environmental restoration and restoration-related compliance for site-specific conditions. In addition to administrative response considerations, *operational elements* including long-range scoping, well thought out statements of work, well-documented work plans, and effective contacting strategies. Together, effective administrative and operational elements can streamline and help accelerate response actions. (Source: DoD. 1995a. BRAC Cleanup Plan Guidebook. Revised BCP Guidebook. Fall.)

competent technical support to provide quickly characterize soil and groundwater conditions and cleanup alternatives, (3) establishing short- and long-term funding to complete the necessary work for cleanup and transfer, (4) ensuring that all BCT members actively take a project forward, communicate with each other and act proactively, and (5) enhancing the ability of the local community to agree on BRAC issues, reuse, and cleanup decisions. Common principles surrounding these success factors that are part of the BRAC process include teamwork, empowerment, shared goals, consensus building, community outreach, coordination of contracts and schedules, efficient procedures, and elimination of impediments (DoD undated). How these and other principles fit into the BRAC process to expedite cleanup and decision making is up to the project team. It is up to the BCT members, individual project members, and community members to uphold these and other principles, since they

U.S. DEPARTMENT OF DEFENSE AND MILITARY BRANCHES

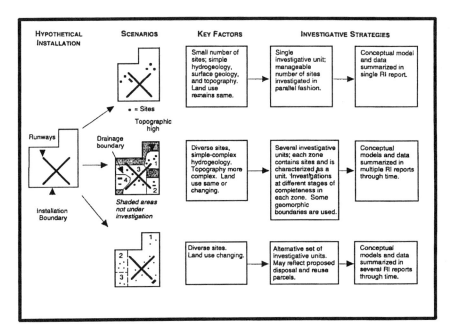

Figure 3-7 Conceptual representation of zone designations useful to manage multiple releases and sites as single investigation units. Zones define the investigation strategy. (Source: DoD. 1995a. BRAC Cleanup Plan Guidebook. Revised BCP Guidebook. Fall.)

are not requirements of the BRAC process, but values and qualities that can be shared between all stakeholders.

The success of accelerating cleanup under BRAC is therefore related to the abilities of the BCT and project team to grasp extremely complex issues related to cleanup and property transfer. The BCT and project team must organize large amounts of information into understandable formats, establish funding to complete actions under the BRAC program, establish a working trust, and work as team players. The ability to communicate and agree on reuse issues is extremely important, and can be a dynamic and volatile process, depending on where the base is located and who the stakeholders are. In summary, the BRAC process offers strategies and approaches to guide project teams in accelerating cleanup and decision making. The use of the process and its success are dependent on many variables, including those listed above, in order to meet the original program goals of fast-tracking cleanup and property transfer. The success of BRAC remains to be seen, since the program was midstream when this book was written. In all likelihood, the program will be a success, and its success will be closely related to the ability of BCT, project teams, and the community to work toward common goals. Lastly, successful implementation of the BRAC process offers other environmental programs a potential prototype to improve fast-tracking cleanup and decision making. One example program with similar goals is EPA's Brownfield Institute.

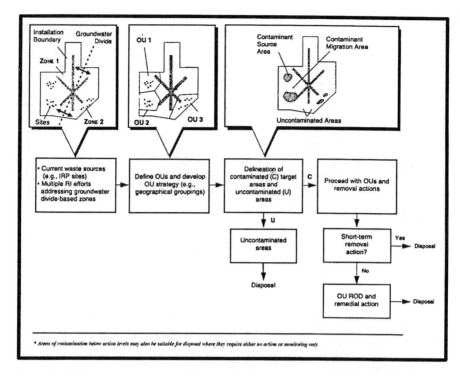

Figure 3-8 OU designations are designated to manage and integrate environmental response actions into comprehensive response efforts. OUs define the remediation strategy. OUs can include geographic, media, chemical, or other site-specific criteria that help manage and more efficiently or effectively conduct remediation efforts. (Source: DoD. 1995a. BRAC Cleanup Plan Guidebook. Revised BCP Guidebook. Fall.)

3.1.6 Establishing Partnerships and Enhancing Community Involvement

As part of several military environmental programs, establishing partnerships at the installation and program levels with regulatory stakeholders is a major component of accelerating efforts at bases. In addition, conducting community outreach programs to encourage and expand community involvement is also part of DoD's approach to running their environmental programs efficiently and effectively.

Partnerships

Partnering with agencies can be achieved either through a formal or informal process. Formal partnerships generally involve various military and regulatory personnel gathering for a 1 to 2 day period. The project team members then learn about each other and their responsibilities, and they identify a common vision or direction for their project. The process is formalized by all attending members signing an agreement that outlines the commitment to work together as a team. Facilitators may be brought in to help the project team members establish a working relationship

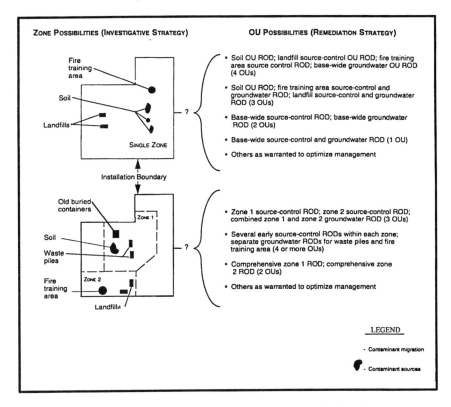

Figure 3-9 The relationship between zones and OUs for two hypothetical sites, assuming the contaminant source and migration pathways are characterized. From a strategic planning perspective, it is important to develop ahead of time an approach by which OUs, and possibly zones, are defined, sequenced, and scheduled up to the ROD, and as far beyond the ROD as possible, so they can be adjusted as investigation proceeds. (Source: DoD. 1995a. BRAC Cleanup Plan Guidebook. Revised BCP Guidebook. Fall.)

and maximize the allotted time. In the informal partnering approach, it is up to the project members to "naturally" establish partnerships through whatever means are available. This may be as simple as meeting after hours at a local establishment to get to know one another better, or reserving a meeting time to plan how their informal partnership will work. There is generally no agreement to be signed by the project team members in an informal partnership. For either of these types of partnerships to be successful, trust, effective communication skills, leadership, and teamwork are critical in order to quickly achieve agreement on difficult issues. The benefits of establishing partnerships are

- Quick identification of stakeholder concerns
- Expedited exchange of ideas for project needs
- Discernment of project direction from different viewpoints
- Fast-tracked decision making through cooperation
- Reduced cost of managing environmental programs
- Reduced likelihood of animosity between stakeholders.

3.1.7 Community Outreach Programs

DoD promotes developing community outreach programs that meet the regulatory aspects of, for example, CERCLA community relations under IRP efforts. DoD generally exceeds community relation requirements by directing project teams to develop an aggressive plan to maximize the opportunity for community participation in environmental programs. The establishment of RABs under BRAC is one innovation DoD developed for closing bases, which may also be applied to nonclosing bases in the future. Community outreach programs that go beyond the regulatory requirements can, for example, involve open houses for local residents to come visit the site, preparation of periodic newsletters on environmental progress and developments, education days where residents can learn more about the environmental program the military is working under, and lists of contracting opportunities for local businesses to bid on. The benefits of utilizing a community outreach program are

- Quick identification of community concerns
- Up front agreement or disagreement with proposed actions and decisions
- Idea generation from a community perspective
- Local economic development through contracting opportunities for businesses.

Utilization of partnering and implementation of a community outreach program does not ensure an environmental program will be expedited. In some cases, one or both are credited for the reason why a program is successful. Conversely, one or both are identified as the reason why environmental progress was impeded. There appears to be no foolproof equation to predict whether the outcome of partnering or community outreach programs will benefit individual projects. Rather, their success appears to be related to complex factors associated with the project team's ability to work together, establish trust, and community aspects that relate to a unified or divergent approach for site restoration, reuse, or transfer. In all cases, a custom partnering program and community outreach program should be developed, with consideration of the project team individuals, and sensitive community concerns, in order to help realize the benefits resulting from a partnership and community outreach program.

3.1.8 Determination of Land Use

DoD considers land use a integral component of decision making at contaminated bases and an avenue to fast-track environmental decision making. For environmental programs where risk is used to propose cleanup actions, future land use can have a significant effect on cleanup decisions and determination of remediation action levels. In the case of residential settings, the most stringent of cleanup criteria are generally necessary to ensure protection of human health and the environment. On the other hand, industrial land use generally has the least stringent of cleanup criteria, since there may be mitigating measures, such as protective clothing and limited exposure time, that allow for workers to be in areas with higher levels of contaminants. Currently, EPA's *Land Use in the CERCLA Remedy Selection Process* (EPA

1995) is the guidance document for making land use determination. Unfortunately, like many guidance documents, there are a variety of interpretations for how land use decisions are made using that guidance. Future DoD policy for land use will likely be developed to help ensure a more consistent approach for DoD facilities. In the interim, future land use decisions are developed using the existing guidance documents and site specific information.

In a base closure situation, the local redevelopment plan is used to identify future land use and focus identification of action levels based on planned or proposed land use, rather than a series of potential land uses where more stringent cleanup criteria may apply. Decisions related to future land use and accelerating cleanup often focus upon more than cleanup levels, such as institutional controls where deed restrictions and easements are attached to a property, limiting uses, or the property use during the cleanup process. In addition, zoning and permitting requirements can be modified to maintain the current or future land use. In both of these cases, less stringent cleanup criteria may be applicable which can expedite remedial decision making, and saving time and money due to less stringent cleanup requirements. However, when making decisions related to institutional controls, it is important to consider the long-term reliability of institutional controls, or if institutional controls will potentially be reinterpreted over time. Achieving the intended goal of institutional controls also depends heavily on cooperation or partnership with the federal cleanup agencies, and state and local governments that control property and zoning laws.

Future land use is a fundamental component of decision making at military bases, where risk-based cleanup is applied. Yet, integrating land use and cleanup decision making is often not inconsistent with the current guidance. Often, an overly conservative residential reuse alternative may be selected by project teams when making remedial decisions, significantly increasing the time and money needed to clean up a site. To counter overly conservative remedial decisions, the project teams must consider input from the community, future land users, and other sources as a way of ensuring realistic land use decisions that may not involve cleaning sites up to pristine or zero-contamination conditions. In addition, the project team can consider other avenues, such as institutional controls, to fast-track transfer or closure of sites, taking into account the inherent limits of these types of controls and their impact to protecting human health and the environment (see also Chapter 8).

3.1.9 Measuring Progress at DoD Installations

DoD understands that accelerating cleanup and environmental decision making can be accomplished through multiple avenues. They also understand that measuring progress and promoting success stories related to accelerating actions is as important as other aspects of their environmental programs. While it may be perfectly clear to the project team that an environmental program is being efficiently and effectively operated, it may not, however, be clear to outside sources, including the public, or the entity funding the work. By the same token, the entity funding the work will likely want to be informed if their funds are being wisely invested to maximize protection of human health and the environment and to minimize the cost of compliance with environmental laws. One example of how DoD measures progress is under the IRP. Section 120 (e)(5) of CERCLA and section 211 of SARA require that federal agencies report to Congress annually on the progress of their CERCLA cleanup programs and provide specific information. The CERCLA progress report includes, but is not limited to, the following information (DoD 1993b):

- A progress report on reaching interagency agreements between DoD and EPA for NPL sites.
- Information on the specific cost estimates and budgetary proposal involved in each interagency agreement.
- A progress report on conducting remedial investigations/feasibility studies (RI/FS) for NPL sites.
- A progress report on conducting remedial actions for NPL sites.
- A progress report on conducting remedial action at facilities that are not on the NPL.
- Identification of the number of sites contaminated with hazardous substances at each installation.
- The status of response actions underway, or contemplated, at each site.
- Cost and budgetary data on response actions.

DoD meets these basic requirements, and continually improves its approach to measuring progress, since it is directly related to funding and meeting program goals. In summary, the IRP is a major component of the military's environmental programs, and substantial funding is allocated each year by Congress to conduct IRP work. As a result, documenting progress is not only important to meet regulatory requirements, but also important to justify continued funding. Congress, the public, and other sources, such as EPA, are interested in knowing how much progress is being made protecting human health and the environment and whether or not environmental programs are being run efficiently and effectively. To date, DoD has been able to show substantial progress in moving sites through the early IRP assessment and study phases, and has been able to identify nearly all potentially contaminated sites (DoD 1993b). For IRPs that follow CERCLA, the measuring approach emphasizes CERCLA milestones, such as preliminary assessments (PA), site inspections (SI), RI/FSs, removal actions, record of decisions (ROD), remedial design/remedial actions (RD/RA), and the number of sites moving through the program.

DoD determined that using the site count to show progress under the IRP has limitations, since early progress under the PA/SI stage is achieved in about 2 years, demonstrating steps forward in a relatively short time period. The remaining CERCLA elements, including the RI/FS, ROD, and RD/RA generally take longer periods of time to complete, which can result in the appearance of limited action. To counter this perception, DoD augmented the information in their reports by adding quantifiable environmental changes to show progress (Figures 3-10 and 3-11). Quantifiable environmental measures can include the number of people relocated or provided with temporary drinking water, number of people for whom risk has been reduced by a cleanup action, the percentage of NPL sites using certain treatment technologies, and volumes of soil, liquid waste, groundwater, surface water, and

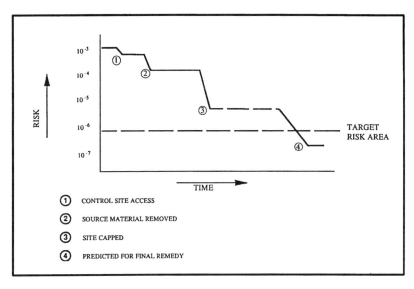

Figure 3-10 Conceptual illustration of risk reduction over time using short-term interim remedial actions, or interim corrective measures. This approach can be used to demonstrate environmental progress over time. Sequentially reducing risk and measuring the change over time is an example of a quantified environmental observations that can be used to show past progress and ongoing efforts are successful (see also Figure 3-11). (Source: DoD. 1993c. Measuring Progress in DoD's Installation Restoration Program. Rpt. 93-S-0625. March.)

sediment remediated at sites. Other measures include counting interim remedial actions (IRA) or other early actions, counting remedial actions that are installed but not complete, describing the environmental significance of procedural milestones, and listing limited or no-action decisions.

Measuring progress is not only important at military bases, but at all environmental sites when planning an accelerated or fast-tracked environmental process, regardless of the environmental regulatory arena. Often, accelerating cleanup involves more up front funding to quickly undertake cleanup actions that would be postponed under normal environmental scheduling. Unless you can document that progress is being made and that the effort accelerated the program, or yielded long-term cost savings, the person or entity funding the effort will likely reduce future funding or find someone else to do the work. How you measure progress is important if you want to go beyond the statutory reporting requirements. Outlining milestone events, decisions, documents, and success stories is critical to documenting an accelerated approach. Not only should the project team relay milestones and success stories to their client, but also to the public as well. The importance of public awareness, celebration of milestones, and communication of success stores from both a conventional and quantifiable approach, as used by DoD, cannot be overlooked in order to fully measure and report the environmental status of investigation, cleanup, and decision making at contaminated sites.

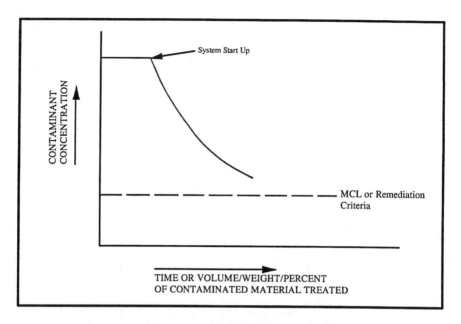

Figure 3-11 Conceptual illustration of chemical concentration reduction over time under long-term *in situ* or on-site treatment. This is an example of an environmentally quantified measure that can be used to show progress at sites undergoing long-term remediation or corrective actions (see also Figure 3-10). (Source: DoD. 1993c. Measuring Progress in DoD's Installation Restoration Program. Rpt. 93-S-0625. March.)

3.1.10 Road to ROD

In 1992, EPA and DoD joined in an interagency effort and prepared Road to ROD to promote effective and timely cleanup of hazardous waste sites under CERCLA (DoD and EPA 1992). The purpose of the guide is to help remedial project managers meet the challenges of selecting appropriate remedies for contaminated sites and to offer insight from past success and share lessons learned. The project was initiated as part of a Total Quality Management (TQM) workgroup with representatives from EPA, DoD, Department of the Air Force, Department of the Army, and Department of the Navy. Overall, the guide is not intended to be national guidance, nor does it modify any national policies or directives in the CERCLA program. However, the guide represents a joint effort between agencies and departments to fast-track decision making related to environmental cleanup. Tips contained in the guide for expediting the ROD process at federal facilities are outlined in this subsection. The ROD process, as a whole, is described in Chapter 2.

Tips to Expedite the ROD Process

The Road to ROD is presented as a pyramid of progress (Figure 3-12). In many respects, the figure is incomplete, since achieving the ROD is based on a decision,

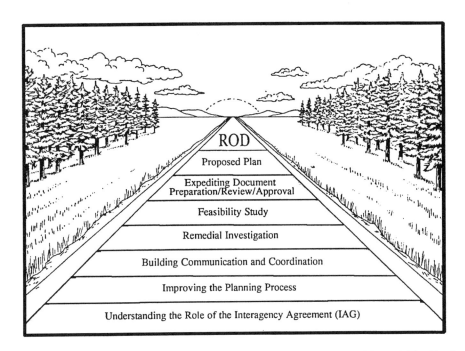

Figure 3-12 *The Road to ROD,* a pyramid of progress. (Source: DoD and EPA. 1992. The Road to ROD Tips for Remedial Project Managers. January.)

rather than cleanup or site closure. Nonetheless, the following tips, when used with other acceleration and fast-tracking measures, can expedite the CERCLA process.

Understanding the Role of the Interagency Agreement (IAG) — Section 1209 (e)(2) of CERCLA requires DoD services to enter into an IAG with EPA for the "expeditious completion" of all necessary remedial action at facilities on the NPL. In cases where an installation is not on the NPL, the installation may enter into an IAG with the state equivalent regulatory agency. Although CERCLA requires an IAG to be signed within 180 days of the completion of the RI/FS, it is more appropriate to enter into an IAG before the beginning of the RI/FS and as close to the beginning of the project as possible, thereby establishing agreement up front on the long-term schedule for the installation components of the proposed investigation and cleanup. Fast-tracking mechanisms, such as those following a SACM approach, could be included in the IAG.

EPA and DoD jointly call the IAG a "Federal Facility Agreement" (FFA), and in cases where the state is the primary regulatory agency often a "Federal Facility Site Remediation Agreement" (FFSRA). Remedial project managers rely on the FFA or FFSRA as a planning document to initiate and control the progress under the IRP, even though it is primarily a legal document. Most often, FFAs and FFSRAs list the maximum time frames needed for site activities, review, and commenting. Therefore, in terms of expediting a project IAGs are important so the project team can identify ways to expedite the project ensuring adherence to the overall schedule, teamwork, and avoidance of dispute resolution. In cases where common ground cannot be found on issues, invoking dispute resolution may be necessary to ensure progress at an installation is not impeded.

Improving the Planning Process — Planning may be the most important factor for timely completion of the ROD (DoD and EPA 1992). In terms of effective planning for the ROD process, DoD lists seven primary aspects:

- Early scoping of the initial investigation tasks through the RI/FS is important, where the scope of work for the site assessment stage is comprehensive enough to approximate health and ecological risk and potential exposure of contaminants. Postponing risk evaluation to close to the decision-making time can potentially delay the ROD preparation.
- Early identification of applicable, relevant, and appropriate regulations (ARAR) and requirements to be considered, streamlines the ROD process. ARARs should be requested early and all ARARs should be identified the first time they are requested. In cases of state ARARs, the ARARs should be requested in writing to document they are officially requested by DoD. Project managers can develop lists of ARARs, and find similar sites that have an existing ROD, and identify which ARARs are likely to apply to their site. If ARARs are not met by the preferred alternative selection, the ROD may be seriously delayed while the RPM conducts additional site investigation activities to meet particular ARARs. A complete set of initial ARARs generally avoids this potential problem. Reviewing the ARARs one last time before releasing the proposed plan (PP) is also important to ensure new requirements have not come into effect.
- Early and accurate updates of completed project milestones help the project team track project progress, and allow them to better coordinate field operations and conduct project reviews.
- Allowing sufficient time to circulate the PP for public, internal, and external review helps account for a potentially lengthy public comment period in response to the PP. Plan on approximately 12 weeks between issuing the PP and the ROD, and consider carefully the comment periods in setting a target date for the ROD to ensure sufficient internal and external coordination.
- Early planning of contractor support is important for developing not only the ROD, but also for conducting the necessary community relation activities, arranging meetings, and ensuring the administrative record and information repositories are up to date. Future contracting needs must also be considered for construction activities supporting the ROD and involvement of the construction contractor up front to assess the PP.
- Early identification of information needs and training needs is important to keep the investigation and planning process going and expedite cleanup decisions. New regulatory approaches or innovative technologies may be important to site remediation, and a thorough understanding of current regulatory information and available technologies may affect a ROD.
- Using project management tools, such as computer planning and tracking programs, is beneficial for relatively complicated sites to keep track of details that could otherwise be overlooked and slow progress. Electronic compatibility with other remedial project managers and the project team is an important factor in order to share information successfully.

Building Communication and Coordination — Fostering good communication promotes effective working relationships. Project members can communicate and coordinate effectively by viewing the ROD process as a team effort, frequently contacting other team members, coordinating closely with other agencies, and using the Technical Review Committee (TRC) as a communication tool. These aspects of communication and coordination are described below:

- Viewing the ROD process as a team approach is important in order to share in the responsibility of developing a ROD that is timely and of high quality. The constraints and goals of individual agencies, or groups, are the constraints and goals of the whole team. Overcoming constraints and reaching goals is much easier as a team. Similarly, a team approach can also enhance the public's perception of the project.
- Frequent contact and face-to-face meetings are important to help establish a team approach, identify concerns related to environmental decision making, and facilitate conversations that generate solutions. Frequent telephone calls, E-mail, facsimiles, and meetings are also important to solve technical issues. Holding conference calls every 2 weeks and monthly internal and external face-to-face meetings are examples of scheduling time with project team members. In regard to external meetings, the location of the meeting should be alternated to bring variety and fairness, and nurture good working relationships.
- Coordination with regulatory agencies is important to build early agreement between the decision makers for the ROD and on technical issues, fast-track ROD completion, and coordinate review of environmental documents. Regional approaches may be available for some technical aspects of site investigation and cleanup, such as regional ecological strategies for projects, and coordination with the applicable agencies or organizations is important to identify these opportunities.
- A TRC is formed for NPL sites, and often for non-NPL sites, consisting of community representatives, environmental groups, and local, state and federal agencies. TRCs generally meet quarterly, although monthly meetings are sometimes held, depending on the level of sensitively related to a project. Although the TRC is not a decision-making group, the TRC provides an opportunity to meet, coordinate, and communicate with all interested parties. In essence, the TRC is a sounding board for issues in anticipation of how the public will react to a PP and ROD. The TRC is therefore important to build consensus before the PP and ROD are issued, and provide early feedback to expedite the ROD processes.

Expediting Document Preparation, and Review and Approval — Project teams can expedite the preparation of the PP and ROD by encouraging EPA and other regulatory agencies to jointly prepare and review these documents. In addition, concurrent preparation and review of the PP and ROD should be used to expedite preparation of these documents.

- Joint preparation of the PP and ROD is important for expediting the decision-making process. Team members may have similar work experience that can be drawn upon to avoid common mistakes and point out innovative avenues or alternatives. Working closely with all agencies when preparing the PP and ROD can also expedite the decision-making process by utilizing the Internet, overnight

delivery, and facsimile to share information and progress between agencies and team members.
- Concurrent preparation of the PP and ROD is important to accelerate the decision-making process, where both the draft PP and draft ROD are reviewed internally simultaneously. In many cases, the nonremedy portions of the ROD can be prepared first when developing the draft PP, which offers flexibility in making final decisions and the opportunity to shorten the time needed to generate the decision-making documents.
- Concurrent external review of the PP and ROD should be used to reduce the time needed to finalize the PP and ROD, and schedule public meetings and review. In addition, a complete list of reviewers should be developed early to pursue parallel review tracks and eliminate unnecessary delay.

3.2 U.S. AIR FORCE

The U.S. Air Force has been extremely active in their environmental programs and has been able to successfully accelerate cleanup at contaminated sites and reduce the overall cost of their environmental programs. This section provides the overall approach the Air Force uses to accomplish this and example programs they operate. While the following passages outline details and actions related to some of their successes, there is one overriding aspect in their approach that is the main reason why the Air Force is a leader in environmental restoration and has been able to move forward quickly and effectively. The Air Force's success can be attributed to empowering their program leaders and RPMs, giving them latitude to aggressively pursue comprehensive actions under their environmental programs. Through the Air Force's tenacity and assertiveness, desire to share responsibility, and ability to maximize their efforts, they have successfully completed their IRP ahead of schedule at bases, reduced costs, and established themselves as a leader in the environmental industry.

3.2.1 The Air Force's Installation Restoration Program and Accelerated Cleanup

Under their IRP, the Air Force generally follows the CERCLA regulatory process at their majority of bases, and less often the RCRA regulatory process, depending on input from federal, state and local regulatory programs. In both cases, the Air Force has taken several steps to streamline the environmental process used in the IRP. Realizing the need for more up-front funding to support accelerated schedules for cleanup and property transfer at closing bases, the Air Force has made a concerted effort to document progress at their bases and identify how cleanup and decision making are accelerated. By doing so, they are able to support their funding needs, and identify new ways of showing future progress. The following approaches used by the Air Force are not only applicable to their IRP sites, but also to private, commercial, and industrial sites (Reid 1996). In addition, their approaches in most cases can be applied to other environmental programs beyond CERCLA and RCRA. Reid (1996) compiled the following approaches the Air Force uses to accelerate actions and reduce cost:

Partnering and fostering an empowered project team comprised of EPA, state representatives, local regulators, consultants, and government RPMs. Air Force partnering, either in a formal or informal sense, is the "glue" that allows a project team to work together efficiently and effectively. Empowered project team members are the building blocks that the partnering "glue" holds together, forming a strong, focused, bonded, and often proactive project team that can accelerate their environmental program and minimize costs. Consider the time it takes to review documents under the CERCLA or RCRA process. In cases where the project team is communicating effectively and working together toward a common goal, review of reports should contain or offer no new material or conclusions that were not already discussed in prior meetings or memorandums. While some details may need clarification, the overall approach is understood by the project team, making it a joint effort. In this one example, the time it takes to review major deliverables, such as an RI/FS, can be significantly reduced compared to the normal review time. Regulatory consensus at all levels before reports are submitted is critical in order for a partnership to flourish. In terms of empowering the project team to act proactively, the concept of failure has less meaning when making decisions, because it is the project team's decision, rather than multiple levels of decision makers. Utilizing this approach, the project team is more likely to consider approaches for investigation and cleanup that will fast-track protection of human health and the environment, save money, and ultimately benefit the local community, because there is incentive and self-satisfaction to be gained in helping the project, the client, and the public. In partnerships where the project team is not empowered to make decisions, members are more likely to be overly conservative in their decisions, since there is no benefit or reward associated with taking risks.

Conduct weekly information transfer meetings or prepare a brief report on project status for the project team. The Air Force encourages frequent project team communication to outline the current status and planned activities for projects. Communication can be in the form of meetings, E-mail, memorandums, as well as other formats. Transfer of pertinent information is encouraged to help the contractor performing the work to be up to date with the Air Force's needs and regulatory concerns, and also to keep the Air Force abreast of field activities and decision-making needs. This critical component is the catalyst for ensuring accelerated actions go smoothly and remain cost-effective.

Tailor risk-based cleanup for known or potential land use. Restoring site conditions to pristine conditions and pursuing a residential reuse scenario as a conservative environmental cleanup strategy is often not practical at some Air Force sites. Making practical decisions related to future land use can have a significant impact on determining cleanup levels. Cleanup levels that are based on the future intended use, planned use, anticipated use, or a needs assessment can be developed, eliminating the need for "how clean is clean" negotiations and developing multiple cleanup levels for multiple potential reuse scenarios. In cases where the location has been under industrial use, such as a mechanics shop for jets at an Air Force base (and the local setting is industrial or there is a large buffer zone between the site and residential areas), the cleanup time and remedial costs can be substantially reduced if industrial cleanup levels are pursued, assuming ecologic exposure to contaminants is not a concern. As part of the development of a plan to investigate sites and make a remedial decision, a comprehensive and concerted effort to evaluate land use is an essential cornerstone in making realistic and practical cleanup decisions.

Standardize cleanup for similar sites. In line with EPA's presumptive remedy process, the Air Force attempts to standardize remedial decisions at sites with similar contaminants and site conditions. This approach reduces the time and effort associated with evaluating multiple remedial alternatives in terms of feasibility studies (FS), design packages, and decision documents. In cases where different sites are on the same schedule, the administrative and construction effort can also be combined to save resources and simplify planning. An example would be the treatment and disposal of three different sludge ponds within an operable unit containing similar wastes. Executing the cleanup effort using a standard Air Force approach would involve using the same treatment system for all three ponds, which would reduce costs, and using a single plan and contractor to conduct the work, which also would reduce costs. If the sites are following different schedules, the planning documents for the initial action could be used in support of the following actions. In this scenario, the preparation and review times are streamlined. Additional information on the use and benefit of presumptive remedies is given in Chapter 2.

Use Interim Remedial Actions (IRAs) under CERCLA or Interim Measures under RCRA to clean up sites before contamination migrates further. In line with EPA's SACM approach, the Air Force encourages project teams to consider IRAs, removal actions, and the equivalent RCRA corrective actions to limit migration of contaminants in air and groundwater. In addition, use of early actions can substantially reduce the time required to initiate and complete restoration at contaminated sites. Postponing actions for migrating contaminants increases the cost of cleanup, because larger areas and more groundwater become impacted. Detailed information on the use and benefit of early actions is also given in Chapter 2.

Consider local weather patterns. Air Force experience has shown that weather should always be considered when developing project schedules. The Air Force's geographic spread in northern and southern latitudes can potentially negatively impact worker productivity, investigation activity, and remedial performance. Linking budgetary needs with sites that experience extreme weather is therefore important. In cases where planning does not consider weather patterns, there may be substantial down time, slow or delayed progress, or cost overruns to accommodate for adverse temperature, precipitation, or wind.

Provide adequate funding at the right time in the project life-cycle. Air Force experience has demonstrated that projects that are completed in an accelerated manner require greater up-front funding to support the effort. Project activities often include decision documents, RCRA permitting aspects, preparation of simple plans and specifications, and construction activities, in addition to the normal characterization of site conditions in the investigation stage. Costs associated with removal actions, interim measures, IRAs, fast-tracked investigative approaches, community relations, materials, services, and other anticipated activities should be included in project cost estimates and scopes of work. In cases where there is a construction contractor that will be retained to conduct the cleanup actions, funding should be requested ahead of time to ensure there is no delay in funding their effort. In turn, a measuring plan that documents progress should be used to justify the additional expenditures and effort, from both a quantifiable (relative risk) and administrative (report or stage) standpoint. Without documenting progress and publishing success stories, funding for accelerated investigation, cleanup, and decision making could be less available in the future because of inaccurate perceptions that progress and cleanup are not being undertaken as quickly as they should be.

Locate OUs by geographical area when possible. In cases where there are sites with similar contaminated media, contaminants, and location proximity, the Air Force groups the sites into an OU to simplify the investigation requirements and remedial decision-making process. In addition, at closing bases, grouping sites in OUs that are in geographic proximity to one another can streamline the transfer of property and reduce the effort required to make remedial decisions at parcels to be transferred. In some cases, the media or type of contaminants may differ significantly from site to site, and before OUs are designated due to location, other considerations related to remediation and scheduling should be considered in order to maximize streamlining and fast-tracking efforts for cleanup and site closure (see also Figures 3-8 and 3-9).

Use standard plans and specifications when appropriate and similar plans and specification for similar projects. This Air Force approach reduces the amount of time the project engineer needs to develop complete plans and specifications to do a job, thus accelerating project completion and saving money. For example, in the case of a relatively simple soil vapor extraction (SVE) or a single recovery well design, a large portion of a previous design package or standard design package is on file, and can be used to expedite development of a new design package. However, as engineering design needs become more complicated due to site conditions, contaminants, safety requirements, and many other factors, the need to customize the design for the job becomes more important. In this case, the available standard design package, or similar design package, may provide the basis for initiating the design, but the project engineer would likely have to add substantial design information to the package to identify all the information the construction contractor needs. In relation to the overall format of design packages, including bidding packages, plans and specifications, cost opinions, and other documents, standardizing each of these deliverables and identifying the minimum information is crucial to ensure all design information is presented in a consistent format. This approach allows engineers the ability to use existing or standard design packages for their work.

Develop a project schedule early in the process. With this approach, the Air Force uses short-term and long-term planning to accelerate investigation, decision making, and cleanup. Planning should be initiated in the beginning of projects to identify the plans and goals for the project team and contractors. Scheduling can also be linked to budget estimation, which will provide a "holistic" approach to project and program management, linking the day-to-day and year-to-year operations with funding estimates and needs. While IAGs establish milestone scheduling, the IAGs schedules are generally the maximum amount of time anticipated to complete the outlined tasks. A proactive project team, however, may opt to be more aggressive with the IAG scheduling. In most cases, a properly administered and managed project can usually be accelerated in comparison to the IAG schedule, assuming funding is available. In cases where the IAG does offer an accelerated schedule, the project team should recognize this and establish internal, nonbinding schedules to accelerate cleanup, if the opportunity arises. An example would be applying EPA's presumptive remedy approach (Chapter 2) through the conventional CERCLA process to shorten the time needed to initiate cleanup. In this example, an internal schedule is developed and the presumptive remedy would be completed prior to the IAG schedule that outlines the typical RD/RA schedule for site cleanup.

Develop a community relations plan early in the process. The Air Force typically develops a community relations plan early in the project to meet statutory requirements, but also to develop community outreach actions that will build trust, understanding, and support for environmental decision making. While the local community is not a decision maker, it is often a stakeholder with a strong voice and political power. Addressing community concerns late in the process can result in unwanted obstacles that slow progress and waste funds. In cases where the local community is supportive of the efforts and proposed plan, plan proposals for accelerated cleanup and cost saving actions can gain momentum, fostering a win–win scenario for all stakeholders. The community relations plan outlines the approach to meet these needs and statutory requirements for the environmental program. In cases where there are sensitive issues related to ecological concerns, base reuse, or off-site contamination, for example, great care must be given to outline how problem resolution will be handled internally and externally to keep the overall process going.

Contract with one company to provide fence-to-fence cleanup from studies through remedial action close-out. The time it takes for one Air Force contractor to understand the work completed by a previous contractor can be significant on projects that are relatively large in scope and have multiple sites. Using a single contractor reduces the review time that is needed for a new contractor to understand previous work. In addition, in cases where there is a perceived liability associated with using data collected by a previous contractor, the new contractor may opt to conduct additional investigation activities to ensure the previous work is satisfactory. For example, contractor A initiates an investigation of a large NPL site. Contractor B replaces the initial contractor A after A recompetes. In many cases, contractor B will duplicate some of contractor A's work because they fear it may be inferior work that could lead to a bad decision. This often over conservative approach is costly and time consuming to clients, yet it happens, and leads to more investigation effort. If the same contractor is used, the institutional knowledge of the site investigation team is more easily passed on to the design team and available to the remedial action team. An example of this type of approach is discussed in the next section regarding the Air Force Center for Environmental Excellence (AFCEE), which provides a broad range of services to its clients. More importantly, however, the topic of fence-to-fence cleanup from study through site closure must consider the potential conflict of interest resulting when a single contractor does all of the work. Investigating sites, proposing a cleanup, and conducting the cleanup at sites has historically been divided between study and action, since the cleanup effort could be artificially enhanced beyond the needed action for self-gain. While this is not typical, nor ethically sound, the potential for conflict of interest must be considered at all times. A thorough review of investigation findings, proposed cleanup, and oversite of construction activities is critical, from both an internal and external perspective, to ensure environmental actions and proposals are reasonable.

Innovative technologies should be used whenever appropriate. The Air Force encourages use of innovative technologies related to investigation and remediation in lieu of traditional technologies to save time and money. In addition, using innovative technologies helps fast-track broad use of these demonstrated innovative technologies. Examples of investigation technologies include immunoassay and field kits to help screen contaminants more quickly, horizontal monitoring wells to sample underneath structures, and direct push/cone penetrometer-related testing. Examples of remedial innovative technologies include proprietary soil washing techniques, integrating SVE and *in situ* bioremediation, heat-enhanced SVE, and microwave induction. Evaluation of innovative technologies should include consideration of site conditions, type of contaminants, the time and cost savings each technology offers, and all other interrelated aspects of a project in order to determine whether an innovate technology has potential at a site. In addition, the level of certainty whether the technology will successfully meet project goals should be considered, since the risk of failure may be greater with some innovative technologies if they are relatively new. The question, "How proven is the technology?" must be asked when considering innovative or new technologies.

Move sites requiring No Action or No Further Action (NFA) off the active site list as soon as possible. The accounting cost, management time, and potential liability of keeping sites posing no risk on an active site list can be expensive. Often, Air Force sites that are considered closed by the project team have to be revisited (which takes time and resources to accomplish) when new stakeholders are introduced to a project. In addition, reducing the number of active sites at an installation (or facility) demonstrates progress and helps simplify the environmental program by limiting the number of sites the project team has to manage. Consequently, formally documenting no action or NFA through the ROD or equivalent decision-making process is useful to document decisions. Documenting no action or NFA is a milestone event and a key measurement, illustrating that the environmental program is working efficiently and effectively. Lastly, formally documenting a no action or NFA decision helps the public participate in environmental decision making through the public meeting and comment process.

Consider conducting third party reviews on project progress, strategy, and technical approach. The Air Force reviews projects from a third-party perspective to ensure the current and proposed actions are prudent and effective. Professionals who are not connected with either the contractors conducting the work or the base itself are contracted to perform this task. This approach results in honest appraisals and can help identify more efficient and effective technical approaches, weaknesses in technical approaches, acceleration opportunities, cost-saving avenues, regulatory issues, and an overall review of the project direction. In some cases, a project team can be too close to a project, and big picture considerations may be missed that can have impact on progress. Undergoing a third party review helps keep a project on track and generates recommendations for time- and cost-saving opportunities.

Use Air Force software to estimate costs for all aspects of the CERCLA and RCRA process. The Air Force funded the development of software to help estimate costs of environmental projects. The software, known as the Remedial Action Cost Engineering and Requirements (RACER) system, estimates the cost of PA/SIs, RI/FSs, RD, RA, operation and maintenance, and long-term monitoring. The system was fielded in 1992 and is currently available in Windows™ format. Analysis of the software shows it to be accurate and quick; for instance, costs related to several hundred sites can be estimated within a few weeks in extensive detail. The Air Force recently allowed the software to be sold to non-DoD customers, such as environmental consultants and industry. As of 1996, there were over 1000 registered users, many of them commercial consultants.

Management Action Plan Guidebook

The preceding section outlines how the Air Force is accelerating the cleanup process and reducing costs. This section outlines how the Air Force transfers information and policies from the program level to the project, or base, level. While all of the military branches are able to transfer information relatively efficiently and effectively, the Air Force has been proactive in developing comprehensive plans and guidebooks for RPMs and project team members to use to develop accelerated cleanup strategies and expand their environmental knowledge and know how. The Management Action Plan (MAP) Guidebook (USAF 1992) was one of the first comprehensive guidebooks developed by a military branch. In many respects, the MAP Guidebook is similar in content to the BRAC Cleanup Plan Guidebook (first released in 1993), which was used, in part, to generate the BRAC guidance. The actual contents and strategy outlined in the MAP Guidebook are similar to those of the BRAC guidance outlined earlier in this chapter. The reason for discussing the MAP Guidebook in this chapter relates directly to the importance of how the Air Force chooses to communicate the goals and objectives of fast-tracked cleanup and decision making to their RPMs and at their installations.

With over 200 bases in the IRP, the Air Force recognizes that by providing environmental leadership from the program level, and by empowering the RPM/project team for each installation to make proactive decisions, the overall goal of expediting the program is realized. In many respects the Air Force has accomplished this goal by communicating at the program level and through the approaches listed in the previous sections. Communicating goals, actions, and examples from the program level is critical in order to foster project leadership, competent technical direction, and meet regulatory requirements at environmental sites. In the area of fast-tracking environmental response actions and decision making, information must be transferred from the program level to the project level efficiently and effectively, mainly because there are more unknowns following an innovative or accelerated approach as compared to traditional approaches for site investigation and cleanup.

Overview of the MAP

The goal of the MAP Guidebook is to help RPMs and project teams develop a comprehensive and consolidated environmental strategy, a master schedule for environmental restoration, and an outline for how the project team should prepare a macro-level environmental status summary and strategy Action Plan. The overall emphasis of the Action Plan is expediting and improving environmental response actions at Air Force installations. The Action Plan is not an enforceable plan, but provides the basis for meeting or modifying deadlines in enforceable agreements, and/or deadlines internal to an installation. Thus, the Air Force recommends establishing a framework in order to develop a comprehensive and consistent approach to plan, execute, and track accelerated response actions at installations, which then benefits the project from an action standpoint and the program from a measuring success standpoint.

Other Air Force guidebooks have been prepared, such as the *Installation Restoration Program Remedial Project Manager's Handbook* (USAF 1993), which reinforces the Air Force's desire to communicate and encourage proactive decision making in their environmental programs. The program communication used by the Air Force to transfer information to the project level can be applied to any large environmental program comprised of hundreds or thousands of sites. More importantly, communicating the policy for accelerating actions and decision making is essential to the widespread success of an environmental program, such as the Air Force's. In addition, communicating available alternative avenues through which the project teams can lead projects is important to assist project teams in working within regulatory constraints for environmental programs and understanding the flexibility within the regulatory programs in order to address site-specific concerns.

3.2.2 Air Force Center For Environmental Excellence

There is an important relationship between accelerating cleanup and decision making, and utilizing comprehensive services that offer leadership, help plan environmental actions, integrate environmental programs, oversee actions, develop project and program-level strategy, design remedial technologies, and understand the environmental regulatory framework, and other aspects associated with acceleration measures. The military as a whole understands this relationship and through its various military branches has several groups and divisions that provide environmental services. Examples include the Army Corp of Engineers and AFCEE, to name a few. AFCEE is pertinent to this discussion because the center represents a concerted effort on the part of the Air Force to offer world-class environmental services that accelerate cleanup and environmental decision making. In the 1980s, the environmental industry experienced the birth of literally thousands of environmental cleanup and investigation businesses. In the 1990s the environmental industry experienced the consolidation of a considerable number of these private environmental consulting and contracting firms as the needs of the industry evolved. This business move orchestrated by environmental executives is in large part related to developing comprehensive services or "one-stop shopping" for their clients. The concept of doing business faster and better is also indirectly related to this consolidation. The military has taken a similar approach by establishing "one-stop shopping centers" such as AFCEE.

AFCEE was formed in 1991 and provides a complete range of environmental, architectural, and landscape design planning and construction management services and products. Over 400 people work for AFCEE in technical, scientific, and support fields. Work can be contracted through AFCEE by utilizing an applicable military contract vehicle and exchanging funding through the military funding interchange process. Their mission is to provide Air Force commanders with a full range of "cradle to grave" services in environmental design, planning, and construction management. Their vision is to provide clients with all the services they need, when they need them, in order to give them the finest services and products at a reasonable cost, and in the process be the easiest to work with. Lastly, AFCEE promises accountability, meaning they work for the client and they listen to the client — with the stipulation that the client is the reason why AFCEE is in business. Although their mission and vision statements do not spell out the goal of accelerating cleanup, their past accomplishments and ever-increasing reputation as a extremely capable agency have proven AFCEE to be extremely successful in reducing costs and accelerating cleanup and decision making. AFCEE has received praise from not only the Air Force, but also DoD and EPA. AFCEE environmental services include

- **Public Affairs**: fact sheets and news releases
- **Construction Management**: products, services, guidance, and reference materials
- **Design Group**: architecture, urban planning, interior design, computer aided design, and geographical information systems
- **Pollution Prevention**: information clearinghouse and transfer, workshops, software, services, and conferences
- **Environmental Conservation**: environmental impact analysis, natural resource management, cultural resource management, and air quality management
- **Investigations and Studies:** preliminary assessments, site investigation, remedial investigation, data validation, technical field oversight, feasibility studies, human health risk assessment, ecological risk assessment, RCRA facility studies, treatability studies, groundwater modeling, and EBSs
- **Design and Cleanup**: design and validation review, remedial action, RCRA corrective action, UST removal and replacement, and low-level radioactive waste
- **Program Support**: program development support, program/peer reviews, plan support, community relations support, administrative record support, and related programmatic support
- **Restoration Technology Research**: research and development, development and demonstration, field demonstration, validation of field results, preparation of protocol and users' manuals, and transfer to field
- **Long-Term-Operation/Monitoring**: wells, air, and remedial equipment
- **Products**: Air Force background concentrations, AFCEE quality assurance project plan (QAPP), AFCEE model field sampling plan (FSP), AFCEE model work plan, and the AFCEE tool kit Bioscreen — Natural Attenuation Decision Support System
- **Mission Support**: Installation Restoration Program Information Management System (IRPIMS)
- **Quality Office**: training, benchmarking, metrics, and surveys award program information
- **Environmental Contracting**: environmental services branch, base restoration branch, and base closure branch
- **Regional Environmental Offices:** eastern region in Atlanta, central region in Dallas, and western region in San Francisco.

The broad spectrum of services AFCEE offers is extremely beneficial for taking a fast-track approach for cleanup and site closure at large and complex sites and bases. Contracting through several government or private entities can be slow and troublesome. Often, scope-of-work issues arise or funding is a problem. In these situations, the entire environmental program can be slowed because of one contact vehicle experiencing problems. For example, a contractor is hired to do the construction phase of a removal action. Due to new information found in the field, the fixed price contract scope-of-work for the construction contractor must be modified. Since the unforeseen conditions are relatively large enough in scope, and costly, the construction contractor must stop work until the funding issue can be resolved. While contractors may do a fine job considering site conditions and identifying cleanup actions, the project can be slowed down because of a contracting problem. In this example, the change in scope could be as simple as increasing the volume of treated soil. Under a comprehensive contract, where the contract is a cost contract, for example, the work is completed by one entity the additional soil volume is more easily dealt with by one entity.

In most cases, with the exception of extremely specialized areas of expertise or when using proprietary products, utilizing a comprehensive service is faster in the long-term to complete efforts compared to a piecemeal contracting approach. AFCEE is one example of an organization that can approach a project from a comprehensive service and product standpoint. The following examples outline some of AFCEE's efforts and technical expertise including intrinsic remediation and environmental decision making, measuring environmental progress under the Air Force IRP, and accelerating cleanup through innovative technologies. These examples include practical application, decision making, program integration, and technical application for fast-tracking the Air Force's environmental programs as well.

Intrinsic Groundwater Remediation

Accelerating cleanup and decision making includes evaluating the need to clean up groundwater contamination. Prior to making a remedial decision or allocating substantial funds to initiate an action, such as pump and treat, the need for such a system must be substantiated. In the past, groundwater that exceeded regulatory criteria or standards was almost always assumed to require cleanup using a remedial technology to reduce or eliminate the threat of groundwater contamination. Through extensive trial and error with groundwater remediation systems, however, it has become clear that systems such as pump and treat are often not as effective as predicted and do not always meet the original goals and objectives agreed by the stockholders. For this reason, alternative and innovative ways of remediating groundwater contamination must be considered to accelerate cleanup and decision making at these types of sites.

AFCEE, in association with EPA, developed technical protocol for data collection, groundwater modeling, and exposure assessment to support intrinsic remediation, sometimes referred to as natural attenuation. AFCEE's technical protocol

(USAF and EPA 1995) provides project teams with methods, strategies, and case studies to evaluate specific site conditions and consider long-term monitoring of fuel-hydrocarbons in groundwater exceeding regulatory criteria or standards. The protocol helps standardize the approach in which a project team determines whether the dissolved-phase fuel of hydrocarbons in groundwater will complete its associated exposure pathway. In many cases, it is possible to show that natural degradation processes will reduce, and sometimes eliminate, concentrations of contaminants in groundwater to levels below regulatory criteria or standards. The technical protocol was primarily developed for the Air Force, its contractors, regulatory personnel, and others working with the Air Force; however, other scientists, consultants, regulatory personnel, etc., that are charged with remediating fuel-hydrocarbons in groundwater have been very active in applying the Air Force's approach for intrinsic remediation. Most importantly, however, the technical protocol does not represent EPA or state guidance, does not replace any existing guidance, and does not apply to any contaminant, or mixtures thereof, other than fuel hydrocarbons in groundwater. Rather, it is another tool developed by AFCEE that allows practitioners to evaluate intrinsic remediation alternatives in feasibility studies.

Intrinsic remediation is achieved when naturally occurring attenuation mechanisms, such as biodegradation (aerobic and anaerobic) bring about a reduction in the total mass of a contaminant dissolved in groundwater. In some cases, intrinsic remediation will reduce dissolved-phase contaminant concentrations to levels below regulatory limits before a plume reaches potential receptors (USAF and EPA 1995). As of 1995, the Air Force had fully or partially implemented the intrinsic remediation alternative at 40 sites, with a cost ranging between $100,000 and $175,000 to implement their protocol, which is now the first option the Air Force evaluates if site conditions warrant an appraisal. Advantages of intrinsic remediation include (1) contaminants are ultimately transformed into innocuous byproducts such as carbon dioxide and water, (2) intrinsic remediation is nonintrusive and allows continuing use of infrastructure during remediation, (3) engineered remediation technologies can pose greater risk to potential receptors than intrinsic remediation because contaminants may be transferred into the atmosphere, making them more available to receptors, (4) intrinsic remediation is less costly than currently available remedial technologies, such as pump and treat, (5) intrinsic remediation is not subject to limitations imposed by the use of mechanized remediation equipment, and (6) fueled hydrocarbons that are most mobile and toxic are generally the most susceptible to biodegradation. Intrinsic remediation is subject to the following limitations: (1) there may be natural and institutionally induced changes in the local hydrogeologic and chemical conditions, and future releases, (2) aquifer heterogeneity may complicate site conditions and performance, and (3) the remedial process is subject to the natural speed at which contaminates are transformed, which may be relatively slow in some cases.

To support implementation of intrinsic remediation, the property owner must scientifically demonstrate that degradation of the site contaminants is occurring at rates sufficient to be protective of human health and the environment. Three lines of evidence can be used to the support intrinsic remediation, including

- **Documented reduction or loss of contaminants at the field scale.** This line of evidence involves the use of statistically significant historical trends in contaminant concentrations of biologically recalcitrant tracers found in conjunction with aquifer hydrogeologic parameters such as seepage velocity and dilution to show that a reduction in the total mass of contaminants is occurring at the site.
- **The use of contaminant and geochemical analytical data.** This line of evidence involves using chemical analytical data in mass balance calculations to show that a decrease in contaminant and electron acceptor concentrations can be directly correlated to increases metabolic byproduct concentrations. This evidence can be used to show that electron acceptor concentrations are sufficient to degrade dissolved-phase contaminants. Solute fate and transport models can be used to aid mass-balance calculations and collate information on degradation.
- **Direct microbiological evidence.** The last line of evidence involves conducting studies of site aquifer material under controlled conditions in the laboratory to show that indigenous biota are capable of degrading site contaminants.

For these three lines of evidence, AFCEE's protocol helps project teams develop a technical course of action that allows the converging lines of evidence to be used to scientifically document the occurrence of and quantify rates of intrinsic remediation. Ideally, the first two lines of evidence should be used to demonstrate intrinsic remediation. To further document intrinsic remediation, direct microbiological evidence can be used. Such a "weight of evidence" approach will greatly increase the likelihood of successfully implementing intrinsic remediation at sites where natural processes are restoring the environmental quality of groundwater contaminated with fuel-hydrocarbons.

Intrinsic remediation results from the integration of several subsurface attenuation mechanisms that are classified as either destructive or nondestructive. Destructive processes include primarily biodegradation, and also abiotic oxidation and hydrolysis. Nondestructive attenuation mechanisms include sorption, dispersion, dilution caused by recharge, and volatilization. Adequately characterizing the site during the investigation phase is an important step for documenting intrinsic remediation and quantifying the attenuation rates. At a minimum, the site characterization phase should provide data on the location and extent of contaminant sources comprised of non-aqueous phase liquids (NAPL) (hydrocarbons present as mobile NAPL and in sufficiently high saturation to drain under the influence of gravity to a well) and residual NAPL (NAPL occurring at immobile residual saturations that are unable to drain to a well by gravity); the location, extent, and concentration of dissolved-phase contamination; groundwater geochemical data; geologic information on the type and distribution of subsurface materials; and hydrogeologic parameters such as hydraulic conductivity, hydraulic gradients, and potential contaminant migration pathway to human or ecological receptors.

The data collected during site characterization can be used to simulate the fate and transport of contaminants in the subsurface and to predict the future extent and concentration of the dissolved-phase plume. Several models, such as Bioplume II (Rifai et al. 1988) have been used successfully to model dissolved-phase contaminant transport and attenuation (see also Chapter 6). Modeling intrinsic remediation has three primary objectives:

- To predict the future extent and concentration of a dissolved contaminant plume by modeling the combined effects of advection, dispersion, sorption, and biodegradation.
- To assess the potential for downgradient receptors to be exposed to contaminant concentrations that exceed regulatory levels intended to be protective of human health and the environment.
- To provide technical support for the intrinsic remediation option at postmodeling regulatory negotiations.

Upon completion of the fate and transport modeling effort, model predictions can be used in an exposure pathways analysis. If intrinsic remediation is sufficiently active to mitigate risks to potential receptors, the project team has a reasonable basis for negotiating this option with regulators. The exposure pathway analysis allows the project team to demonstrate that potential exposure pathways to receptors will not be completed.

Detailed information on intrinsic remediation is presented in AFCEE's technical protocol (USAF and EPA 1995), which includes methods of site characterization for documenting intrinsic remediation, processes affecting contaminant concentrations and the use of fate and transport models, and general data interpretation and calculations. In addition, the technical protocol contains information on negotiations with regulators and monitoring and verification, and presents detailed cases studies. Other comprehensive information on intrinsic remediation includes *Symposium on Intrinsic Bioremediation of Ground Water* (EPA 1994). The following is a brief summary of the Air Force's protocol for implementing intrinsic remediation.

Protocol for Implementing Intrinsic Remediation

The primary objective of the intrinsic remediation investigation is to show that natural processes are reducing contaminant concentrations in groundwater, which will eventually meet regulatory standards or criteria before exposure pathways are completed (USAF and EPA 1995). This requires projecting the potential extent and concentration of the contaminant plume in time and space. The projection should be based on historic variations in, and the current extent and concentration of, the contaminant plume, as well as the measured rates of contaminant attenuation. Since there is an inherent uncertainty associated with such predictions, it is the responsibility of the project team to provide sufficient evidence to demonstrate that the mechanisms of intrinsic remediation will reduce contaminant concentrations to acceptable levels before potential receptors are reached. This requires the use of conservative input parameters and numerous sensitivity analyses so that consideration is given to all plausible contaminant migration scenarios. When possible, both historic data and modeling results should be used to provide information that collectively and consistently supports the natural reduction and removal of the dissolved contaminant plume.

Predicting the future extent of a contaminant plume requires quantifying groundwater flow and solute transport and transformation processes, including the rate of natural attenuation. Quantification of contaminant migration and attenuation rates and successful implementation of the intrinsic remediation option requires the implementation of eight steps, outlined below and in Figure 3-13. For detailed information

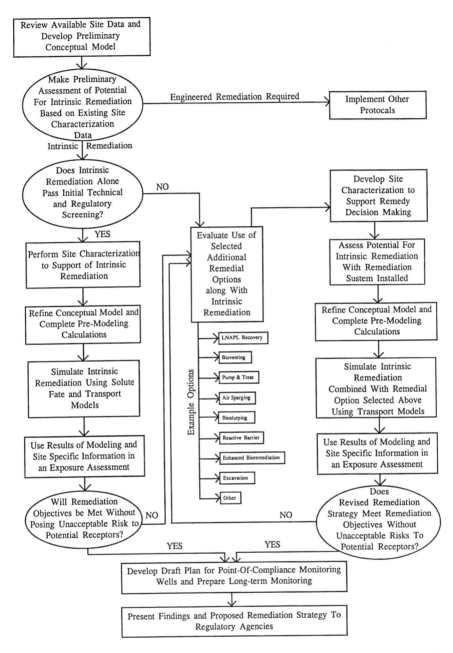

Figure 3-13 Intrinsic remediation flow chart. (Source: USAF et al. 1995. Technical Protocol for Implementing Intrinsic Remediation with Long-Term Monitoring Option for Natural Attenuation of Fuel Contamination Dissolved in ground water. Revision 0. Vol. I and II. August.)

on implementing intrinsic remediation, the reader should refer to the Air Force's intrinsic remediation protocol (USAF and EPA 1995).

1. **Review existing site data.** Data review should include evaluation of soil and groundwater quality data from a horizontal and vertical perspective in time and space. As part of this review, all geologic and hydrogeologic data and locations of potential receptors for groundwater and surface water should be evaluated. In cases where there are no existing data, future site characterization activities should include collecting the data necessary to support intrinsic remediation. Presentation of the data for evaluation and reporting purposes should also be presented in graphical displays outlining both the horizontal and vertical perspectives.
2. **Develop a preliminary conceptual model and assess potential for intrinsic remediation.** After reviewing existing site characterization data, a conceptual model should be developed and a preliminary assessment of the potential for intrinsic remediation should be made. Developing a model involves defining the problem from typical conceptual model perspective using site conditions, but also includes integration of the location of contaminant sources, release mechanisms, transport pathways, exposure points, and receptors. Successful development of conceptual models typically involve defining the problem to be solved, presenting and integrating of all site data (see 1 above), and determining of specific data requirements (boreholes, sampling and analysis, etc.) to support assessment of intrinsic remediation. An assessment of the potential for intrinsic remediation is made after the conceptual model is developed. This is achieved by estimating the migration and future extent of the plume based on site conditions and consideration of biodegradability, aquifer properties, groundwater velocity, and the location of the plume's leading edge and contaminant source location relative to potential receptors. Consideration of the exposure pathway, existing data, and potential for biodegradation are the primary components of the assessment. If intrinsic remediation is determined to be a significant factor in contaminant reduction, site characterization activities should be performed to support intrinsic remediation. If exposure pathways have been completed (or are likely to be completed) and contaminant levels exceed regulatory levels, other remedial alternatives should be considered. However, as a way reducing the overall cost and duration of the comprehensive remedial plan (such as pump and treat), the added benefits of intrinsic remediation should be evaluated in the overall plan.
3. **Perform site characterization in support of intrinsic remediation.** Detailed site characterization is necessary to document the potential for intrinsic remediation and should be based on recommendations for additional data needs (see 2 above). The data collected should be used to determine if the intrinsic remediation processes are occurring at rates sufficient to meet regulatory criteria and protect human health and the environment. In addition, the data should allow for the prediction of the future extent and concentration of a contaminant plume using solute fate and transport modeling techniques. Collection of specific data should at a minimum include (1) the extent and type of soil and groundwater contamination, (2) the location and extent of contaminant source areas with mobile and residual NAPL, (3) potential for ongoing leaching of contaminants into groundwater, (4) aquifer geochemical parameters, (5) regional hydrogeology in relation to drinking water and confining units, (6) local hydrogeology in relation to drinking water, water wells, aquifer use, stratigraphy, grain size, aquifer hydraulics, groundwater flow

direction, local surface water bodies, and local recharge and discharge areas, and (7) identification of potential exposure pathways and receptors. The list of soil, groundwater, and aquifer data that should be collected during the site characterization state must include a full range of both physical and chemical data that support not only site characterization, but also intrinsic remediation (see the Air Force's protocol for a list of data needs). More importantly, a detailed evaluation of the processes that affect the fate and transport of fuel hydrocarbons must be completed in the characterization step, which includes assessment of advection, hydrodynamic dispersion, sorption, biodegradation, volatilization, and infiltration. The detailed evaluation can also include conducting a field dehydrogenase test to qualify the ability of aerobic bacteria to biodegrade fuel hydrocarbons, microcosm studies to provide further evidence that microorganisms necessary for biodegradation are present, and evaluation of the presence of volatile fatty acids which indicate biodegradation is occurring. In all, these data, and evaluation of these data, are used to support the intrinsic remediation alternative. Specific information on how to document the rates of intrinsic remediation processes are outlined in the Air Forces protocol.

4. **Refine the site conceptual model based on site characterization data, complete premodeling calculations, and document indicators of intrinsic remediation.** This step involves refining the conceptual model with data from geologic logs, cone penetrometer logs, cross sections, potentiometric and contaminant contour maps, and electron acceptor/metabolic byproduct and alkalinity contour maps. In addition, premodeling calculations are needed prior to implementation of the solute fate and transport modeling effort, which generally includes sorption and retardation calculations, fuel/water partitioning calculations, groundwater flow velocity estimates, and anaerobic biodegradation rate calculations. Consideration of the extent and distribution of contamination and electron acceptor and metabolic byproduct concentrations and distributions are of paramount importance in the documentation of biodegradation occurrence and simulation of the fate and transport of fuel hydrocarbons. This, for example, includes identification of below background concentrations of oxygen, nitrate, and sulfate, presence of elevated concentrations of iron(II) and methane, and elevated total alkalinity concentrations, all of which generally indicate the likelihood that processes promoting intrinsic remediation are active at a site.

5. **Model intrinsic remediation using numerical fate and transport models.** Simulating intrinsic remediation allows prediction of the migration and attenuation of the contaminant plume through time. While the modeling effort is a site-specific exercise related to actual site data, by themselves the results of the modeling effort are not sufficient proof that intrinsic remediation is occurring at a given site; they must be augmented through extensive site characterization to increase the validity of the modeling effort. In other words, the modeling effort is only as good as the data used in the model. Under this step, the model selected should be a well-documented and widely accepted model that can simulate the influence of advection, dispersion, sorption, and biodegradation. See also information on analytical and numerical models in Chapter 6.

6. **Conduct an exposure assessment.** After the rates of natural attenuation have been documented and predictions of the future extent and concentrations of the plume have been made using the solute fate and transport model, the data and information collected should be compiled to negotiate the intrinsic remediation alternative. To support this alternative, an exposure pathway analysis is generally needed. This

analysis includes identifying potential human and ecological receptors at points of exposure under current and future land and groundwater use scenarios. The results of the contaminant fate and transport modeling effort are central to the exposure pathway analysis, where conservative input parameters provide conservative estimates of plume migration. From this information the potential impacts to human health and the environment can be estimated.

7. **Prepare a long-term monitoring plan.** This step involves developing a long-term monitoring plan to locate monitoring wells and propose a sampling and analysis strategy. The plan is used to monitor plume migration over time and verify that intrinsic remediation is occurring at rates sufficient to protect potential downgradient receptors. The plan should be developed based on site characterization data, the results of solute fate and transport modeling, and the results of the exposure pathways analysis. The plan should identify long-term monitoring wells, installed to determine if the behavior of the plume is changing, and point-of-compliance monitoring wells, which are generally installed to detect movement of a plume outside of the negotiated perimeter of containment that triggers an action to manage the contamination. Figure 3-14 provides an example illustration of a hypothetical monitoring strategy. Location of the long-term monitoring wells are based on the behavior of the plume, and the number and location may very depending on site characterization data. Point-of-compliance wells are placed a distance of 500 ft downgradient from the leading edge of the plume or the distance traveled by the groundwater in 2 years, whichever is greater. If the property line is less than 500 ft downgradient, the point-of-compliance wells are placed near and upgradient from the property line. Long-term monitoring analyses should be limited to determining, for example, benzene, ethylbenzene, toluene, and xylene (BETX), dissolved oxygen, nitrate, sulfate, and methane concentrations, unless other requirements need to be addressed. Also, water level and NAPL thickness should be measured during each sampling event. Quarterly sampling of long-term monitoring wells is recommended during the first year, and based on these findings, the sampling frequency may be reduced to annual sampling in the quarter showing the greatest extent of the plume (assuming the intrinsic remediation processes have stabilized the plume).

8. **Conduct regulatory negotiations.** This step involves presentation of scientific documentation that supports intrinsic remediation as the most appropriate remedial option for the site. All information gathered and evaluated in the preceding steps should be integrated into a consistent and complementary presentation to conduct the regulatory negotiations. Of particular importance is presentation of scientific documentation that intrinsic remediation is occurring at a rate sufficient to meet regulatory compliance levels at the point of compliance and that the preferred alternative will be protective of human health and the environment. In addition, the regulators must be presented with a "weight of evidence" argument in support of the intrinsic remediation option to gain consensus. Last, presenting the long-term monitoring plan is also important in order to demonstrate the commitment to tracking the effectiveness of intrinsic remediation at the site.

Measuring Progress with the Installation Restoration Program Information Management System

Measuring progress and developing success stories are critical for implementing a comprehensive program that fast-tracks cleanup and environmental decision making.

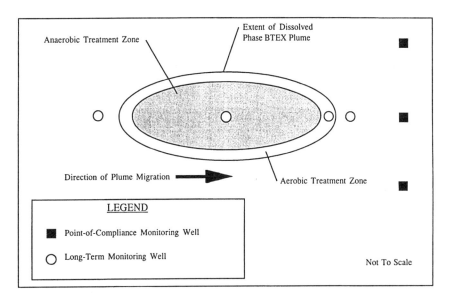

Figure 3-14 Hypothetical long-term monitoring strategy under the intrinsic remediation alternative. (Source: USAF et al. 1995. Technical Protocol for Implementing Intrinsic Remediation with Long-Term Monitoring Option for Natural Attenuation of Fuel Contamination Dissolved in ground water. Revision 0. Vol. I and II. August.)

The Installation Restoration Program Information Management System (IRPIMS) is one example of how the military, and specifically AFCEE, is accomplishing this task. The importance of measuring progress is outlined in the DoD section of this chapter. In general, measuring progress is critical because it communicates to the public, congress, and DoD that progress is being made and that funds are being wisely spent. The following text outlines a computer database and process used to track Air-Force-wide progress in the IRP. During this process, the Air Force identifies, quantifies, and compiles environmental contamination data for Air Force installations and facilities. The IRPIMS is a relational database used to store, analyze, and report information under the Air Force IRP. IRP projects generate technical reports containing generally large volumes of hydrogeological and chemical data that are difficult to manage with manually maintained systems. Mere storage and availability of this type of hard-copy data does not support information transfer needs in today's world without the ready access and computational capability of a mainframe computer equipped with the query tools of a relational database. With these factors in mind, the IRPIMS was designed by the Air Force using a multidisciplinary team of professionals consisting of hydrogeologists, chemists, applied statisticians, system analysts, and IRP project managers. The major impetus in designing the system back in 1986 was to provide an application tool to assist technical personnel, contract administrators, and program managers. The design of the system took approximately 1 year, and the first-generation system was operational in 1987. Major changes in system architecture were made in 1988, and the second-generation system was developed within a year. Since 1989, there have been relatively minor changes made in the data structure, and more attention has been given to technical applications. The following is a summary of how AFCEE manages large amounts of data from an environmental program. The summary is useful in providing an example of how other measurement programs planned or under development could be designed.

The system's hardware consists of a Digital Equipment Corporation VAX 6620 computer. Data is entered, stored, and managed with ORACLE, a commercially available relational database management system. Other application software, existing both in the VAX and personal computer (PC) environment, support the system relative to data entry, graphics, statistics, reporting, and groundwater modeling. The system and hardware are maintained by AFCEE, where information is stored in separate tables and record relationships have common fields within tables, and all IRPIMS tables are interrelated. Air Force IRP contractors must submit data (historical and current) in a format compatible with IRPIMS. The data generated from studies comes from field investigations, laboratory analyses, site characterizations, risk assessments, remediation feasibility studies, remedial designs, and remedial actions.

There are two methods through which a contractor can submit data. If a contractor has an Air Force scope of work to remediate an Air Force installation, the data obtained from this remediation effort would then be considered current data and should be submitted in text or "ASCII" format. AFCEE provides the contractor software to facilitate the data entry. Contractors have the choice of using IRPTools or their own software for downloading the electronic data into IRPIMS format. IRPTools is software that provides a relational database, has the look and feel of other Windows™-based products, allows for the import and export of data, and has a validation command on the menu. All data submissions must be processed through the validation module of IRPTools or through another software (QCTool) before submission to AFCEE. The validation software incorporates a number of automatic error checking routines to identify duplicate record sets, incorrect date/time/number formats, invalid codes, failure to complete key and required data fields essential for file integrity, and field, record, and submission level validation. IRPTools can be used to prepare the following types of IRPIMS data: (1) location definition information (LDI), (2) site and location cross reference information (SLI), (3) well completion information (WCI), (4) lithologic description information (LTD), (5) groundwater level data (GWD), (6) calculated hydrologic parameters (CALC), (7) environmental sampling information (SAMP), (8) analytical test information (TEST), and (9) test results information (RES).

IRPTools generates three groups of data for submission to AFCEE:

- Group I — LDI, SLI, and WCI
- Group II — LTD, GWD, and CALC
- Group III — SAMP, TEST, and RES

Project information is included with each submission file. In addition, IRPTools/PC can generate text files containing the contents of a table for local use by the contractor.

A contractor can also be directed by the Air Force to perform data entry functions to load or enter pre-IRPIMS reports into the IRPIMS database. Historical reports, which were generally prepared before the IRPIMS database was created, pertain to remediation efforts and are as important to the IRPIMS database as the data currently being collected, since they establish a benchmark for site conditions current data can be compared with. Requirements for submitting data to the IRPIMS are generally specified in a contractor's statement of work. When questions arise concerning data submitted to the IRPIMS, the RPM for the contract is the point of contact for coordinating contractual determinations. The submitter is responsible for the accuracy and completeness of all data submitted. Any errors in the submission identified by AFCEE must be corrected by the submitter. If a submission is not in the proper IRPIMS data format or does not pass AFCEE VAX validation, it will not be processed and loaded into the central IRPIMS database and will be returned to the submitter for correction. This rigid format and approach for the IRPIMS is essential in order to ensure accurate data tracking and measuring progress at Air Force bases.

In its effort to manage IRP data, the Air Force has also documented how funding is spent and can estimate funding needs. Expanding the Air Force's accelerated cleanup program depends heavily upon importing IRP data into and out of a central database in order to measure progress and demonstrate success. Through a comprehensive data collection and processing system such as the IRPIMS, documenting progress, milestones, and success stories is much easier, since data from all sites are compiled and are available to generate reports and comparisons to historic information. Successful sites in the Air Force environmental program are promoted as success stories to DoD and Congress, helping ensure future funding by effectively managing and reporting large amounts of information.

Innovative Technologies

Under the AFCEE Environmental Restoration Directorate, the Technology Transfer Division operates the Air Force's leading innovative restoration technologies evaluation, transfer, and application for the Air Force. The Technology Transfer Division program includes research and development of new technologies, demonstration of new technologies, validation of field results, preparation of protocol and users' manuals, and transfer of applications to the field. As of March 1995, there were 82 bases, including Army National Guard, Navy, Army, and Air Force installations, where technology transfer programs were operated in the U.S. A list of some of the more innovative technologies that the AFCEE Technology Transfer Division have demonstrated includes

- Natural attenuation through intrinsic remediation (see intrinsic remediation discussed in this chapter)
- Cone penetrometers fitted with a laser fluorimeter for on-site and real time data collection at contaminated sites
- Bioslurping to remove and separate free product using a vacuum application and enhanced bioremediation through increased oxygen flow in soil through bioventing

- Bioventing to enhance the oxygen supply to naturally occurring soil microorganisms, supporting *in situ* degradation of petroleum hydrocarbons
- Internal combustion engines (ICE) coupled with SVE to extract and destroy fuel hydrocarbons
- Controlled blasting to form trenches in bedrock to maximize collection of contaminated groundwater in bedrock.

In addition to innovative technologies, AFCEE's Environmental Restoration directorate is involved with application of demonstrated and traditional remedial technologies such as SVE, extraction wells, and air sparging. Detailed information on AFCEE's Technology Transfer Division is available from Technology Transfer Division, telephone number (210) 562-4329. As discussed in Chapter 2 on EPA's Superfund Innovative Technology Evaluation (SITE) program, development and demonstration of innovative technologies is an important component for accelerating cleanup. New technologies can offer substantial time and cost saving over existing technologies, and offer cleanup opportunities when traditional methodologies fail. Innovative technology programs, such as AFCEE's Technology Transfer Division, help identify, evaluate, and demonstrate innovative technologies, and help keep the environmental industry and accelerated cleanup viable in the future.

Proactive Communication in Air Force Facility Cleanup

AFCEE personnel use fact sheets and publications to promote and transfer environmental information related to accelerating cleanup and using innovative technologies. This type of communication has helped AFCEE become an environmental leader in the military. As an example, the following is a paper adapted from an AFCEE fact sheet paper that was written by Gil Dominguez (1997) of AFCEE:

> As I see it, innovation in Air Force cleanup technology is driven by the service's desire to do things faster, more effectively, and cheaper. Innovation may help us expedite remediation on closing bases so that we can turn them over to their respective communities for reuse as quickly as possible. Of course, we also need innovative approaches to remediate our active installations, assuring that our military and civilian personnel are working in a safe and clean environment. The Air Force has an organization that is dedicated to introducing innovative technology to the cleanup process. The Technology Transfer Division at the Air Force Center for Environmental Excellence, Brooks Air Force Base, Texas, evaluates, demonstrates, and applies existing and innovative technologies to Air Force environmental problem areas. The division is currently pretesting some form of new technology at 84 bases in the United States and some overseas locations.
>
> The bioslurping system is just one of several technologies being tried out. I think many of you already know what bioslurping is. If you want to impress people, you can call it vacuum-mediated free product recovery/bioremediation. As the name implies, it is a process that draws fuel-contaminated water out of the ground, much like sipping soda through a straw. The liquid that comes out of the ground is treated and then discharged. As the liquid is being extracted, air is flowing into the ground, stimulating the growth of helpful microorganisms that "feed" on the contamination.

Compared to other recovery technologies, the bioslurper system promises to be more cost-effective and possibly superior. A bioslurper pilot test was completed last April on the island of Diego Garcia in the Indian Ocean. During the war in the Persian Gulf the island was a serving base for Air Force B-52 bombers. In 1992 it was discovered that part of an 18-inch pipeline in the underground refueling system had broken, leaking an estimated 60,000 gallons of JP5 jet fuel. The bioslurper system removed more than 2,000 gallons of aircraft fuel during the first month of operation and expanded use of the technology is planned beginning in January 1997.

The Air Force is also looking into natural attenuation. The idea behind this type of remediation is that Mother Nature can clean up contamination just as well — if not better — than any procedure people can devise. She can also do it a lot more cheaply. Oxygen and other "electron acceptors" in water, such as nitrate, iron, sulfate, and carbon dioxide, stimulate the naturally occurring microorganisms which degrade dissolved fuel components. For chlorinated solvents such as trichloroethene, or TCE, the microorganisms usually require a food source such as natural carbon or spilled fuel in order to cause the solvents to degrade. Together with volatilization and adsorption, the "bugs" eventually stop the contamination plume dead in its tracks. But the bugs won't tell us if they are doing the job. To make sure they are, our researchers use computer models that simulate the contamination's migration and predict how far the plume will go before natural attenuation stops it. We verify that fact by installing sentry wells in the plume's path and along its center line. Periodic samples taken from the wells helps us determine exactly how fast the contamination is going away naturally.

Sometimes innovation is born of necessity. Travis Air Force Base, California, was installing a hydrant system for refueling aircrafts, but the pipes needed to be laid in an area found to have groundwater contaminated with TCE. The presence of the solvent so close to the surface posed a potential health hazard to the construction people. It also meant a possible work delay while the contamination was cleaned up. However, the base and the Air Force contractor came up with an innovative approach that allowed both projects to be completed at the same time. Most water-extraction systems use vertical wells drilled straight into the ground, but that would have required drilling about 20 to 30 wells in the area. Plus, keeping the wells active and accessible while the new construction continued presented a logistical nightmare. The idea the base and contractor officials came up with was to install two horizontal wells to do the same work as 30. The wells are about 500 feet long and act as drains. Water flows into them and then into two vertical casings, each about 40 feet deep. The water is pumped from the casings to holding tanks and from there goes to other tanks for treatment. This treated water is not wasted; the base has received permission from the California Water Quality Control Board to use the liquid in a landscape sprinkler system. Using these wells, more than three million gallons of water and 40 pounds of TCE were removed in the first four weeks of operation. Additionally, the water table was lowered to about 10 feet below the surface.

Innovation does not always mean technology. For example, when the capital city of Austin, Texas, came up with a very ambitious plan to create a new municipal airport at the closing Bergstrom Air Force Base, we were able to cut in half what would normally be an eight-year environmental cleanup program. Was this accomplished using some fancy new technology? No: Just plain, low-tech human communication.

The fast-track process at Bergstrom was made possible by the creation of a six-member executive team composed of city, contractor and Air Force officials. The committee meets on a regular basis to iron out problems and consider new procedures. Everyone agrees that increased communication has allowed all the parties to come to closure on decisions so that the work can proceed in a fast pace. Incidentally, Bergstrom was also the site of a treatability study involving the new technology called air sparging, which is used to clean up groundwater and soil contaminated with jet fuel. The process involves pumping air into the ground and again letting the microbes there do their work. But because the process releases volatile gasses into the soil, another technology, soil vapor extraction, was used to pull the gasses so they could be treated and rendered harmless.

Innovation also comes from team work. Throughout DoD, the emphasis is now on joint operations and projects. Working together allows the military services to pool their scarce resources, share ideas and make better use of the tremendous talent and expertise available in the Defense scientific and engineering community. We also save time by working together. If one of us already has the solution to an environmental problem, the others don't have to spend time, money and effort coming up with the very same solution. The Air Force, Army and Navy have been holding a series of meetings, held alternately at each environmental service center, in which they have discussed inter-service technology development and transfer. The services have talked about cooperative efforts to develop environmental technologies and assess the technologies that private industry is offering to each service individually.

Currently there is a partnering effort to standardize contract laboratory protocols. It involves the Omaha District of the Army Corps of Engineers, the Naval Facilities Engineering Services Command and the Environmental Protection Agency. The Quality Assurance Project Plan, or QAPP, developed by the Air Force Center for Environmental Excellence standardizes lab protocols and sets quality parameters to improve the accuracy of the data that results from the lab analysis. It will also standardize the contract laboratory evaluation/audit programs. The QAPP has the potential of being the first Tri-Service Quality Assurance Project Plan that will ultimately cut costs by eliminating the need for each service to have different analytical requirements. The three services have agreed to use the plan and it has already been accepted by two EPA regions, with positive comments from others.

3.3 U.S. NAVY

As with the Air Force, the U.S. Navy has been extremely active in its environmental programs and has been able to successfully accelerate cleanup at contaminated sites and reduce the overall cost of its environmental programs. This section provides several example programs and approaches the Navy uses to accomplished this. While the following text outlines some of the detail and actions for their success, one overriding aspect of its approach stands out as the primary reason why the Navy is a leader in environmental restoration and has been able to move forward quickly and effectively. The Navy's success can be attributed to building partnerships with their regulatory counterparts in environmental programs and emphasizing teamwork in order to pursue cleanup actions. As a direct result of the Navy's ability to establish working relationships with stakeholders, it has successfully completed some IRP activities ahead of schedule, reduced costs, and established itself as a leader in military environmental programs.

3.3.1 Navy Environmental Leadership Program

The Navy Environmental Leadership Program (NELP) is a relatively new environmental program established in 1993 to evaluate and test innovative treatment technologies. The program is focused on developing new management procedures to run the Navy's environmental programs more efficiently and effectively. The goal of the program is to be the Navy's primary evaluation program for new and innovative technologies, to develop focused management approaches that address the full spectrum of environmental challenges facing the Navy, and to export successful technologies and approaches throughout the Navy. Under NELP, innovative technologies are organized into cleanup, compliance, conservation, and pollution prevention (P2) programs (U.S. Navy 1996). Technologies that are shown to be effective and reduce costs are exported Navy-wide to accelerate cleanup and improve environmental management techniques. Naval Air Station (NAS) North Island at San Diego, CA and Naval Station (NAVSTA) Mayport in Florida are NELP bases where the majority of innovative technologies and management approaches are demonstrated. The connection between accelerating cleanup and decision making and NELP is strong, because demonstration of innovative technologies is crucial in order to identify more efficient and effective investigation and remediation technologies and ways to better manage environmental problems. Implementation of the NELP program also includes building partnerships with all stakeholders, including base personnel, the NELP management team, the regulatory agencies, and the community. NELP is an excellent example of how the military is overcoming environmental challenges using innovative solutions from technological, management, and partnering aspects. The following is a summary of the NELP management system, which provides an excellent example of how the Navy is operating a program that promotes new technologies and partnerships, which has the effect of accelerating cleanup and decision making.

NELP Management Team

The key to successful operation of a complex, broad-scoped program such as NELP is the effectiveness of NELP's management team. NELP has selected a simple and efficient organizational design that maximizes functional capabilities while minimizing unnecessary controls and administrative costs. Through the management team, 28 NELP partners from various projects have been established, which substantially broadens the scope of the program and allows the Navy to demonstrate and transfer information related to innovative technologies. Importantly, the NELP partners include agencies and associations outside of the Navy, which helps to build partnerships and unified direction for program-wide approaches to utilize innovative technologies and accelerate actions. Identification of problems and ideas for innovative solutions often comes from the local community, which deals with these issues on a daily basis. In recognition of this, the NELP management team established Process Action Teams (PAT) in the areas of cleanup, compliance, conservation, and P2. These four focus areas are the cornerstones of the overall DoD environmental security program. Each PAT includes representatives from the tenant Navy command that NELP is active in and is sponsored by a member of the NELP management team. The PATs are charged with identification and evaluation of candidate NELP projects.

The Cleanup PAT identifies and prioritizes remediation problem areas and implements innovative projects and methodologies to achieve quicker, more cost-effective cleanups. The Compliance PAT focuses on development of methods to streamline regulatory compliance and to coordinate requirements between regulatory programs and agencies. The Conservation PAT is responsible for identifying innovative methods to conserve and protect natural, cultural, and energy resources. The P2 PAT strives to implement innovative technologies that rapidly reduce existing sources of pollution or minimize the potential for accidental release of pollutants. Projects that pass the PAT evaluation are forwarded to the NELP management team, which makes the ultimate decision as to which projects are implemented. Using this management approach that involves all stakeholders, the Navy is able to efficiently and effectively demonstrate innovative technologies and management approaches. Information for any of the NELP projects is available in the NELP Program Guide (U.S. Navy 1996) or from the NELP Management Team identified in Appendix A. An example of NELP innovative technology demonstration project is included in Chapter 6, Section 6.4.

3.3.2 Integrated Project Schedules

Accelerating cleanup and decision making and working toward a common goal often depends heavily upon the ability of different military and regulatory counterparts to work together. In addition, multiple contractors are often used to complete field tasks. In some situations, it may be possible to take a project from beginning to end using one comprehensive service, as described in the previous section, and tie the various project needs together under the direction of one entity. However, in most situations within the military, a partnership must be established between contractors and the military to accomplish similar results when using multiple contractors. To achieve this goal, the Navy's Engineering Field Activity (EFA) West in San Bruno, CA developed a comprehensive protocol (PRC EMI and IT Corp. 1996) for integrated scheduling at selected Navy installations conducting IRP cleanup. The purpose of the protocol is to streamline the transition between contractors by creating a schedule outlining key objectives, milestone dates, funding requirements, and transitions points for contractors. This cradle-to-grave scheduling approach allows for improved project management and revision of project schedules in response to changes in priority, funding, and other impacts. Using an integrated scheduling approach results in

U.S. DEPARTMENT OF DEFENSE AND MILITARY BRANCHES

- A more effective and successful environmental program, measured by more timely and cost-effective cleanup solutions.
- Improved project development facilitated by taking advantage of the skills and knowledge base of all team members in order to recommend appropriate solutions and provide a better, more cost-effective, jointly developed project.
- Maximization of the Navy's opportunities for getting a project into the cleanup phase and completed in a timely and cost-effective manner.
- Elimination of situations wherein projects are delayed or funding is lost due to a lack of coordination or delayed action dates.

The key elements of EFA West's integrated scheduling approach include the following four points:

1. **Tie project prioritization to funding.** The key components of starting the Navy's integrated scheduling approach are prioritization of projects and identification of funding. Based upon priorities and funding, the various contractors can work jointly to expedite selected projects and ensure both program and contractual goals are met. Projects lacking firm funding commitments can also go through the integrated scheduling process, but this effort will generally result in a generic schedule, since construction and cleanup cannot be started without a construction award, the timing of which cannot always be determined. Therefore, funding must be available in a timely manner to avoid schedule delays. In addition, prioritization and schedule integration of projects will also allow selection of the level of design early in the environmental process and will allow greater flexibility in determining which contract vehicle to use.
2. **Conduct meetings to develop an integrated schedule for each installation and project.** From the installation level, the Navy installation and/or project-level RPMs conduct meetings with their contractor staff and determine the overall program priorities and milestones for their particular installation and contractors. This process is based on using realistic budget programming and funding availability information. Prioritized projects with definite funding are first to undergo integrated scheduling, followed by projects with probable funding, and finally, all other projects. This approach results in all project team members having a clear understanding of installation priorities and the ability to identify high-priority tasks, which has an overall effect of accelerating the environmental program.
3. **Assign target dates for major milestones and identify critical integration points.** Once the overall installation priorities are established, individual projects are discussed and the expected time frames for each project milestone are established. Agreeing upon project milestones enables the framework for integration and coordination to be determined. In turn, the Navy's contractors have a clear understanding of their responsibilities and their commitment to the whole team in making sure their time frames for deliverables are met. Based upon project prioritization and complexity, along with other factors, the integrated scheduling effort results in identification of necessary integration points and the planned transition points in projects, such as a transition from the design contractor to the construction contractor to conduct actual cleanup actions.
4. **Keep data current and keep communication open.** Integrating schedules is a dynamic process, and as such, the Navy's contractors must communicate project milestones and any problems, such as delays in the regulatory review process, to the

Navy RPMs in order for the Navy to take appropriate action. As part of this process, the Navy's legal council is included in the communication loop to assist RPMs in enforcing the Navy's lead agency role and help maintain project schedules. The RPMs are also part of the communication loop and must communicate funding developments and other pertinent information to the contractors in order to convey information that may accelerate or delay projects. This integrated level of communication fosters teamwork between all parties; however, it must be maintained throughout the project due to the large number of internal and external stakeholders on projects.

Integration points where study and design work are being conducted include, for example, the engineering evaluation/cost analysis (EE/CA) and FS stage, the conceptual design stage for a remedial action, or the definitive design stage. In all these examples, the Navy would request the services of an investigation contractor and a remediation contractor for one project, who in turn, must work together. In this same scenario, earlier transition or integration of efforts between contractors is possible (including the project discovery stage or RI stage) when site conditions warrant a transition to, for example, a removal action phase. More importantly, the point of transition should be decided upon early when integrating schedules, and should be based upon contracting and project information. Flow diagrams and narratives should be used to present the integration and transition points and project milestones. In terms of project milestones, each project generally has a study, design, and construction phase, assuming cleanup is warranted at a site. Each phase should be evaluated to identify primary and secondary milestones. In turn, the Navy reviews the milestones semi-annually or quarterly at the installation level with the lead RPM to receive updated information and identify accomplishments.

Project Coordination

Coordination, cooperation, teamwork, and schedule adherence between all parties is critical for successful integrated scheduling. A collaborative approach built upon trust, understanding of the value each party adds to the relationship, and a positive nonthreatening coordination environment builds and encourages teamwork. Teamwork, in turn, creates a synergy that benefits the Navy's environmental program and achieves greater success than any single party working alone or functioning within a nonteamwork setting. Meetings and teleconferences are the primary means to achieve teamwork. Therefore, meeting schedules have to be developed for each project based upon project-specific requirements and milestones. A key element to improved teamwork is increased informal communication, where contractors are able to directly communicate on coordination and technical issues. However, contractor discussions outside of these subjects are not permitted, such as discussions involving money, since there may be a conflict of interest. Meetings and teleconferences include regular program-level coordination, prescheduled design-phase meetings and construction-phase meetings, and regular meetings related to integrated scheduling. These meetings and teleconferences act as the checks and balances with integrated schedules, helping nourish partnerships between the Navy, regulatory agencies, and major contractors.

Contract Vehicle Selection and Planning

In cases where there are multiple contract vehicles available, the overall objective of timely and cost-effective cleanup is directly dependent upon the type of contract used to accomplish cleanup. In general, there is an inverse relationship between timeliness and costs, where cost of project (COP) contracts are more expensive than firm fixed price (FFP) contracts, yet COP contracts can be utilized in a more timely manner. With FFP contracts it generally takes longer to plan and develop deliverables, since they have to be more definitive in order for a contractor to accurately bid on the work; however, construction costs may be less compared to a COP contract. Integrated scheduling will help identify candidate projects for the available contracting vehicles and help plan the selection of cost-effective contracting vehicles to meet the stakeholder's needs for operating and closing bases. For example, in cases, where there is a time-critical removal action proposed because the site poses a serious threat to human health or the environment, using a COP contract may be most appropriate, since it usually takes less time to get cleanup actions underway.

Early in the integrated scheduling process, the Navy project team should meet to evaluate all projects based on future contractor needs. Following EFA WEST guidance (WESTNAVFACENGCOM ROICC/ACO Remedial Action Contract Administrative Plan dated 24 June, 1994, section 1.1) the RPM, based on contracting officer approval, should determine which contracting mechanism is appropriate based on contractor recommendations, program goals and schedules, and current command requirements and project specific factors. Evaluation factors that are part of the acquisition planning process used to determine if a project should be a fixed price or a cost contract include

- **Urgency of action.** Unless the action is time-critical or urgent in order to meet program objectives and master schedules, the project is a candidate for FFP/invitation for bid (IFB) contracting since this will help keep costs low. However, time-critical actions should use the COP contract, since they typically can be initiated more quickly.
- **Technical difficulty.** Cleanup projects that are straightforward in technical difficulty and utilize common technology should be candidates for firm priced remedial action contract (FRAC) IFB contracting. Conversely, technically challenging projects with difficult processes should be contracted for using COP contracts, since there may be advantages in having flexibility in the services rendered.
- **Construction difficulties or risk.** Projects that have above-average risk or construction difficulty for the contractor should be cost contracts. Construction difficulty may be due to the integration of several different technologies or processes within the project. Projects that are straightforward and use routine construction methods and operations, and projects with little risk, however, are candidates for the FRAC/IFB vehicle.
- **Project conditions.** Both IFB/FRAC and cost contracts can make allowances for unknown site conditions. If a site is adequately characterized and utility locations are known, the project typically meets FRAC/IFB requirements. However, unforeseen utilities or subsurface conditions can be expected on virtually every project and should not be the sole criteria for selection of a cost contract.

- **Complexity.** Complex projects with extensive process equipment and control requirements, a high degree of coordination or integration between process systems or technologies, or a high risk associated with the complexity of managing the overall project should use a cost-plus contract. Projects involving the normal complexity encountered in the field and during construction are candidates for the FRAC/IFB process.
- **Project size (cost).** Projects with high estimated costs should be considered for FFP/IFB contracting. The government will achieve most cost savings and benefit on competitively bidding large-scale or high-value projects. Cost benefits may accrue not only from direct pricing competition obtained from a competitive bid, but also from avoidance of program costs associated with cost contracts. Low-value projects should be executed under the expedited process of COP contracts in order to reduce engineering costs. Potential savings from competitive bids do not offset the design costs on low-value projects. High-value projects should be evaluated to determine if IFB contracting can minimize construction costs.

In addition to the above project-specific factors, the contracting officer must take into consideration program factors such as support of small and small disadvantaged business goals, existing contract capabilities, and the need for continuity of services in making the decision as to the most appropriate contract to accomplish the action. Procedures, and requirements and goals of the Federal Acquisition Regulation (FAR), must also be followed and met.

From this perspective, each project is evaluated against the above criteria in order to select which contract vehicle and project approach will be used to complete the desired work tasks. To a large extent, the selection process will also help identify the type and level of design effort that will be necessary for a project. In addition, identification of projects for FFP/IFB or FRAC contracting does not always mean they will be less timely. Typical design actions for a remedial action may take 2 to 6 months, excluding regulatory review and approvals, which are generally required for actions. Effective integrated scheduling and proactive planning will allow the team members to move to FFP/IFB contracting efforts without delays. Certain types of projects with easily definable requirements should be reviewed as a group, since these projects are ideal candidates for a FRAC/IFB contract.

Level of Effort for Design Projects

The effort and level of design needed to conduct a cleanup action is directly related to how fast project decisions are made and how long it will take to mobilize a construction contractor into the field. In these terms, the level of effort is directly related to how fast cleanup and decision making are accelerated. For projects requiring cleanup and implementation of a remedial technology, some level of design must be completed, whether conceptual or definitive. The ultimate goal of the level of effort used to develop a project design is to provide the contractor with the minimum necessary construction details and requirements (either written or in other forms) to properly complete the work in an efficient and effective manner. Therefore, the level of design should be appropriate for the complexity and size of the project, where the adage "pay me now or pay me later" is appropriate for environmental construction projects. The level of design effort needed is based upon

- **Remedial objectives.** The more widely encompassing or long-term a cleanup action, the more design and planning are required for overall construction cost-effectiveness and life-cycle cost considerations. Site-specific, short duration projects may be more easily accomplished with minimal design effort. Projects with a broader scope, a long period of operation, or with complicated physical site conditions will require more intensive design and planning. The remediation or cleanup objectives play an important role in determining the level of effort required and the type of design documents produced (plans and specifications vs. sketches and performance specifications).
- **Type of contract vehicles available.** Selection of the contract vehicle will help determine the level of effort required to successfully accomplish the project in a cost-effective manner. Projects planned for execution under a cost contract may require only minimal design effort, taking advantage of the knowledge and experience of the project team. Projects selected for fixed price contracting will require a greater level of design detail, and are dependent upon contract requirements. Projects for IFB will require traditional design documents (100% design packages) including plans and specifications. FRAC projects may be developed to a lesser extent, perhaps at the 35% or 65% design level. Table 3-2 provides a listing of common remediation technologies and requirements and their suitability for various contract vehicles.
- **Technical complexity and the preferred technology to be applied at the site.** Projects that have a complex technology or multiple interacting components will require an increased design effort. Using standard original equipment manufacturer (OEM) items and readily available off-the-shelf process equipment may not require detailed analysis or design. However, sizing calculations, overall process integration, code requirements, and performance still have to be evaluated. Projects using more than one type of physical, chemical, or biological process, having specific flow control parameters and requiring specific fabrication in order to meet project needs, require added design effort. The more components or delivery points required to meet cleanup objectives, the more complex a system becomes. Long-term maintenance and operations are also considerations requiring additional design evaluation.
- **Regulatory and community interaction and needs.** Active regulatory bodies and community reuse groups may influence the level of design required. Some projects may need project plans and specifications prepared to allow state regulatory agencies to review and approve the project. Work plans prepared by a contractor may be adequate for more routine projects where agency and public groups are less involved. The record of involvement at an installation by these groups may help decide how future projects are handled and designed. More sophisticated community groups and regulatory agencies are interested not only in the details of the construction, but also in construction quality control and long-term operations and maintenance.
- **Site considerations.** Site considerations may play a part in determining design and planning levels. Availability of electricity and other utilities, existing subsurface utilities, site lithology, site elevations, and site constraints all have implications for the design effort. Careful and detailed incorporation of site factors into the design phase can simplify and speed construction.

Table 3-2 Remediation Technologies And Contracting Suitability Matrix

	RAC (Performance)	RAC/FRAC (Conceptional)	FRAC Definitive (65% FRAC)	IFB/FFP or FRAC (Full Design)
Bioremediation	X	X		
Soil removals	X	X	X	X
Sewer cleanout	X	X	X	
Sewer renovation	X	X		X
GW collection and treatment	X	X	X	
Pump and treat	X	X		
Well point extraction system	X	X	X	X
Plume containment	X	X	X	X
WW treatment	X	X		
Process water treatment	X	X	X	X
Filtration/RO	X	X		
Wetlands restoration		X	X	X
Process or industrial sump cleanout	X	X	X	X
Erosion control		X	X	X
Landfill cap/restoration/closure		X	X	X
UST removal	X	X	X	
Soil/GW petroleum	X	X		
PCB contamination	X	X		
Sheet pile containment	X	X	X	
CAMU	X	X	X	X
UG pipeline removal		X	X	
Pesticide remediation	X	X	X	X
Slurry wall	X	X		

Notes: RAC = Remedial Action Contract; FRAC = Firm Price Remedial Action Contract; IFB = Invitation for Bid; FFP = Firm Fixed Price; PCB = Polychlorinated Biphenosis; CAMU = Corrective Action Management Unit; GW = Groundwater; WW = Waste water; UST = Underground storage tank; RO = Reverse osmosis; UG = Underground.
Source: PRC and IT Corp. 1996. CLEAN/RA Integration and Coordination Plan. Prepared for EFA West Naval Facilities Engineering Command, San Bruno, CA. August.

- **Value added for project completion.** The potential added value of the design effort to a project must be evaluated and considered. Additional designs will typically reduce construction cost as unknowns, ambiguities, or assumptions are reduced or eliminated. Design activities allow for the selection and sizing of equipment, utilities, and conveyance systems. More detailed designs allow for greater consideration of innovative alternatives for projects and can potentially lead to significant reduction of project costs. An example of this is a landfill project. A typical prescriptive cap may cost $500,000 per acre, while an engineered alternative may cost less than 20% of this or less than $100,000 per acre. More detailed design may allow for elimination of equipment, reduction of rework efforts, and optimization of the technology.

- **Navy requirements.** Navy program and financial objectives and requirements may factor heavily in the determination of the level of effort for design as well as which contractor will accomplish the design. In addition, several of the available contractors may have similar design capabilities. In all cases, however, congressional guidance as of 1996 requires 80% of environmental funding to be spent on cleanup (construction) activities for DoD environmental contracts.

Design Considerations

In a less than perfect industry, it is easy to under- or over-design projects; both errors result in lost time and additional costs to projects. Achieving middle ground is more difficult; however, by utilizing the Navy's integrated scheduling approach it becomes much easier through improved communication via meetings and establishment of partnerships to meet this goal. As noted above, there are many factors and considerations that influence the appropriate level of design for a particular project. Some factors carry more weight than others; for instance, outside influences or politics may change an otherwise rational or scientific approach to the evaluation process to a baffling, more costly, and less timely final selection. Ideally, the rational approach for evaluating the level of design necessary for a project should focus around open discussions in integrated scheduling meetings, so all parties can discuss project priorities, financial considerations, schedules, and project milestones. From this perspective, the Navy considers command goals, contracting objectives and limitations, and program scheduling to decide what level of design it will require for each project.

The following definitions for design work are provided to outline the expected level of design effort, and the final product resulting from the design effort for field implementation. Understanding each of these design efforts is important to reduce confusion and misunderstanding, reduce costs and unnecessary or redundant work, and expedite the process.

Performance Specifications — Performance specifications are synonymous with the "enhanced" EE/CA. The enhanced label is added to show that additional detail not usually found in an EE/CA must be provided so the contractor can develop an implementation work plan. The minimum information provided in an enhanced EE/CA includes

- Project boundaries (drawing)
- Site conditions, improvements, and utilities (drawing)
- Preconstruction contamination levels or influent concentrations for all contaminants of concern (narrative and lab results)
- Contaminant locations or influent flows (narrative/drawing)
- Required postremediation contaminant levels or effluent concentrations for all contaminants of concern (narrative)
- Area/volume/quantity for cleanup (narrative/drawing)
- Screened technologies and selection of preferred technology/method (narrative)
- Siting locations if appropriate (drawing)
- Discharge and disposal requirements of contaminants, waste, effluents.

Conceptual Design — A conceptual design provides more detail and engineering than a performance-based specification such as is found in an enhanced EE/CA. In addition to all of the requirements noted in the performance specification section above, a conceptual design has the following components and requirements:

- Process flow sheets (drawing)
- Site layout and conceptual layout for integration of equipment (drawing)
- Project scope and requirements statement (narrative)
- Identification of utility pick-up and connection points
- Process and instrumentation diagram (P&ID) drawings
- Hydraulic profiles
- Site-specific data

Definitive Design — The definitive design provides the next level of design and detail and should be used for technically complex or challenging projects or when different technologies or equipment types are used on the same project. Greater design and detail is required to ensure integration of components or multiple systems. In addition to the requirements noted for both conceptual designs and performance specifications, the following requirements exist:

- Determination of O&M requirements (narrative)
- Site characterization (drawing)
- Equipment specifications and construction details
- Electrical drawings and panel layouts (drawing)
- Equipment schedules, sizing, ratings (table)
- Connection details and mechanical piping sizing (drawing)
- Specific construction details not available as a RA contractor standard construction detail or specification (narrative)
- Narrative specifications/CSI specifications describing work requirements
- Grading plans or finish plans (drawing)
- Demolition drawings (drawing)
- Laydown and access areas (drawing)

The above-listed requirements for drawings may not apply to all projects. Each requirement is not a drawing in itself, and multiple information may be placed on one drawing. In some cases, it is possible to minimize the design effort and accelerate cleanup. The following are examples of minimal design efforts that accelerate cleanup.

- **CERCLA process at EE/CA stage.** If screening criteria show that a cleanup project can be accomplished with minimal design, the EE/CA can provide the basis for conducting the field effort. The EE/CA defines the problem, selects the technical remedy, compares costs, and defines the cleanup objectives. Based upon the EE/CA, an implementation work plan is prepared for Navy and regulatory review. The work plan is also reviewed by the contractor who developed the EE/CA to ensure all EE/CA requirements and concerns are clearly understood by the cleanup contractor and will be addressed in the construction effort. Overall, the design and planning work is minimal and is provided as part of the work plan to allow the remediation contractor crews to successfully complete the project. At a minimum the EE/CA provides information on the project scope (objectives), contaminants of concern, preconstruction contaminant levels and postconstruction contaminant levels to be met, the area/volume of contamination to be remediated, and the selected remedial technology. Confirmation sampling, oversight, and a final remediation implementation report are completed as part of the project. In this case, using the EE/CA as a conceptual design document streamlines engineering needs in cases that warrant a minimal design.

- **RI/FS with interim removal action.** Based upon the complexity of a project and the screening criteria results, a conceptual design or definitive design may be adequate for a construction job. The remediation contractor, for example, develops the implementation work plan from an FS and supplements the work plan with any design effort or drawings required to efficiently construct the project. The FS contains the selected technology alternative, cleanup requirements, discharge requirements, utility- and public-owned treatment works (POTW) information, and definition of influent streams and/or contaminated area boundaries. The implementation work plan is submitted to the Navy and regulatory agencies for review, and allows the contractor who prepared the FS to review the work plan and ensure the scope and intent of the FS is met. Utilizing this approach, the FS can be used as a resource to develop design supplements for a remedial work plan; as a result, the design effort is streamlined and standard design packages are avoided. A typical definitive design example would be a landfill closure. An FS and work plan design package incorporate regulatory requirements issued by the regulatory stakeholders and undergo extensive regulatory agency review. In addition, the design package provides specific details on site topography, cap structure and construction requirements, grading and drainage, and control of landfill gases for construction activities to proceed without delay.

- **Full design.** A full design effort would be accomplished for sealed bid and firm fixed price contracts and is likely the most common or standard design package engineers develop. Based upon acquisition planning and the screening criteria identified, projects are selected for the sealed bid process. In addition, there are a wide variety of technologies that are suitable for a firm fixed price contract. A full design effort provides complete and comprehensive plans and specifications of all work items, construction materials or equipment, and construction details to allow a contractor to prepare a fixed price bid to complete the work. Since a full design package is generally the most complete of the design packages outlined, it generally takes the longest to prepare.

3.3.3 Developing a Comprehensive Scope of Work

Most large government contracts require negotiating costs and the technical approach to be used for each individual scope of work before a contractor initiates the work (Payne 1994). An example of this type of contract is the Navy's Comprehensive Long-Term Environmental Action Navy (CLEAN) contract, where the Navy issues Contract Task Orders for contractors to perform under the CLEAN contract. Limiting the number of Contract Task Orders can help accelerate a project, because the amount of time and resources needed to develop a new scope of work, negotiate the cost and technical approach, and obtain funding are reduced. In addition, new task orders tend to expend contract dollars on administrative duties rather than cleanup. This is especially evident when unplanned work is required, such as conducting an additional investigation at a site that was not previously identified.

Under large contracts such as CLEAN, task orders can be issued in sequence to perform work, or they can be multi-tasked and combined into a limited number of comprehensive task orders. Simplistically speaking, a multi-tasked scope of work could include expanding an effort to prepare a decision document after the final investigation is completed. In this case, the project team knows that a decision document must be prepared and plans ahead by including the effort in the current scope of work being developed. From a complex perspective, multi-tasked task orders can include a scope of work for conducting contingency activities, such as collecting additional samples that are later determined to be necessary based on the first round of sampling, or pilot testing a remedial technology early in the regulatory process, such as the RI stage. These are potential actions or "semi-quantified" tasks the project team could plan for based on using assumptions, available site data, and past experience.

Other opportunities to accelerate cleanup arise from new information or unanticipated changes in a project. Using the comprehensive scope of work approach, it is possible to help the project team take advantage of new opportunities to accelerate cleanup when the opportunity arises or unanticipated changes occur. Flexibility in the scope of work in order to address acceleration opportunities is therefore important in task orders. An example of an unanticipated change would be identification of buried drums in an area undergoing a site inspection. For the sake of argument, assume the area had a prior history of buried drums and the newly identified drums can be easily excavated under a removal action. Developing a new task order to physically remove the drums is necessary, since the cost of such an action is likely beyond the hypothetical scope of work, and likely an action to be undertaken by a remediation contractor. However, the action memorandum and other supporting documents required to initiate and plan the removal action could be prepared, if included in the scope of work. Using this approach, planning for a potential removal action is possible by preplanning removal action document preparation in a scope of work, which has the effect of accelerating the overall removal process by as much as 3 to 6 months, or the time it would take to develop a new scope of work, reserve funding, and begin preparing the removal action documentation.

In developing a comprehensive scope of work, the scope of work should include not only required tasks, but also tasks that are prudent and go beyond the minimum requirements. In the above example, the additional effort of preparing an action memorandum, EE/CA, or other documentation for an unanticipated removal action would need to be part of, for example, an investigation scope of work. Other examples would be inclusion of tasks to collect additional samples in newly discovered contaminated areas, or conducting additional public relations efforts to ensure project information is transferred to the local community quickly and efficiently when it is necessary. A comprehensive scope of work that includes contingency efforts is difficult to negotiate because speculation may be required to develop cost for some tasks, and because some tasks may be based on what is needed to maximize the effort vs. what are the minimum requirements to the job. These costs are sometimes hard to accurately define. Therefore, any supporting information in favor of the additional effort, such as similar site conditions or past experience, is needed to negotiate the additional task(s) or proposed contingency actions. In addition, determining the cost of contingency items is difficult, since the size of the effort may not be known. However, if the scope of work allows flexibility for the project team to make field decisions or take advantage of new data, it will result in accelerating cleanup and reducing costs.

3.3.4 Accelerated Scheduling

Developing a strategy to accelerate a project schedule for investigation, design, and remediation of contaminated media, or closing sites posing little or no risk, can substantially streamline projects and reduce costs. At several Naval facilities, the Navy and its contractors have applied DoD's aggressive nonlinear strategy (described earlier in this chapter) to manage and schedule multiple tasks and actions simultaneously. This approach, which is also used by other military branches, expands the step-by-step, one-dimensional approach, or conventional process into a multi-tasked, two-dimensional or three-dimensional approach for early site closure, which is closely related to accelerating a project schedule.

Project schedules can be accelerated to either a small or great degree depending on site conditions and the desire to expedite project activities. A project team that is given plenty of time to complete a project will most likely take the allotted time rather than getting the job done early. Developing schedules with realistic time frames built into them helps move a project quickly from start to finish. However, overaccelerating a project schedule can jeopardize the quality of work performed and should always be avoided. In general, investigation, design, and cleanup actions at relatively uncomplicated sites are more easily accelerated than actions at complicated sites. However, complicated sites with multiple contaminants and impacted media may also be accelerated by conducting multiple tasks, such as simultaneous investigation and cleanup actions. For example, removal actions can be conducted while performing an SI or RI. However, as more tasks are managed simultaneously, the need for clear communication with the project team grows and management becomes more difficult from both an administrative and technical standpoint.

The ability of a project team to accelerate a project schedule is closely tied to the entire team working toward a common goal. For example, in the case where a regulatory agency and the Navy have established a partnership and are communicating well, less time is needed to review documents, which accelerates the project schedule. Rather than utilizing the standard 60 days, for example, to review large deliverables, only 30 days may be needed to complete a thorough review, accelerating the schedule by 30 days. In this example, 30 days is adequate because the information that is presented in the document was already discussed in detail by the project team in prior meetings. There are no "surprises" or unresolved issues when deliverables are reviewed by regulatory counterparts. In some cases, however, it may be advantageous to "agree to disagree" on some unresolved issues for the sake of continual progress. In these situations, future information or resources will be made available to address the unresolved issue(s).

In terms of scheduling activities and remediation, the project teams can accelerate a project schedule by conducting investigation activities, removal actions, risk assessments, FSs, and conceptual designs simultaneously, rather than one after another at multiple sites. Data and information for each of the tasks must be shared quickly, efficiently, and openly with the project team in order to address issues. Utilizing a case-by-case approach, the project team can evaluate and identify which activities can be completed simultaneously vs. those activities that must follow other actions because they are dependent on forthcoming information or decisions. One example of how the Navy accelerated a project schedule is the IRP for Fleet and Industrial Supply Center, Oakland, CA. During completion of the phase II RI/FS, the project team agreed to prepare a draft Proposed Plan and preliminary ROD for submittal with the draft phase II RI/FS report. Information including risk assessment data, future land use, and community input significantly limited the number of possible remedial alternatives that needed to be considered. Using a combined approach to preparing a final report with a preliminary decision document, the Navy accelerated the schedule by 6 to 12 months. This scenario is unique to sites where potential remedial alternatives are easily determined or in cases where presumptive remedies are applicable. Under other conditions, the opportunity to develop a decision document ahead of schedule is questionable, and should not be undertaken until pertinent information is available in order to make a confident remedial decision.

3.4 U.S. ARMY

Similar to other branches of the military, the U.S. Army has taken proactive steps to accelerate and streamline cleanup and decision making at toxic waste sites. Although primarily state-of-art pollution prevention programs have been implemented by the Army using comprehensive geographical information systems (GIS), acceleration programs have also been developed that include administrative, technical, and measurement initiatives to help meet Army and DoD program goals. As in other military branch programs, the Army administers programs such as CERCLA, RCRA, TSCA, and others. Since the Army's environmental programs share similar approaches with other military branches, detailing their program approaches is unnecessary. However, one environmental program the Army administers that should be mentioned is the U.S. Army Environmental Center (USAEC) and the associated programs for UXO.

While UXO is not new to the military, it is relatively new in terms of evaluating risk, decision making, and conducting cleanup actions on active, closing, and inactive bases. From military ranges, and public land to historic battle fields, UXO are a significant concern to military personnel and civilians because of high explosive (HE) material associated with spent or partially spent rounds, bombs, and rockets. Contact with UXO can result in immediate injury or death if the UXO is improperly handled. Safe and effective investigation, characterization, and cleanup of UXO is under development by the military, universities, and private environmental industry. From a contamination standpoint, UXO represents a substantial challenge to the environmental industry in terms of protecting human health and the environment.

With the quantity of UXO around the world, not including landmines, cleanup of and/or protection from UXO is expected to last long into the future, making it a relatively new industry for environmental professionals and a significant task for the military to address. In order to make the UXO program efficient and effective from an investigative, risk evaluation, and cleanup perspective, the Army is helping start the UXO program on the right foot, and learning from mistakes made under other environmental programs in the 1970s and 1980s.

Only limited information on UXO has been written that ties the entire UXO program together from a regulatory, investigation, decision making, technology, or cleanup standpoint. The DoD Military Range Rule (which was not final at the time this book was written) is part of the military's attempt to standardize UXO efforts for bases that are closed prior to BRAC, or transferred or closed under BRAC. In light of these needs, the USAEC has been very active in UXO research and development with focus on improving technology, demonstrating innovative technologies, and program approaches. The USAEC, in association with the Naval Explosive Ordnance Disposal Technology Division (NAVEODTECHDIV), have taken a leadership role with UXO, and new approaches and technology related to UXO are developed and demonstrated for application in the field. The risks posed by property containing UXO could be great depending on the amount of UXO present and how the property is or may be used in the future. Under the UXO Clearance Technology Program, the USAEC and the NAVEODTECHDIV are working together to enhance and demonstrate systems that can detect, identify, characterize, and remediate a vast array of UXO in a variety of environmental settings. Not only must these systems be reliable and accurate, but they must also be efficient, safe, and cost-effective. UXO information, including points of contacts and information on UXO risks and remediation prepared by the USAEC and the NAVEODTECHDIV, is available from USAEC for those environmental professionals that have little or no familiarity or experience with UXO. In the future, investigation and cleanup of UXO may become much more common, and the information generated by the USAEC is and will continue to be extremely valuable for those who work in this field.

3.5 REFERENCES

Dominguez, G. 1997. Testing Technologies at the Air Force Center For Environmental Excellence. Internal AFCEE Fact Sheet, March.

Payne, S.M. 1994. 1994 Spring Workshop: Fast-Track to CLEAN. Proceedings from the Department of the Navy Engineering Field Activity West Naval Facilities Engineering Command Spring Workshop, April.

Payne, S.M. and Ocampo, L.A. 1994. Implementing Accelerated Cleanup on Large, Multiple-Site Projects. Proceeding of the 8th National Outdoor Action Conference and Exposition, Minneapolis, May 23-25. pp. 17-26.

Perry, W.J. 1995. An Annual Report from DoD to the President and the Congress of the U.S. on Environmental Security. February.

PRC Environmental Management, Inc. (PRC) and IT Corporation. 1996. CLEAN/RA Integration and Coordination Plan. Prepared for EFA West Naval Facilities Engineering Command, San Bruno, CA. August.

Reid, M.E. 1996. Cutting Cost Becomes Big Business. Pollution Engineering Environmental Technology for DoD/DOE Sites. Special Report. Vol 28, No. 8. August, pp. 13-19.

Rifai, W.R., Bedient, P.B., Wilson. J.T., Miller, K.M., and Armstrong, J.M. 1988. Biodegradation Modeling at Aviation Fuel Spill Site. Journal of Environmental Engineering, Vol. 114, No. 5. pp. 1007-1029.

U.S. Air Force (USAF). 1992. United States Air Force Environmental Restoration Program Management Action Plan Guidebook. May.

USAF. 1993. Installation Restoration Program Remedial Project Manager's Handbook.

USAF, U.S. Environmental Protection Agency (EPA). Engineering Science. 1995. Technical Protocol for Implementing Intrinsic Remediation with Long-Term Monitoring Option for Natural Attenuation of Fuel Contamination Dissolved in Ground Water. Revision 0. Vol. I and II. August.

U.S. Department of Defense (DoD). Undated. Key to Opening the Door to BCT Success. Prepared by the Fast-Track Cleanup Implementation Work Group of the Defense Environmental Response Task Force.

DoD. 1993a. Department of Defense Strategy for Environmental Security. Secretary Aspens DoD Strategy for Environmental Security.

DoD. 1993b. Management Guidance for Execution of the FY94/95 and Development of the FY96 Defense Environmental Restoration Program. Memorandum from Ms. Sherri W. Goodman, Deputy Under Secretary of Defense, to the Assistant Secretaries of the Army, Navy, Air Force, and Directors of the Defense Logistics Agency and Defense Nuclear Agency.

DoD. 1993c. Measuring Progress in DoD's Installation Restoration Program. Rpt. 93-S-0625. March.

DoD. 1995a. BRAC Cleanup Plan Guidebook. Revised BCP Guidebook. Fall.

DoD. 1995b. Base Realignment and Closure Cleanup Plan Handbook. Office of the Deputy Under Secretary of Defense Environmental Security. Fall.

DoD. 1995c. Fast Tract to FOST. A Guide to Determining if Property is Environmentally Suitable for Transfer. Office of the Deputy Under Secretary of Defense Environmental Security. February.

DoD. 1995d. BRAC 1995 Quick Reference: Community and Environmental Activities. Office of the Deputy Under Secretary of Defense Environmental Security.

DoD and EPA. 1992. The Road to ROD Tips for Remedial Project Managers. January.

DoD and EPA. 1994. Restoration Advisory Board Workshop Guidebook. Summer.

U.S. Environmental Protection Agency (EPA). 1994. Symposium on Intrinsic Bioremediation of Ground Water. EPA/540/R-94/515. August.

EPA. 1995. Land Use in the CERCLA Remedy Selection Process. OSWER Dir. No. 9355.7-04. May.

U.S. Navy. 1996. Naval Air Station North Island Navy Environmental Leadership Program Guide. Edition 3. July.

CHAPTER 4

U.S. Department of Energy

The U.S. Department of Energy (DOE) is a major player in accelerating cleanup and decision making, and is especially active in the research and development area of environmental technologies. From a contamination standpoint, DOE faces environmental remediation and waste management problems resulting from over 50 years of nuclear weapons and energy production. Figure 4-1 shows DOE's overall strategic plan to reduce risk and impacts to the environment and expedite cleanup under its environmental programs. The volume, extent, broad distribution, and complexity of DOE's contaminated soils and groundwater pose a unique and formidable challenge due to nuclear contamination and other non-typical contaminants. Development of scientifically sound characterization, remediation, performance assessment, and long-term monitoring technologies for nuclear wastes and other contaminants that are cost-effective and result in acceptable risk to human health and the environment are important components of DOE environmental programs.

At most DOE sites, *in situ* approaches for site restoration and containment of wastes are desirable (over excavation and pump-and-treat technologies), because much of the contamination is difficult to handle after it is removed. Contaminants may also be widely dispersed in the environment, exist at relatively dilute concentrations, or may be otherwise inaccessible because of depth or location beneath structures. Environmental restoration at DOE facilities is complicated by diverse subsurface environments which include arid, wet, hot, and cold climates, and a diversity of geological settings with unique and complex conditions, as well as exotic mixtures of compounds.

Classes of chemical contaminants present in sediments and groundwater at more than half of all DOE facilities include fuel hydrocarbons, chlorinated hydrocarbons, metals, and radionuclides. Compound classes most frequently detected at DOE sites are metals and chlorinated hydrocarbons, radionuclides, anions, fuel hydrocarbons, and ketones. Organic acids, phthalates, explosives, alkyl phosphates, complexing agents (i.e., EDTA, DTPA, and NTA), and pesticides are also frequent contaminants at individual sites. More than half of the DOE sites evaluated by Riley et al. (1992) were contaminated with mixtures of two or more compound classes. The most common mixtures reported were metals and radionuclides. Co-disposal mixtures of compounds have resulted in modified transport and toxicity properties that can

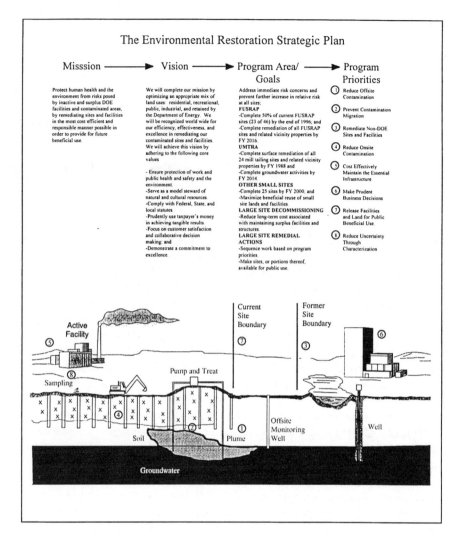

Figure 4-1 DOE's illustrated environmental restoration strategic plan. (Source: DOE. 1995b. Environmental Restoration Strategic Plan: Remediation the Nuclear Weapons Complex. DOE/EM-0257.)

increase ecological or health risks. For example, complexation with ligands enhances the mobility of metals and radionuclides, alters behavior at mineral surfaces, and affects availability to microorganisms.

In many cases at DOE facilities, contamination is a result of direct disposal of wastes into cribs and trenches or by deep injection, or indirect contamination following the loss of integrity of landfills and leaking underground storage tanks (Figure 4-2). The actual costs of remediating DOE's largest plumes of contaminated groundwater and sediments have not yet been determined, but estimates for cleanup of DOE's contaminant soils, sediments, and groundwater range from tens to hundreds of billions of dollars (DOE 1995a).

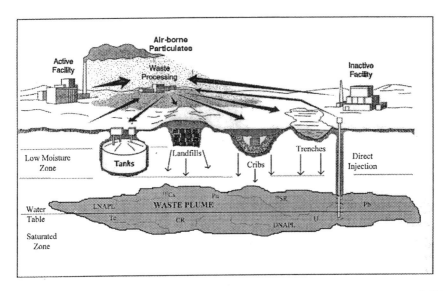

Figure 4-2 DOE's legacy from the Cold War and unique set of remediation challenges. DOE estimates there is 2500 billion liters of contaminated groundwater and 200 million cubic meters of contaminated media. (Source: DOE. 1995c. Natural and Accelerated Bioremediation Research Program Plan. DOE/ER-0659T UC-402.)

4.1 OFFICE OF ENVIRONMENTAL MANAGEMENT

DOE created the Office of Environmental Management (OEM) in response to legal and ethical requirements to deal with the environmental contamination at about 130 DOE sites. Collectively, these facilities contain more than 2500 billion liters of contaminated groundwater and more than 200 million cubic meters of contaminated soil. More than half of the DOE sites are contaminated with mixtures of two or more compound classes. In addition, soil at approximately 5000 other properties is contaminated with uranium tailings (DOE 1995a). OEM's major responsibilities include waste management, environmental restoration, and relevant technology development. DOE must overcome a myriad of technical challenges to ensure efficient waste management and environmental restoration at its sites. Scientific breakthroughs and technology development are important to enable implementation of cost-effective technologies that can meet performance standards. OEM has instituted a new management approach that is designed to ensure that DOE's environmental research and development (R&D) programs remain focused on OEM's most pressing remediation and waste management needs (DOE 1994). DOE has identified five major R&D remediation and waste management focus areas:

- Contaminant plume containment and remediation
- Mixed-waste characterization, treatment, and disposal
- High-level waste tank remediation

- Landfill stabilization
- Facility transitioning, decommissioning, and final disposition.

The Department has made considerable progress in accelerating the pace of cleanup over the past few years. For example, the 1995 Baseline Environmental Management Report (BEMR) (DOE 1995a) estimated that cleanup would take 75 years to complete at DOE facilities, stretching ahead to the year 2065. The most current estimates, however, show that 80% of the Environmental Management site cleanups will be completed by the year 2021. These changes reflect a more focused technical approach, oriented toward results and fueled by performance-based contracts, but also a fundamental change in the strategy of the OEM. This strategy involves addressing urgent risks first, stabilizing sites, investing in technology development and basic science, reducing mortgage costs, and basing decisions on future land use considerations. Utilizing this approach, DOE's OEM program expects to succeed in completing cleanup at most sites within the next 20 years. The approach used to increase the pace of acceleration is focused on risk reduction and lower fixed management costs, leaving only monitoring, surveillance, and maintenance tasks after 20 years. Cost reduction is achieved through shortening cleanup schedules, which not only reduces the opportunity for the spread of contamination in the environment, and reduces fixed management costs by optimizing schedules, but also minimizes the impact of inflation by completing the work earlier. Section 4.2 provides additional details on DOE's acceleration efforts as described in the Environmental Restoration Acceleration Report (DOE 1996).

4.2 OEM ENVIRONMENTAL RESTORATION PROGRAM

The Environmental Restoration Program under OEM develops and implements innovative and effective management processes for expediting DOE's environmental programs (DOE 1996). OEM's accelerated approach will likely result in substantial benefits, including avoidance of unnecessary health risk exposure, quicker turnover time for DOE facility reuse, and cost reductions. The following information provides DOE's strategy and accomplishments related to their accelerated approach. More detailed information is available in their Environmental Restoration Acceleration Report (DOE 1996).

4.2.1 OEM's Environmental Restoration Strategy and Accomplishments

In 1993, Assistant Secretary Thomas P. Grumbly directed the OEM program to focus on reducing risk and lowering mortgage costs. He further emphasized an orientation toward cleanup fueled by performance-based contracts. Since that time, the OEM program has made significant progress toward these goals. Current estimates indicate that the total cost of running the program will be reduced due to accelerating environmental restoration activities and rescheduling OEM projects based on DOE acceleration initiatives. The potential savings derived from accelerated programs and projects will continue to drive OEM toward reducing risks and lowering mortgage costs. The Office of Environmental Restoration, which is within OEM, developed an outcome-oriented national strategy to accelerate the pace of site restoration, produce tangible cleanup results, and reduce risk and long-term costs associated with the cleanup of the nation's nuclear defense production and testing sites and facilities. The strategy is articulated in the Environmental Restoration Strategic Plan (Figure 4-2) and is being implemented through both national and site-specific efforts (DOE 1995b). The strategy achieves several immediate benefits, including addressing urgent risks, stabilizing sites, investing in technology development and research, reducing mortgage costs, and basing environmental decisions on future land use considerations.

Congress demonstrated its support of this strategy by funding an acceleration program and encouraged the Department to take further steps in accelerating the cleanup at DOE facilities (U.S House of Representatives 1995). In addition, the National Defense Authorization Act for fiscal year (FY) 1996 directed the Secretary of Energy to accelerate environmental restoration and waste management activities at nuclear facilities. Specifically, section 3156, Accelerated Schedule for Environmental Restoration and Waste Management Activities, provided the following mandate:

> The Secretary of Energy shall accelerate the schedule and projects for a site at a Department of Energy defense nuclear facility if the Secretary determines that such an accelerated schedule will achieve meaningful, long-term cost savings to the Federal Government and could substantially accelerate the release of land for local reuse. Section 3156 (b) also provided that in determining how to invest these resources, the Secretary shall consider the following: (1) the cost savings achievable by the Federal Government; (2) the amount of time for completion of environmental restoration and waste management activities and projects at the site that can be reduced from the time specified for completion of such activities and projects in the baseline environmental management report . . .; (3) the potential for reuse of the site; (4) the risks that the site poses to local health and safety; and (5) the proximity of the site to populated areas.

Finally, the Authorization required the Secretary to provide a report to Congress by May 1, 1996, discussing each site where environmental restoration and waste management activities and projects have been accelerated. The following is DOE's response to that requirement, describing the strategy and benefits gained from the program which have been or will be realized.

Acceleration Strategy

This section provides a brief overview of the strategies which the Environmental Restoration Program is using to execute its assigned mission. It also includes a discussion of the program's efforts to streamline cleanup activities and increase its program cost-effectiveness. The Environmental Restoration Program is undergoing completion by means of a cleanup strategy predicated on the following set of core values: ensure protection of worker, public health and safety, and the environment; serve as a model steward of natural and cultural resources; comply with federal, state, and local statutes; prudently use taxpayers' money in achieving tangible results;

focus on customer satisfaction and collaborative decision making; and demonstrate a commitment to excellence (DOE 1995b). In order to effectively accomplish its mission, the Environmental Restoration Program has been divided into five program areas based on logical groupings of similar site or facility characteristics. The five program areas are: (1) Formerly Utilized Sites Remedial Action Program (FUSRAP), (2) Uranium Mill Tailings Remedial Action (UMTRA) Project, (3) other small sites, (4) large site decommissioning, and (5) large site remedial actions. This categorization of sites and facilities allows for a subdivision of priorities and budgets, a refining of program focus and direction, and the development of common program strategies.

In all, there are 132 geographic sites in 31 states, one territory (Puerto Rico), and two Indian tribal lands for which the Office of Environmental Restoration has (or had) responsibility for characterization, cleanup, and/or long-term monitoring/surveillance, and maintenance. Table 4-1 shows the breakdown of the Environmental Restoration Program by geographic site grouping and by program area with a brief discussion of scope and status. In some cases the 46 formerly utilized sites (or FUSRAP sites) and 24 uranium mill tailings sites (or UMTRA sites) have associated nearby residential/commercial properties that also require characterization and remediation. The 47 other small sites and 15 large sites are typically comprised of a number of individual release sites and/or facilities, each of which requires characterization and subsequent cleanup if determined to be necessary. In this case, each site represents a unique location at which a release has occurred or is suspected to have occurred and a facility is a uniquely identified building or structure. Through FY 1995, 21 FUSRAP sites, 15 UMTRA surface cleanups, and 8 other small sites have reached completion (excluding long-term surveillance and monitoring).

The strategy for streamlining and accelerating the Environmental Restoration Program involves striking a balance between short- and long-term program accomplishments and program costs. Work is sequenced to address high-risk activities first, while optimizing the schedule for effective completion of the program's technical scope. Where possible, this includes a cost-effective and responsible integration of long-term commitments and accelerated activities. The combined effect of these strategies accelerates the pace of the Environmental Restoration Program, reduces costs, and improves efficiency. The strategies also provide for maximum on-the-ground progress, while protecting both human health and the environment. In addition, the strategies have been developed to leverage and enhance the cost-effectiveness of budgets received from Congress. Numerous initiatives influence each stage of budget formulation, program execution, and performance evaluation. The Environmental Restoration Program strategies include

1. **Demonstrating outcomes and performance.** The Environmental Restoration Program measures overall progress toward mission completion, primarily by tracking the number of site and facility closures. The Program also tracks the number of closed sites within geographic areas. Figure 4-3 shows an example chart of DOE's Environment Restoration Program status at the close of FY1995. Tables 4-2 and 4-3 show example tracking tables of percentage complete by release site and by problem type. Mission completion measures, like the number of closed sites and

U.S. DEPARTMENT OF ENERGY   151

Table 4-1 Summary Of Environmental Restoration Geographic Sites And Program Areas

Group	Program	Summary
Small sites	FUSRAP	• Includes 46 formerly unitized sites in 14 states • Cleanup at 21 sites completed at the end of FY 1995 • 309 vicinity properties identified • 180 vicinity properties closed at the end of FY 1995
Small sites	UMTRA	• Includes 24 sites requiring both surface remediation and potential groundwater compliance issues • 15 surface remediation projects completed at the site at end of FY 1995 • 5,294 vicinity properties identified • 5,138 vicinity properties closed
Small sites	Other sites	• 47 sites classified as small site based on cost of completion • 8 sites completed at the end of FY 1995, not including long-term surveillance and monitoring • The 47 sites are comprised of 1,356 release sites and 199 facilities
Large sites	Decommissioning	• Includes 9 of the 15 large sites undergoing decommissioning, specifically focused on facilities and structures • 68 of 621 facilities at large site decommissioning projects are completed at the end of FY 1995
Large sites	Remedial Action	• 15 sites classified as large sites based on estimated cost of completion or complexity • 7,364 release sites and 257 facilities were active in the large site remedial action grouping as of 1996

Notes: FUSRAP = Formally Utilized Sites Remedial Action Program; UMTRA = Uranium Mill Tailings Remedial Action; FY = Fiscal Year.
Source: DOE. 1996. Environmental Restoration Acceleration Report. DOE/S-0116.

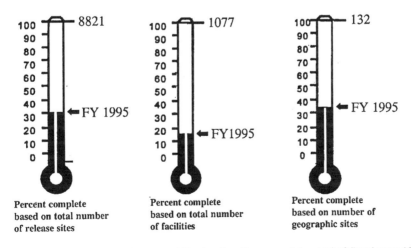

Figure 4-3 Status of the Environmental Restoration Program at the end of fiscal year 1995. (Source: DOE. 1996. Environmental Restoration Acceleration Report. DOE/S-0116.)

Table 4-2 Release Sites and Facilities by Problem Type

Problem Type	Total Release Sites and Facilities	Number Completed	Percentage Complete
Surface or subsurface material or waste	2257	644	29%
Building and equipment	1882	464	25%
Spills and leaks	1274	440	35%
Tanks (hazardous and radioactive)	1212	558	46%
Underground test areas	906	3	0.3%
Above-ground material or debris	803	360	45%
Liquid surface contamination	764	150	20%
Dispersed surface contamination	331	63	19%
Surface water or groundwater	204	19	9%
Miscellaneous	208	65	31%
Mill tailings piles	57	16	28%
Total	9898	2782	28%

Source: DOE. 1996. Environmental Restoration Acceleration Report. DOE/S-0116.

facilities, are incorporated into the program's performance goals and serve as the benchmark for future performance measurement in accordance with the Government Performance and Results Act of 1993 (PL 103-62 August 1993). Annual increase in the number of closed sites, facilities, and vicinity properties is a critical component of the program's ability to demonstrate its productivity and cost-effectiveness.

2. **Emphasizing risk reduction, compliance, and cost-effectiveness as part of the budget formulation process.** DOE personnel work in a collaborative manner with the regulators, stakeholders, public officials, community, and others to establish funding priorities at each site. The Environmental Restoration Program and DOE headquarters place high priority on achieving relative risk reduction for both human health and the environment, complying with site regulatory agreements, investing dollars in the most cost-effective manner, and reducing mortgage and out-year costs of their environmental programs. As part of the budget review process, a determination is made as to the criteria that drive an activity for funding consideration. Consideration is also given to funding activities that are driven by other criteria, such as potential for reuse. If the Department decides an activity should be considered for funding, the funding is evaluated from a performance perspective where benchmarking is used to evaluate the performance. In addition, the Department considers the contracting strategies, utilization of technology, permitting status, milestone completion, and site-specific factors. Using this approach, the Environmental Restoration Program emphasizes risk reduction, compliance, mortgage reduction, and cost-effectiveness, all of which are an integral part of DOE's budget formulation process.

3. **Allocating resources to cleanup vs. studies.** The Environmental Restoration Program focuses on "on-the-ground" performance. This strategy is traditionally measured by the distribution of assessment dollars vs. cleanup dollars. The percentage allocation of the Environmental Restoration budget between the functions of studies, cleanup, and "other" (i.e., program management, landlord, monitoring/surveillance, and maintenance) is one of the most important performance measures tracked within the program. As Figure 4-4 illustrates, more DOE funds were expended on

Table 4-3 Release Sites and Facilities by State

State	Total Release Sites and Facilities	Number Completed	Percentage Complete
Alaska	3	1	33%
Arizona	4	2	50%
California	441	248	56%
Colorado	219	25	11%
Connecticut	2	1	50%
Hawaii	3	3	100%
Idaho	569	374	66%
Illinois	534	386	72%
Indiana	11	10	91%
Kentucky	159	0	0%
Maryland	1	0	0%
Massachusetts	3	1	33%
Michigan	1	1	100%
Mississippi	26	0	0%
Missouri	120	71	59%
Nebraska	1	1	100%
Nevada	2373	331	14%
New Jersey	14	2	14%
New Mexico	2377	1048	44%
New York	81	12	15%
North Dakota	4	0	0%
Ohio	343	22	6%
Oregon	3	2	67%
Pennsylvania	4	3	75%
Puerto Rico	1	0	0%
South Carolina	374	35	9%
Tennessee	629	25	4%
Texas	241	137	57%
Utah	22	3	14%
Washington	1331	36	3%
Wyoming	4	2	50%
Total	**9898**	**2782**	**28%**

Source: DOE. 1996. Environmental Restoration Acceleration Report. DOE/S-6116.

actual cleanup than on studies for the first time in FY1995. In the future, DOE acceleration strategy is anticipated to expand this contrast, providing further evidence of the program's ability to achieve more cleanup.

4. **Streamlining cleanup strategies**. At all DOE facilities, stakeholders, regulators, and DOE personnel are working together to reduce costs and accelerate cleanup. Some sites (ORNL, Hanford, Savannah River, Mound) utilize DOE's Streamlined Approach for Environmental Restoration (SAFER), which allows projects to move through the characterization phase more efficiently and on to the actual cleanup phase. SAFER accomplishes this through streamlined data collection efforts and encourages the participation of regulators, the public, and other stakeholders in decision-making, similar to integrated assessments under the U.S. Environmental Protection Agency's (EPA) Superfund Accelerated Cleanup Model (SACM). In

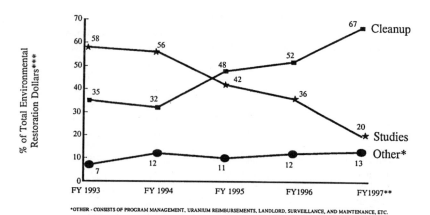

Figure 4-4 Trends in DOE's Environmental Restoration Program studies and cleanup. (Source: DOE. 1996. Environmental Restoration Acceleration Report. DOE/S-0116.)

addition, SAFER integrates the data quality objectives process with the observational approach (refer to Chapter 6 for more information on the observational approach). Overall, SAFER provides a consistent approach to focus data collection, reduce significant uncertainties, and manage remaining uncertainty during environmental restoration. The major streamlining component of the SAFER approach is accelerated site investigations and studies, which expedites decision making and when remediation is undertaken. Coupled with accelerated site investigations and studies is development of plans for managing uncertainty during the remedial process using the observational approach. DOE's SAFER approach provides an effective means for accelerating actions and decision making. Other sites, like the Idaho National Engineering Laboratory, the Laboratory for Energy-Related Health Research, and the Mound Plant, have implemented custom facility-wide strategies to streamline cleanup. As a result, DOE facilities have reduced costs utilizing a variety of tools, such as limited field investigations and removal actions, SAFER, and innovative technologies.

5. **Minimizing the program's uncosted balances**. Extensive focus has been placed on the effective management of uncosted carryover balances to both reduce and optimize these balances through sound management practices. The Environmental Restoration balances for FY1993–FY1995 are shown in Figure 4-5. This figure shows the progress that has been made in reducing the uncosted balance as a percentage of the overall environmental restoration program budget from 32% in FY1993 to 15% in FY1995. The program has set an uncosted balances target of 15% for FY1996. These levels are considered necessary to ensure that the program's progress continues unimpeded. To encourage high performance, each facility is required to fully justify and account for uncosted balances at the end of the fiscal year.

6. **Implementing "Workout" sessions with regulators and stakeholders**. The OEM program has instituted a management process called "Workouts" to achieve consensus between and among program stakeholders with respect to site priorities. The goal of the Workouts is to develop integrated, accelerated site cleanup plans that can be supported by DOE, regulators, and other stakeholders. Workouts have

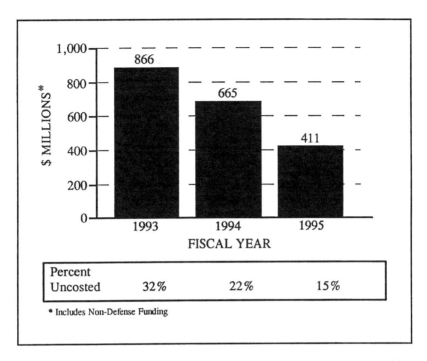

Figure 4-5 DOE's Environmental Restoration uncosted balances trend. (Source: DOE. 1996. Environmental Restoration Acceleration Report. DOE/S-0116.)

been held at some of the larger sites including Hanford, Rocky Flats, Oak Ridge, and Savannah River in conjunction with federal and state regulators and other stakeholders. At the Workouts, all of the principal decision-makers are brought together to develop specific solutions for accomplishing critical site activities and milestones within restricted budgets. Workouts address fundamental program management issues such as priority setting, funding requirements, opportunity for cost savings, and waste treatment, storage, and disposal assumptions. Representatives of the Waste Management, Environmental Restoration, and Nuclear Materials and Facility Stabilization programs participate in the Workouts to facilitate the development of comprehensive and integrated site plans. These sessions often result in the development of a long-term vision for each site, with interim accomplishments or goals to be met at various points in the future.

7. **Streamlining program requirements and roles and responsibilities.** The multiplicity of local, state, and federal regulators with jurisdiction over the DOE's cleanup activities can result in difficulties in achieving agreement on program objectives and milestones. The Department is addressing this challenge by combining regulator requirements and streamlining regulator roles and responsibilities. The purpose in combining requirements is to ensure little or no duplication of procedures and documentation, and to institutionalize a site-wide approach to remediation. In addition, DOE is promoting the "lead regulator" concept in its process for negotiating compliance agreements. Under the lead regulator concept, a single regulatory agency will generally be involved in the day-to-day oversight and decision making on specific environmental management activities. This concept is being utilized for the execution of the Tri-Party Agreement at Hanford and

was proposed in a draft Rocky Flats Agreement. At both sites, the concept promises to significantly improve program performance and accelerate cleanup by streamlining requirements, reducing redundant paperwork requirements, and minimizing inconsistent, and potentially conflicting, direction from regulatory agencies. Using this approach, congressionally approved funding levels can be applied to activities with more direct results, accelerating the rate of progress at DOE sites. In another example, the DOE, EPA, and the Defense Nuclear Facilities Safety Board (DNFSB) have entered into a Memorandum of Understanding (MOU) at Rocky Flats. The memorandum outlines respective roles regarding oversight authority, which will result in a clearer definition of roles and responsibilities.

8. **Linking compliance agreement milestones to the budget.** The "rolling milestone" approach to compliance is a process to help ensure that the establishment of regulatory-driven program milestones reflects the balance between progress on regulated activities and reality given federal funding uncertainties. Under this approach, near-term activities within the Department's 3-year program planning, budget formulation, and program execution horizon are generally designated as enforceable milestones subject to penalties if the Department fails to meet these milestones. Out-year activities, beyond the 3-year budget planning horizon, are generally designated as nonenforceable deadlines in recognition of potential technical and funding uncertainties associated with these dates. Each year, milestones are evaluated with the regulators and stakeholders in concert with the budget formulation process to determine if changes are warranted based on funding availability, new priorities, new technologies, or other factors. This approach has shown great promise for DOE in terms of facilitating teamwork. While regulators are not bound to keep milestones within appropriated funding levels, they have agreed to consider these levels in good faith in establishing and reviewing milestones. The process promotes a more cooperative working relationship with regulators, resulting in a more results-oriented program. DOE requests that regulators incorporate the "rolling milestone" approach in all new cleanup and compliance agreements that contain long-term milestones with significant funding commitments.

9. **Improving contracting strategies.** The Environmental Restoration Program also tracks the value of contracts awarded within the program by contract type. In addition, a formal team is established to monitor and enhance program-wide contracting strategies. The Environmental Restoration Program places great importance on transitioning from cost plus fixed fee contracts to performance-based contracting in accordance with the Department's contract reform initiative. In addition, the program evaluates and implements innovative contracting mechanisms within new and existing contracts, encouraging contractors with incentives to reduce costs and improve performance. The Environmental Restoration Program continues to evaluate the use of additional contract controls and flexible contract options, such as basic ordering agreements, to ensure it receives the best possible service and products. While dramatic changes seldom occur quickly, DOE's enhanced contracting strategies can result in significant cost reductions, fostering the overall acceleration of the program.

10. **Developing performance-based specifications rather than technology standards during the remedy selection process.** The Office of Environmental Restoration has developed a remedial technology deployment strategy in conjunction with the Office of Science and Technology. The initiative outlines improving the procurement process by emphasizing performance specification-based criteria, rather than preselecting specific technologies and proceeding with the traditional

design/construction/operation approach. DOE issues performance-based requests for proposals and chooses a vendor/technology by comparison of alternatives provided by the marketplace.

11. **Applying cost-effective technologies.** Another goal of the Environmental Restoration Program is to ensure that appropriate, efficient, and cost-effective technologies are utilized in its cleanup program. The Environmental Restoration Program uses a technology screening process, coupled with the performance specifications previously discussed, to optimize the selection and deployment of cleanup technologies. Technical experts periodically conduct peer reviews of the technologies that have been selected for each of the field site cleanup projects in an effort to share lessons learned, and discuss issues and problems. DOE headquarters personnel have developed a database of preferred technology alternatives for common cleanup requirements, and use this information in order to assist environmental restoration of field sites by screening available technologies and identifying the appropriate candidates for specific cleanup challenges. Technology screening and peer review processes promote the use of appropriate and effective technologies in the Environmental Restoration Program, which increases DOE's program efficiency.

12. **Implementing privatization.** Where cost-effective, the Environmental Restoration Program is actively pursuing the privatization of activities and functions that were formerly performed by the Department's Management and Operating (M&O) contractors. A major privatization initiative has been implemented for the Environmental Restoration Program's Pit 9 Comprehensive Demonstration Project at the Idaho National Engineering Laboratory (INEL). This innovative, cost-effective approach to remediation of a waste disposal pit, containing approximately 150,000 ft^2 of transuranic waste, is one of several at INEL, and is being performed as an Interim Action under the Comprehensive Environmental Response, Compensation and Liability Act (CERCLA). The effort is being accomplished under an innovative service subcontract, with payment based, in part, on fixed unit prices for the treatment of waste. Under this privatization effort, the subcontractor owns the facility and assumes the associated financial risk. Substantial savings can result from privatization and is anticipated to be completed in 1999. The Environmental Restoration Program is actively exploring other privatization initiatives to further improve the cost-effectiveness of the program.

13. **Basing cleanup decisions on future use.** Making land and facilities available for future beneficial use has a significant impact on DOE's planning and program execution process. The Environmental Restoration Strategic Plan, and specifically the small site strategy, emphasizes the need to accelerate site cleanup and return land and facilities to other beneficial uses. DOE's Management Action Process includes considering future land use for each major site. The Department's strategy for future site use is outlined in the 1996 Environmental Management report (DOE 1996). The process for remediating each site and making it available for beneficial reuse ultimately must address the objective of protecting the safety and health of the public and the environment. However, the extent to which this objective is achieved, and the cost of conducting remediation activities, depends largely on the intended future uses of remediated land. The Department's overall approach includes working with stakeholders and regulators to link future use considerations with realistic risk-based cleanup decisions. Developing optimal remedial action solutions, based on realistic and responsible land use planning, enables the Department to reduce the time and cost of remediation, while accelerating the cleanup of the sites. The Office of Environmental Restoration anticipate a significant cost

savings relative to prior approaches and to develop cleanup levels which use conservative assumptions with respect to risk, exposure, and future use.
14. **Implementing program cost reductions.** Among the more significant cost reduction measures implemented over the past few years in DOE are the reduction of reliance on DOE headquarters administrative and technical support, reductions in travel and administrative costs, and use of technological advances such as teleconferencing and video calls. DOE reports these measures have resulted in reducing the Environmental Restoration Program's reliance on administrative and support service contractors by more than 50% between FY1993 and FY1995. Established cost and schedule controls have reduced the Office of Environmental Restoration's primary support contractor labor costs by $7 million and DOE headquarters staff travel by more than 20% in FY1996. In addition, DOE reorganized, eliminating a level of management and organized headquarter's activities into integrated team efforts. These measures have reduced the overall Environmental Restoration Program management cost to only 3% of the total program costs. These corporate management costs are substantially lower than the management cost of other similar government programs (8–9%) and are comparable to private industry (1–3%) (DOE 1996). Taken collectively, these cost-saving measures and efficiencies have permitted the redirection of limited resources from DOE headquarters to the field, where accelerated site cleanup is occurring.

OEM Case Studies — Program Acceleration Accomplishments

The Department developed and implemented a multiphase process to distribute acceleration funding to DOE sites and facilities and implement their acceleration strategy. Resource distribution consisted of (1) developing selection criteria that incorporated the congressional guidance for accelerating cleanup at sites and facilities; (2) communicating this guidance to all field organizations, with a solicitation for acceleration funding proposals; (3) evaluating acceleration funding proposals received in response to the solicitation; (4) meeting collectively with DOE program representatives for all field facilities to reach a consensus on the projects and activities to be accelerated; and (5) allocating the resources and monitoring progress. In turn, DOE's acceleration strategy could be implemented program-wide. A total of $60 million was allocated to 40 environmental restoration activities to augment budgets for the purpose of accelerating actions. The total project costs ranged in size from $1 million to over $100 million and comprised both remedial action and decommissioning projects. The acceleration funding allocated between the projects was used to enhance performance and implement the acceleration strategy. A brief summary of select projects is provided below.

Energy Technology Engineering Center, California — The Oakland Operations Office received $2.5 million in funding for accelerating the decommissioning of the Rockwell Hot Lab at the Energy Technology Engineering Center (ETEC). The additional funding, coupled with $1.8 million budgeted for the project, permits early demolition of the Hot Lab and construction rubble disposal, decommissioning waste disposal, independent verification, and final site grading and project closeout, resulting in a estimated cost savings of $1.5 million. The accelerated cleanup activities reduced the time for project completion and site restoration by 3 years. The site will likely be returned to the owner, Rockwell, for unrestricted use. The completion of this accelerated activity will eliminate all potential risks to the public and environment.

Rocky Flats Environmental Technology Site, Colorado — The Rocky Flats Field Office received $5.0 million in funding to accelerate the excavation and treatment of contaminated soils in the T3 and T4 trenches at the Rocky Flats Environmental Technology Site. Cleanup of these trenches was scheduled in the scope of work of the Interagency Agreement (IAG) and was included in the closure of Operable Unit (OU) 2, along with several other projects. The cleanup of OU 2 was delayed due to budget constraints, and the project was not progressing. However, because of available acceleration funding, the cleanup of the trenches was accelerated under a removal action, resulting in a reduction in time for project completion of almost 3 years, and an estimated cost saving of $2.0 million. Fast-tracking the cleanup activities under the removal action greatly reduced the risk to workers, the public, and the environment by removing the source of groundwater contamination.

Pinellas Plant, Florida — The Albuquerque Operations Office was allocated $1.4 million in funding for accelerating treatment of groundwater at the Northeast Site at the Pinellas Plant, Florida. The accelerated cleanup permitted focus on early introduction of innovative technologies, such as rotary steam stripping and bioremediation, to enhance groundwater cleanup. Use of these innovative technologies at Pinellas Plant are estimated to accelerate groundwater cleanup by more than 10 years. The estimated savings from the acceleration of cleanup is $3 million.

Shoal Test Area, Nevada — The Nevada Operations Office received $0.35 million in funding to accelerate cleanup of the Shoal Test Area. This activity involves the isolation and containment of contamination associated with a device emplacement shaft which is open at the surface by (1) fencing off the area to limit access, (2) filling the area to the ground surface with a combination of grout and clean fill, and (3) installing two wells to monitor soil gases and groundwater. Remedial activities will be completed 6 months early and will realize an estimated cost savings of $0.35 million. The Department will retain subsurface rights at the Shoal Test Area to prevent inadvertent intrusion into the nuclear shot cavity; however, the surface will be released to the public for unrestricted use.

Sandia National Laboratories, New Mexico — The Albuquerque Operations Office was allocated $4.1 million in FY1996 acceleration funding to address potential contamination at 12 sites at Sandia National Laboratories, New Mexico. The accelerated site work included assessing cleanup activities at six high-risk Solid Waste Management Units and two groundwater plumes, implementing an interim action at one medium-risk site and remediating two high-risk sites and one medium-risk site. The accelerated cleanup activities at these 12 sites resulted in an estimated cost savings of $1.3 million and reduced the time for project completion by 3 months for each of the sites.

Battelle Columbus Laboratories, Ohio — The Ohio Field Office received $6.8 million in FY1996 acceleration funding to complete the decommissioning at Battelle Columbus Laboratories' King Avenue site. Nine buildings at the King Avenue site were contaminated primarily with uranium and thorium. Four buildings have been decontaminated and released for use without radiological restrictions. Decontamination of the remaining five buildings was completed in 1996, 7 months earlier than planned, with an estimated savings due to acceleration of $4.8 million. The early completion of the King Avenue site contributes to mortgage reduction by relieving DOE of liability for an entire privately owned site and makes the site available to the facility owner for reuse without radiological restrictions.

Mound Plant, Ohio — The Ohio Field Office received $2.0 million in funding to accelerate mitigation of contaminant migration through the removal, storage, and disposal of approximately 6000 yd^3 of contaminated soils at the Mound Plant, Ohio. The accelerated cleanup permitted the removal of the Thorium-contaminated soil to begin earlier than planned. The estimated savings from the accelerated program amount to $12 million, and are attributable to adjustments in the amount of soil required to be excavated by using a new geographical information system (GIS) that helped reduce soil excavation amounts from 18,000 to 6000 yd^3. The completion of this project will help expedite the availability of the entire site for release to the local community. The removal of the contamination will reduce potential worker exposure and mitigate the further potential spread of contaminants.

OEM Summary

The above accelerated cleanup accomplishments are examples of how the Environmental Restoration Program produces tangible results that have both immediate and long-term benefits. DOE's overall efforts to improve their program activities, reduce costs through benchmarking, and focus on tangible outcomes through performance plans and metrics demonstrated in these examples illustrates its commitment to their vision and to teamwork. Previously, DOE's efforts were focused primarily on understanding and managing the program's scope and reducing costs. DOE's new focus has shifted to encompass a more strategic view, one with clear and measurable endpoints. This perspective, coupled with demands from stakeholders and Congress for greater progress and value, has allowed the DOE to improve its approach in accomplishing environmental work. Today, embedded within all national and site-specific planning initiatives are efforts to complete site cleanup, reduce overall costs, and accelerate progress where ever possible. The legacy of this change in perspective is the development of an effective management infrastructure that enables the Office of Environmental Restoration, OEM, and DOE as a whole to demonstrate immediate results from utilizing their accelerated cleanup strategy. The use and appropriation of acceleration funding is an important example of how the Department was able to complete 40 activities an average of 1 year ahead of the original schedule and save the department $48.5 million of an initial investment of $60 million.

Overall, OEM is an agency that is committed to accelerated cleanup whenever possible. Congressional authorization for DOE accelerating cleanup activities was an important investment that has paid dividends. In addition, the 1996 congressional appropriations to DOE represented an important endorsement of OEM's strategy to accelerate program accomplishments that can be used or adapted by other environmental programs. Equally important to this endorsement is the explicit understanding that a multiyear commitment is necessary to realize the anticipated savings. Budget uncertainties and/or reductions can severely affect the ultimate success of DOE's accelerated cleanup program. With ongoing support, DOE will likely continue to demonstrate significant benefits from their accelerated program.

4.3 OFFICE OF HEALTH AND ENVIRONMENTAL RESEARCH

DOE has been extremely active in research and development, and is an important clearinghouse of new technologies for investigation and cleanup of contaminated sites. Under the Office of Health and Environmental Research (OHER), the Natural and Accelerated Bioremediation Research Program (NABIR) is an example program DOE sponsors to conduct R&D in the environmental industry. In terms of accelerating cleanup, continued funding of technical research programs such as NABIR is critical for the advancement of the investigation and cleanup technologies that expedite cleanup and increase the efficiency of remedial actions. Other research programs related to bioremediation and other technologies are supported by DOE funding, making DOE one of the leading single entities for research and development of new technologies. The following information is a summary of DOE's Natural and Accelerated Bioremediation Research document (DOE 1995c), which provides an overview of the NABIR program. The NABIR program is presented here because it is an excellent example of how new research programs can be utilized to help meet society's needs and expectations for environmental programs to develop a way of that fast-tracking cleanup and streamlining decision making.

4.3.1 Overview of the NABIR Program

OHER, with advice and assistance from OEM, established a 10-year program to improve the scientific understanding needed to harness and develop natural and enhanced biogeochemical processes in relation to bioremediation of contaminated soils, sediments, and groundwater at DOE facilities. The program builds on OHER's tradition of sponsoring fundamental research in the life and environmental sciences and was motivated by OHER's and the Office of Energy Research's (OER) joint commitment to supporting DOE's environmental management mission and the belief that bioremediation is an important part of the solution to DOE's environmental problems.

Bioremediation is defined by the American Academy of Microbiology as "the use of living organisms to reduce or eliminate environmental hazards resulting from accumulations of toxic chemicals or other hazardous wastes" (Gibson and Sayler 1992). The NABIR program addresses both natural bioremediation, which relies on naturally occurring microbial and plant processes, and accelerated bioremediation, which seeks to accelerate desirable processes through the addition of amendments (e.g., nutrients, electron acceptors) or microorganisms, or by manipulating physical, chemical, or hydrological processes. Bioremediation has been implemented successfully for degradation of petroleum hydrocarbons and, to a limited extent, degradation of explosives and chlorinated hydrocarbons, as well as for immobilization of toxic trace metals. However, the effectiveness of bioremediation cannot always be predicted reliably, due to numerous factors ranging from lack of basic scientific knowledge to engineering limitations. More importantly, few if any investigations have addressed bioremediation of the contaminants present at DOE sites, where mixtures containing chlorinated hydrocarbons, metals, radionuclides, polychlorinated biphenyls (PCB), and inorganic contaminants are common. Finally, there is general agreement among the research community that field-based research is needed to realize the full potential of bioremediation.

Summary of Bioremediation Processes

Transformation and degradation processes differ, depending on physical environment conditions, microbial communities, and contaminants, but the fate of many contaminants, from petroleum hydrocarbons to radionuclides, is influenced by microbial activity. Over the past 2 decades, it has become widely accepted that microorganisms, and to a lesser extent plants, have the ability to transform and degrade many types of contaminants. These processes form the foundation for both natural bioremediation, which relies on intrinsic rates and processes, and accelerated bioremediation, which seeks to enhance desirable processes through the addition of amendments.

Microorganisms degrade or transform contaminants by a variety of mechanisms. Petroleum hydrocarbons, for example, are converted to carbon dioxide and water or are used in generating new cells by aerobic bacteria. In this case, microorganisms use the petroleum hydrocarbons as a primary food source. Chlorinated hydrocarbons can be degraded, but the degradation takes place as a secondary or cometabolic process. Enzymes created during aerobic utilization of carbon sources such as methane fortuitously degrade the chlorinated solvents. Under anaerobic conditions, chlorinated solvents such as trichloroethylene (TCE) are degraded through a sequence of steps, where some of the intermediary by-products may be more hazardous than the parent compound (e.g., vinyl chloride). Inorganic contaminants such as nitrate can also be converted by microbial activity and plants. Depending on environmental conditions, nitrate can be converted to nitrogen gas or used as a nutrient to support cell production.

While metals and radionuclides cannot be degraded by biological activity, they can be transformed from one chemical form to another or transported from soils to above- and below-ground plant tissues. Fungi, for example, can convert dissolved arsenic and selenium to gaseous forms through methylation. Bacteria similarly have been shown to reduce mercury to its volatile elemental form. Bacteria have been shown to change the oxidation state of some heavy metals (e.g., chromium, selenium, and mercury) and radionuclides (e.g., uranium) by using them as electron donors or acceptors. In some cases, the solubility of the altered species will increase and consequently the altered species will more easily be flushed from the geologic host material. In other cases, the opposite will occur, and the contaminant will be immobilized *in situ*. The immobilization (or sequestration) of chemical and radioactive contaminants through *in situ* biological processes can make an important contribution to site remediation.

In light of the impressive capabilities of microorganisms and plants to degrade and transform contaminants, systematic attempts to understand and harness these abilities should provide tremendous benefits. In fact, some of these remediation ideas are not new and have provided the foundation for many *ex situ* waste treatment processes (including sewage treatment) and a host of *in situ* bioremediation methods that are in practice today (Hinchee et al. 1994). For example, bioremediation technologies such as bioventing and land farming have become accepted practice for remediating petroleum-hydrocarbon-contaminated soils and sediments (CISB 1993). In addition, several projects, including one at DOE's Savannah River Site, have demonstrated that TCE can be cometabolically degraded *in situ* by providing methane as a carbon source. Explosives and PCBs have been biodegraded in field-scale experiments (Harkness et al. 1993). Bacterial transformation and immobilization of trace metals such as selenium also have been implemented on a limited basis (Benson et al. 1993). Technologies to promote volatilization and dissipation of selenium have been demonstrated in surface water and soils (Frankenberger and Benson, 1994).

NABIR Background Information

NABIR was shaped by a team of scientists after reviewing recommendations in published research (DOE 1989; 1990) and an analysis of related programs in DOE and other federal agencies. The mission of the NABIR program is to provide the scientific understanding needed to harness natural processes and to develop methods to accelerate these processes for the bioremediation of contaminated soils, sediments, and groundwater at DOE facilities. The program focuses on *in situ* bioremediation of complex mixtures of contaminants present at DOE facilities. Primary contaminants of concern include mixtures of halogenated compounds, organic acids, chelating agents, metals, and radionuclides. Over the course of the program, scientific understanding is gained by performing fundamental laboratory and field research on biotransformation and biodegradation processes, community dynamics and microbial ecology, biomolecular science and engineering, biogeochemical dynamics, and innovative methods for accelerating and assessing *in situ* biogeochemical processes. Field research centers and the supporting infrastructure established as part of the program facilitate long-term, interdisciplinary research. Several key themes guide the scientific approach:

- Fundamental research is required to advance our understanding of the biological, chemical, and physical processes important for natural and accelerated bioremediation.
- Integration of scientific ideas across disciplines is essential for the development of the knowledge needed to predict and optimize bioremediation rates and processes.

- Field research centers are the best vehicles for integrating research, identifying crucial research needs, and focusing the program on DOE's most significant problems.
- Access to research and development shared infrastructure is required to advance measurement and diagnostic techniques.
- To realize the enormous potential that this program has to contribute to remediating DOE's sites, the goals must be identified clearly and used to focus the program.
- Ethical, legal, and societal issues associated with bioremediation, especially with regard to genetically engineered organisms, must be identified and addressed immediately.
- Linkages to other related programs must be established and maintained.

DOE has a legacy of environmental problems to deal with that results from production of nuclear weapons and energy. Among the most serious are widespread contamination of soil, sediment, and groundwater. The total life-cycle cost of remediating these sites is not known (DOE 1995a). Moreover, many of the contaminated soils, sediments, and groundwater are believed to be impossible to remediate with existing technology. Examples of such intractable problems include the Snake River Aquifer in Idaho, contaminated groundwater at the "100" and "200" areas at Hanford, Washington, contaminated sediments in the Columbia River, and groundwater at the Nevada Test Site (DOE 1995a). The huge cost, long duration, and technical challenges associated with remediating DOE facilities present a significant opportunity for science to contribute cost-effective solutions.

While some of DOE's environmental remediation problems are shared by other federal agencies and the private sector, DOE faces a unique set of challenges associated with complex mixtures of contaminants, especially those mixtures that contain radioactive elements. OEM's Office of Technology Development has embarked on an ambitious path to develop innovative technologies for solving many of its problems. However, they and others have recognized that, in many cases, the fundamental scientific information needed to develop effective technologies is lacking. Natural and accelerated bioremediation, which have potential to play an important role in DOE's environmental restoration program, represent one area where advances in scientific understanding could make a significant contribution.

NABIR Goals

The NABIR program goals to improve natural biogeochemical processes and develop methods to accelerate these processes include

- Develop interdisciplinary teams and focus them on gaining the fundamental knowledge necessary to overcome the obstacles facing current technologies for remediating complex contaminant mixtures in natural environments.
- Establish field research centers for long-term research on the scientific foundations of bioremediation.
- Develop the scientific knowledge, computational methods, and monitoring techniques needed to implement and predict the effectiveness of bioremediation of contaminant mixtures in a wide variety of natural environments.
- Train a new generation of scientists and engineers to address interdisciplinary problems related to biogeochemical processes in complex environments.

- Identify opportunities for using knowledge gained from this program for other applications such as *ex situ* waste treatment, pollution prevention, and ecosystem monitoring.
- Develop effective partnerships to address ethical, legal, and social issues as well as to use and share the knowledge acquired from the program for optimal application of bioremediation at DOE sites. Partners include intra- and interagency representatives, state and federal regulators, the public, and the research and development community.

These goals are realized through implementation of seven interrelated program elements described below:

1. **Biotransformation and biodegradation**. Fundamental research in microbiology to evaluate the mechanisms of biotransformation and biodegradation of complex contaminant mixtures.
2. **Community dynamics and microbial ecology**. Fundamental research in ecological processes and interactions of biotic and abiotic components of ecosystems to understand their influence on the degradation, persistence, and toxicity of mixed contaminants.
3. **Biomolecular science and engineering**. Fundamental research in molecular and structural biology to enhance our understanding of bioremediation and improve the efficacy of bioremedial organisms and identify novel remedial genes.
4. **Biogeochemical dynamics**. Fundamental research in the dynamic relationships among *in situ* geochemical, geological, hydrological, and microbial processes.
5. **Assessment**. Fundamental research in measuring and validating the biological and geochemical processes of bioremediation.
6. **Acceleration**. Fundamental interdisciplinary research in flow and transport of nutrients and microorganisms, focused on developing effective methods for accelerating and optimizing bioremediation rates.
7. **System integration, prediction, and optimization**. Fundamental research to develop conceptual and quantitative methods for describing community dynamics, biotransformation, biodegradation, and biogeochemical dynamic processes in complex geologic systems.

The goals and program elements of the NABIR program are important, because they represent a holistic approach for researching bioremediation in terms of improving the effectiveness and efficiency of *in situ* and *ex situ* cleanup technologies (Figure 4-6). Consideration of NABIR goals and program elements by other research programs, whether or not they are researching bioremediation, can aid in developing a holistic research program. Through the development of a holistic program, avenues can be explored, extending beyond the typical technical aspects of R&D, and expand into the ethical, educational, societal integration, and communication components that differentiate average research programs from excellent research programs.

Rationale for DOE Researching Bioremediation

At the present time, bioremediation is often the preferred method for remediation of petroleum hydrocarbons because it is cost-effective, and it converts petroleum

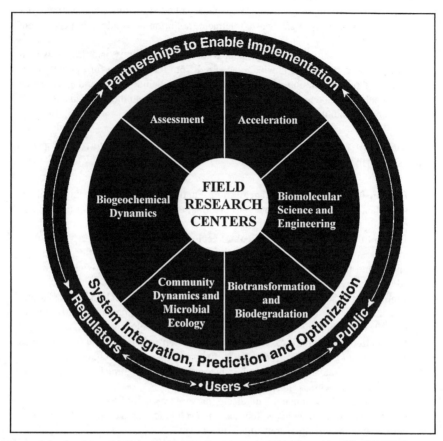

Figure 4-6 A schematic diagram showing the seven program elements of the NABIR program, integration of the field research centers, and partnerships used to implement new bioremediation technologies. (Source: DOE. 1995c. Natural and Accelerated Bioremediation Research Program Plan. DOE/ER-0659T UC-402.)

hydrocarbons into harmless by-products such as carbon dioxide and water. Over the past decade, opportunities have been identified for applying bioremediation to a much broader set of contaminants. Indigenous and enhanced organisms have been shown to degrade industrial solvents, polychlorinated biphenyls (PCB), explosives, and many different agricultural chemicals. Pilot, demonstration, and full-scale applications of bioremediation have been carried out on a limited basis. Equally important, microorganisms that transform and sequester metals and radionucleides have been identified and employed, to a limited extent, for *in situ* bioremediation. However, the full benefits of bioremediation have not been realized, because processes and organisms that are effective in controlled laboratory tests are not always equally effective in full-scale applications. Failure to perform optimally in the field setting stems from a lack of predictability due, in part, to inadequacies in the fundamental scientific understanding of how and why these bioremediation processes work.

The advantages of *in situ* bioremediation compared to, or in combination with, other remediation technologies include, but are not limited to, the following:

- *In situ* bioremediation can be used to completely degrade and detoxify some organic contaminants, thereby permanently removing liability for the contaminants.
- For deep, widely dispersed plumes of metals and radionuclides, *in situ* biosequestration and immobilization may be the only viable solution for addressing DOE's remediation needs.
- For some types of contaminants, physical and chemical methods of remediation may not completely remove the contaminants, leaving residual concentrations that are above regulatory guidelines. Bioremediation can be used as a cost-effective secondary treatment scheme to decrease the concentration of contaminants to acceptable levels. In other cases, bioremediation can be the primary treatment method, and followed by physical or chemical methods for final site closure.
- In some cases, natural attenuation, including natural bioremediation, of the contaminant plumes may be the only cost-effective solution (DOE 1995a). Natural biogeochemical processes that degrade organic contaminants, convert nitrate to nitrogen gas, and sequester metals and radionuclides will play major roles in natural attenuation. These biogeochemical processes must be adequately understood before regulatory agencies and the public accept natural attenuation as an alternative to more aggressive, less-construction oriented remediation methods.
- For radionuclides and metals broadly dispersed in surface soils, phytoremediation (bioremediation using plants) may be the only practical way to concentrate and collect the contaminants. Alternatively, plants that do not accumulate radionuclides or heavy metals can be planted to prevent wildlife exposure to contaminated sources of food.
- In highly heterogeneous geologic environments, physical or chemical methods that rely on advective transport of remediation agents to the contaminants may be ineffective. However, in such environments, *in situ* remediation schemes that rely on diffusive transport of remediation agents (e.g., nutrients) to indigenous microorganisms for degrading or transforming the contaminants may be more effective.
- For complex mixtures of contaminants requiring a combination or sequence of physical and chemical remediation methods, bioremediation techniques that use microbial consortia to concurrently address all contaminants may be faster and more cost-effective.

Results of NABIR research are designed to reduce the cost and improve the effectiveness of remediation of DOE's contaminated sites. Although this program emphasizes fundamental research on *in situ* bioremediation of soils and groundwater, knowledge applicable to *ex situ* waste treatment is also gained and transferred. Similarly, this program focuses on microbial bioremediation, compared to phytoremediation, because the majority of DOE's contaminants are below the rhizosphere. However, when appropriate, phytoremediation research is also sponsored.

NABIR Participants

Participants in the NABIR program include the national research community, OEM, and federal officials with management and oversight responsibilities. The national research community is engaged in the broadest sense, including academia, DOE laboratories, other federal agencies, and industrial researchers. Research projects are carried out by teams of laboratory scientists and engineers, academic researchers, postdoctoral fellows, graduate students, and field technicians. The scientific teams are selected through a competitive process, and research projects are formally peer-reviewed biannually. Periodic scientific forums are scheduled to stimulate the transfer of ideas and information between researchers and to help guide the direction of the scientific program. Programmatic peer reviews are conducted prior to critical decision points in the program. The scientific direction of the program and its projects is adjusted in response to the two types of peer review and broader input from the scientific community. In addition, a multi-agency steering committee provides coordination between complementary research and technology development programs.

Field research centers are a critical component of the NABIR program to ensure that the knowledge gained from NABIR is used to help solve DOE's environmental management problems. Locating the field research centers at DOE facilities provides ongoing opportunities for a two-way transfer of information between the OEM problem holders and the research community. In addition, researchers learn more about site-specific needs for fundamental and applied research. Site personnel responsible for the cleanup also keep abreast of the latest scientific developments and new opportunities for applying bioremediation. The concept for operating these field research centers is illustrated in Figure 4-7. The centers support four types of activities:

1. Small-scale, investigator-driven experiments related to community dynamics of soil microbiota and ecology, biotransformation and biodegradation processes, the survival and effectiveness of bioengineered organisms, biogeochemical dynamics, new methods of assessment, and acceleration strategies.
2. Large-scale interdisciplinary assessment of the rates and processes influencing natural bioremediation.
3. Large-scale manipulative experiments where methods of accelerating bioremediation and the underlying scientific foundations can be evaluated.
4. Pilot-scale evaluation of manipulative methods of accelerating bioremediation.

Strategies for Scientific Integration

One of the major scientific challenges facing bioremediation (and other remedial technologies) is that solution development for complex environmental problems requires an interdisciplinary approach. Successful bioremediation requires information from technical specialists such as microbiologists, hydrogeologists, geochemists, and engineers and nontechnical individuals such as the community members, land-use planners, and manufacturers. Together they represent important aspects of conducting a holistic research program vs. conducting a solely technical research program. Overcoming the traditional boundaries between these disciplines and integrating nontechnical elements are key to the success of NABIR. The NABIR program employs several strategies to achieve the integration of their research activities:

- The field research centers are a major vehicle for integrating research activities and promoting cooperation among research teams. Databases containing a wide variety of site-specific information are shared, and large-scale experiments are jointly conceived, designed, and implemented by multiple investigators.

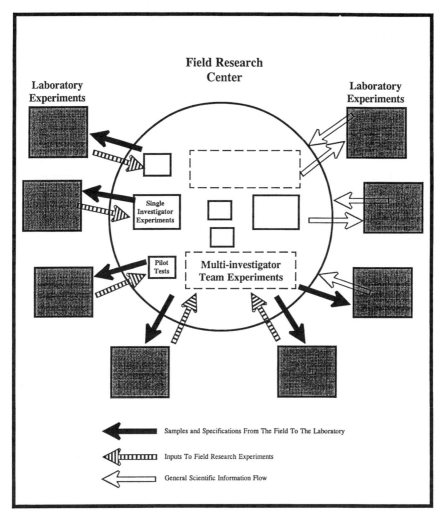

Figure 4-7 DOE's field research center concept for the NABIR program. (Source: DOE. 1995c. Natural and Accelerated Bioremediation Research Program Plan. DOE/ER-0659T UC-402.)

- Scientific project teams that include a variety of disciplines are encouraged to submit proposals to the program. Proposals submitted by individual investigators are integrated into larger programs involving a number of investigators. The peer-review process is designed to facilitate formation of these teams.
- The system integration, prediction, and optimization program element enables project teams to develop computational models that describe the dynamic interaction among microbial, geochemical, and hydrogeological processes. Participation of scientists from a variety of disciplines and research projects is required to achieve this goal. For example, engineers are involved in scaling up these models to develop tools that accurately incorporate these processes, but also can assist in the design, optimization, and evaluation of full-scale bioremediation efforts.

- Through the use of shared R&D infrastructure, information flow occurs rapidly as new experimental data is gathered and interpreted.
- Research projects that include investigators from more than one discipline are favored over those involving only one discipline.
- Regular scientific and public forums are held wherein the results of research in progress are shared among the research community including technology developers and those involved in technology transfer. These forums stimulate the rapid exchange of ideas and help keep the project focused on the most important scientific issues.

Partnerships and Education to Enable Implementation

Too often, scientific development, research, engineering implementation, and societal expectations are poorly synchronized. The focus of scientific research sometimes does not reflect engineering and societal needs and thus contributes little to the mainstream of higher education and/or the solution of immediate and important problems. At other times, scientific knowledge that could contribute to solutions is overlooked because of poor communication between problem holders, the scientific community, regulatory agencies, and the public. The underlying intention of the NABIR program is to help overcome these barriers by establishing proactive partnerships and educational vehicles to enable implementation of the scientific knowledge. Transferring the scientific knowledge gained through NABIR is crucial to the success of the program.

To achieve a seamless, integrated research program in which new knowledge is acted upon quickly and effectively in the field, as well as utilized for training scientists and engineers of the future, the NABIR program provides two types of activities:

1. **Linkage with the OEM problem holders**. Identifying those problems that are most important to solve is the critical key to funding meaningful research. An ongoing, systematic approach of working with DOE's OEM is used for continued program success, including ongoing interaction with DOE headquarters staff, field offices, site problem holders, and identifying high-priority problems that may be solved through bioremediation. OEM staff participate in programmatic peer reviews and act as advisors to the program. At least one person from OEM acts as a liaison between NABIR and OEM, and helps bridge the gap between NABIR's fundamental research and OEM's technology development programs.

    NABIR also seeks input on directions and priorities from within OHER, from other divisions of DOE, and from other government agencies and bioremediation programs. Program managers performing related research in OHER and DOE's Basic Energy Sciences Program act as advisors to the program. In addition, communication is established and maintained with other federal, state, and private bioremediation research programs to coordinate complementary programs and avoid unnecessary duplication. Managers of related programs are also invited to serve as advisors.

    Transfer of scientific knowledge is critical in the NABIR program. Through peer review and open forums with the scientific community, results of the program are communicated and the most promising avenues of research are identified on an ongoing basis. The field research centers play a major role in transferring

scientific knowledge to problem holders and students alike. Opportunities for other programs and universities to use the sites help accomplish this goal, as well as short courses and summer stipend support for higher education provided throughout the DOE program and through scientific societies and other appropriate venues.
2. **Information and education outreach programs.** Implementing technologies based on new scientific knowledge requires effective communication with problem holders, regulatory agencies, and the public. The program establishes and maintains an effective information outreach and education program to accomplish this objective. Several different methods of disseminating information are used: the World Wide Web to publish program opportunities and accomplishments, a newsletter aimed at DOE sites and regulatory agencies, a curriculum program for higher education, and timely publications to publicize successful programmatic products.

NABIR Summary

Over the last decade, progress has been made in expanding the number and type of contaminants to which bioremediation can be applied. Significant progress has also been made in the number of practical methods for implementing *in situ* bioremediation. For example, alternative strategies have been developed for delivering chemical additives such as oxygen. Chemical additives for increasing the bioavailability of recalcitrant organics have been identified. Techniques such as hydrofracturing have been developed for improved delivery of nutrients or microorganisms in low-permeability geologic media. In addition, methods have been developed for creating passive treatment systems such as biofilters (Taylor et al. 1993). Innovative concepts for using microbially produced biopolymers as *in situ* plugging agents have also been explored (Li et al. 1994). These advances are important steps toward establishing bioremediation as one of the viable solutions for *in situ* remediation of contaminant mixtures in a wide range of complex environments. And of great importance, data are now available that can demonstrate the cost-effectiveness of bioremediation in comparison to physical and chemical remediation methods (Saaty and Booth 1994; Wijesinghe et al. 1992; Atlas 1995).

NABIR builds on all recent activities and advances in bioremediation and focuses on filling the existing knowledge gaps. However, significant work must be done to build a more extensive scientific foundation to support widespread application of bioremediation, especially of contaminant mixtures, metals, and radionuclides. Future NABIR efforts will identify the ethical, legal, and social impacts, and the natural and accelerated bioremediation processes of bioremediation to implement cleanup at DOE facilities that is acceptable to the public and meets regulatory requirements.

4.4 REFERENCES

Atlas, R.M. 1995. Bioremediation. Chemical and Engineering News, April, p 32-42.

Benson, S.M., Delamore, M., and Hoffman, S. 1993. Kesterson Crises. Irrigation and Drainage Engineering 119, pp. 471-483.

Committee on *In Situ* Bioremediation (CISB). 1993. *In Situ* Bioremediation: When Does It Work? Water Science and Technology Board, Commission on Engineering and Technical Systems, National Research Council. National Academy Press, Washington, D.C.

Frankenberger, W.T., Jr. and Benson, S.M. 1994. Selenium in the Environment. Marcel Dekker, New York.

Gibson, D.T. and Sayler, G.S. 1992. Scientific Foundations of Bioremediation: Current Status and Future Needs. American Academy of Microbiology, Washington, D.C.

Harkness, M.R., McDermott, J.B., and Abramowicz, D.A. 1993. *In Situ* Stimulation of Aerobic Degradation of PCB Biodegradation in Hudson River Sediments. Science 259, pp. 503-507.

Hinchee, R.E., Anderson, D.B., Metting, F.B., Jr. and Sayles, G.D., eds. 1994. Applied Biotechnology for Site Remediation. Lewis Publishers, New York.

Li, Y., Yang, C.-Y., Lee, V.-I., and Yen, T.F. 1994. *In Situ* Biological Encapsulation: Biopolymer Shields, in Applied Biotechnology for Site Remediation (R.E. Hinchee et al., eds.). Lewis Publishers, New York, pp. 275-286.

Riley, R.G., Zachara, J.M., and Wobber, F.J. 1992. Chemical Contaminants on DOE Lands and Selection of Contaminant Mixtures for Subsurface Science Research. Department of Energy, DOE/ER-0547T.

Saaty, R.P. and Booth, S.R. 1994. *In Situ* Bioremediation: Cost Effectiveness of a Remediation Technology Field Tested at the Savannah River Integrated Demonstration Site. LA-UR-94-1714. Los Alamos National Laboratory, Los Alamos, NM.

Taylor, R.T., Hanna, M.L., Shah, N.N., Shonnard, D.R., Duba, A.G., Durham, W.B., Jackson, K.J., Knapp, R.B., Wijesinghe, A.M., Knezovich, J.P., and Jovanovich, M.C. 1993. *In Situ* Bioremediation of Trichloroethylene-Contaminated Water by a Resting-Cell Methanotrophic Microbial Filter. Hydrological Science Journal 38, pp. 323-342.

U.S. Department of Energy (DOE). 1989. Evaluation of Mid- to Long-Term Basic Research for Environmental Restoration. DOE/ER-0419.

DOE. 1990. Basic Research for Environmental Restoration. DOE/ER-0482T.

DOE. 1994. EM's Action Plan for a New Approach to Environmental Restoration in DOE. Unpublished report.

DOE. 1995a. Estimating the Cold War Mortgage: The 1995 Baseline Environmental Management Report. Vol. 2, DOE/EM-0230.

DOE. 1995b. Environmental Restoration Strategic Plan: Remediation the Nuclear Weapons Complex. DOE/EM-0257.

DOE. 1995c. Natural and Accelerated Bioremediation Research Program Plan. DOE/ER-0659T UC-402.

DOE. 1996. Environmental Restoration Acceleration Report. DOE/S-0116.

U.S. House of Representatives. 1995. 104th Congress First Session. Conference Report to Accompany H.R. 1905. Rpt., October, pp. 104-293.

Wijesinghe, A.M., Knapp, R.B., Taylor, R.T., and Carman, L.M. 1992. Preliminary Feasibility and Cost Analysis of the In-Situ Microbial Filter Concept. UCRL-ID-111021. Lawrence Livermore National Laboratory, Livermore, CA.

CHAPTER 5

State and Local Environmental Programs

A vast amount of environmental work at toxic waste sites is conducted at the state and local levels under numerous environmental programs. In most cases, these programs follow federal guidance, and less often, more stringent state criteria or state cleanup requirements that go beyond federal mandates. Such programs are relevant because many state programs are successful in accelerating investigation and cleanup of contaminated sites, and offer ideas and approaches for other states and programs to follow. This chapter describes the research, concepts, approaches, and integration of state environmental programs to expedite environmental action. In order to improve the efficiency and effectiveness of state environmental programs, states have joined together to demonstrate and approve innovative cleanup technologies, develop policy to fast-track and streamline site closures, developed state-funded programs that not only meet the requirements of federal programs, but also streamline the regulatory process, and accelerated actions and reduced costs working within federal guidance.

This chapter provides a relatively brief summary of how a select number of states have been able to fast-track environmental programs and pursue opportunities to act proactively and reduce costs. In several cases, federal and state agencies have joined forces to open communication channels between one another and the community to ensure local concerns related to environmental decisions are met. Unfortunately, review of all states programs and agencies is beyond the scope of this book. To illustrate this point, take, for example, state underground storage tank (UST) programs, or the federal equivalent state programs that follow the Comprehensive Environmental Response, Compensation and Liability Act (CERCLA) as amended by Superfund Amendments and Reauthorization Act (SARA). In a broad sense, a comparison of UST or CERCLA programs from state to state shows that they are relatively similar. However, when examined closely, differences emerge between state programs with regard to nomenclature, flexibility, funding, and other aspects, presenting confusing contrasts between programs, regulations, and opportunities to accelerate actions. In order to adequately summarize the vast amount of information n state environmental programs, each state would require its own chapter outlining the basic regulatory framework, the pros and cons of each environmental program, and sections or chapters tieing those findings to other state programs, which unfortunately may be outdated relatively quickly.

To remedy this, selected environmental programs from California, North Carolina, Texas, Iowa, Nevada, and Montana are used as examples in this chapter, along with pertinent information from other states, to illustrate how some states conduct environmental business. The purpose of this chapter, and other chapters, is to describe the foundation, background information, strategies, and resources needed to accelerate actions, no matter which state or program you are working under. Tailoring these ideas, strategies, and approaches to the state or program in question is the first step toward adopting or proposing a more streamlined or fast-tracked investigation, cleanup action, or accelerated decision making process, whether approaching a problem from a regulatory or private sector perspective. Many states in addition to the states outlined in this chapter offer streamlined approaches or encourage expedited actions under their environmental programs, and it will be necessary to research these programs and to develop a plan, either internally or externally, for implementing more efficient and effective actions.

5.1 RESEARCH PROGRAMS

A substantial amount of research is conducted at state universities, higher learning and research facilities, private colleges, and in institutions in other countries around the world. The number of environmental programs advertised on the World Wide Web, for example, totaled about 275 programs offering environmental education opportunities in the U.S. alone (at the time this book was written). In addition, some environmental programs may not be listed on the World Wide Web, or are listed under other education programs such as forestry, agriculture, geology, engineering, and others. The total number of these programs and environmental programs are a major source of environmental research and development and help improve the industry through the generation of new ideas, technical breakthroughs, and innovative solutions. Professors, research specialists, and students publish research and development work in professional papers, books, theses, dissertations, conferences, and other media. Cumulatively, these results and findings provide substantial benefit to the environmental industry in a variety of areas, including areas of innovative investigation, remedial, or analytical methodologies or technologies; new and improved remedial technologies; improved engineering designs; innovative environmental planning; policy development; and other topics. At many learning and research institutions there are federal, state, and private grant monies available to conduct environmental research and test environmental technologies that have a direct effect on how quickly human health and the environment are protected and improve the way environmental programs are run. Consequently, research and development at state universities, research facilities, private colleges, and worldwide learning institutions are important for meeting long-term environmental goals, future technology needs, and assisting the environmental industry as a whole to be successful.

STATE AND LOCAL ENVIRONMENTAL PROGRAMS 175

A large percentage of research and development conducted at learning institutions directly or indirectly relates to accelerating environmental actions, improving decision making, reducing costs, and other topics discussed in this book. The following are examples of research projects useful to illustrate the variety of work generated by learning and research institutions at the state level. The examples presented directly or indirectly relate to accelerating cleanup, investigation, or streamlining environmental decision making. The examples include a variety of projects, publication sources, and research institutions that produced the work. In each project summary, a direct or indirect link to accelerated cleanup, investigation, or decision making is outlined. The indirect links to accelerated actions are as important as the direct links, because they are often the regulations, science, and sometimes the technologies that allow models for implementing accelerated investigation and cleanup to be recognized and developed.

Work: Groundwater Remediation Using "Smart Pump and Treat"
Author: Fredrick Hoffman
Institutions: University of California, Environmental Protection Department, Environmental Restoration Division, Lawrence Livermore National Laboratory, Livermore, CA.
Source: Ground Water, Vol. 31, No. 1, January–February 1993
Summary: This paper outlines nine techniques that can be applied, all or in part, to reduce time and costs of achieving cleanup goals using pump and treat groundwater remediation systems. Technical and decision-making techniques, along with regulatory considerations, are presented to provide a holistic approach to fast-track groundwater remediation. The paper is directly related to accelerating cleanup and implementing cost reduction actions, because it focuses on streamlining the decision-making process and technical evaluation of pump and treat remediation at contaminated sites.

Work: Oxygen Pellets Spike Bioremediation
Authors: Steve Vesper[1], Larry Murdoch[1], and Wendy Davis-Hoover[2]
Institutions: (1) University of Cincinnati, Engineering Research Division, Center for GeoEnvironmental Science and Technology, Cincinnati, OH; (2) Environmental Protection Agency, Risk Reduction Engineering Laboratory, Cincinnati, OH.
Source: Soils, May 1994
Summary: This paper outlines the success of adding a slow release solid oxygen source to contaminated groundwater and using a hydraulic fracturing technique to improve bioremediation performance and stimulate increased microbiological activity. Their work is useful for evaluating the feasibility of bioremediation in cases where above-ground pumps and equipment are unsuitable, or in cases where relatively low levels of contamination are present. The work is directly related to accelerated cleanup and cost reduction. However, use of their approach is generally restricted to sites where specific site conditions apply, which limits its overall applicability. In cases where site conditions do warrant this alternative, this approach can fast-track evaluation of remedial alternatives and enhance bioremediation because other alternatives may be less applicable.

Work: Hazardous Waste Site Remediation
Author: Domenic Grasso
Institution: University of Connecticut, Department of Civil and Chemical Engineering

Source: Lewis Publishers
Summary: This book is a comprehensive compilation of remedial technologies, processes that govern the remedial technologies, engineering principles related to environmental cleanup, advantages and disadvantages of cleanup technologies, background regulatory information, fundamental risk assessment methods, and other related information that environmental professionals need to properly evaluate cleanup alternatives and collect data needed to support remedial decisions. The book is directly related to accelerated cleanup or decision making; because it is a compilation of the regulatory and technical baseline information that environmental professionals need to develop or adopt technological, administrative, or regulatory approaches that will accelerate environmental actions. Professional judgment and experience play critical roles in developing a proactive stance or an acceleration plan. Similarly, having a fundamental understanding of environmental technologies, engineering needs, regulations, field techniques, and other baseline information is also a critical for making informed acceleration decisions. Grasso's book, and other books like his, contain the baseline information environmental professionals need to make informed acceleration decisions.

Work: *Exploiting Opportunities for Pollution Prevention in EPA Enforcement Agreements*
Authors: Monica Becker[1] and Nicholas Ashford[2]
Institutions: (1) University of Massachusetts, Massachusetts Toxic Use Reduction Institute, Lowell, MA; (2) Massachusetts Institute of Technology, Cambridge, MA
Source: American Chemical Society, Vol. 29, No. 5, 1995
Summary: This paper offers policy analysis related to two relatively new U.S. Environmental Protection Agency (EPA) policies related to pollution prevention in regulatory enforcement settlements. In addition, it is an example of how research is conducted outside of technical aspects of the environmental industry to improve environmental performance and identify cost reduction opportunities. This work is indirectly related to accelerated cleanup because it relates to pollution prevention, which helps preclude the need to accelerate cleanup. However, integrating the aspects of pollution prevention into other environmental programs is sometimes the catalyst for accelerating cleanup and decision making. For example, money saved through pollution prevention actions can be applied to cleanup programs to address historic contamination, which can help fast-track cleanup through supplemental funding.

Work: *Design Factors for Improving the Efficiency of Free Product Recovery System in Unconfined Aquifers*
Authors: Jagath J. Kaluarachchi[1] and Robert T. Elliott[1]
Institutions: (1) Utah State University, Utah Water Research Laboratory, College of Engineering, Logan, UT; (2) Hill Air Force Base, UT
Source: Ground Water, Vol. 33, No. 6, November–December 1995
Summary: This paper evaluates recovery of floating free product using a vertically integrated computer flow model where the effects of steady state, multiple-stage pumping (variable time and rate), delayed start-up times, and residual oil saturation variables are tested to determine how they affect recovery systems. Their results showed multiple-stage pumping provides optimal design conditions, if cost and product recovery are primary project goals. However, if containment and recovery are the primary goals, steady-state pumping tends to provide optimal design

conditions. This work is an example of how design aspects and remedial performance can be evaluated to increase the efficiency and effectiveness of cleanup actions, which directly relates to acceleration opportunities because it can be used to hasten free product recovery.

Work: Environmental Restoration and Cleanup
Author: David Ziberman
Institution: University of California Berkeley, Department of Agricultural and Resources Economics
Source: World Wide Web
Summary: This work presents optimal cleanup level graphs, illustrating how complete cleanup to zero contamination may not be optimal. Land use is introduced as the heterogeneity that may cause the optimal level of cleanup to range, depending on the land use designation. Graphs are used to illustrate how increased money vs. increased cleanup levels have a direct relationship to environmental cleanup decisions. This work directly relates to accelerating cleanup, because decisions for future land use have a direct effect on how fast, or possible, cleanup is at a contaminated site. Depending on funding limitations, proposed land use, and the desired cleanup level, it is possible to develop optimal cleanup curves to foster an outcome-based decision-making process, instead of "worst case" decision-making processes. Using an outcome-based decision-making process, cleanup can be expedited by focusing funding on optimal future land use that is constrained by environmental cleanup limitations.

Work: Use of a Drive Point Sampling Device for Detailed Characterization of a PCE Plume in a Sand Aquifer at a Dry Cleaning Facility
Authors: Seth Pitkin, Robert A. Ingleton, and John Cherry
Institution: University of Waterloo, Waterloo Centre for Groundwater Research, Waterloo, Ontario, Canada
Source: Presentation and paper for the Eighth Annual Outdoor Action Conference, May 23–25, 1994
Summary: This paper outlines an innovative and patent-pending sampling device developed at the University of Waterloo, Canada to reduce costs, decrease volume of investigation-derived waste, and streamline the investigation process at contaminated sites. A case study is presented to show the success of their equipment in the field. This effort is directly related to accelerating investigation activities at known or suspected contaminated sites because their technology, and technologies similar to theirs, reduces the time it takes to investigate sites. In addition, their direct push technology reduces investigation costs compared to utilization of conventional drilling technologies for investigation.

The examples show both direct and indirect links that environmental research has on accelerating environmental actions, developing new cleanup technologies, and improving decision making, policy development, cost-reduction measures, and innovative approaches. While limited in scope and size, these examples represent a wide array of research conducted under many types of research and educational programs. Also, the examples demonstrate integration between state education and research programs and state to federal research programs. Note that many of the papers are authored by individuals from more than one state or environmental program. In addition, note that while most of the examples are published in periodicals, conferences, or books, the Internet is now also a source of environmental information that will likely continue to grow and serve as an information highway environmental researchers can use to learn about other research and publish new work. In summary, state learning and research institutions are a significant source of new environmental information. Continued environmental research, technology development, and policy analysis at universities and research institutions are needed for new improvements in the environmental industry and identification of new acceleration opportunities.

5.2 VOLUNTARY CLEANUP AND NEGOTIATING ACTIONS UNDER STATE PROGRAMS

State and local governments handle a variety of cleanup actions under environmental programs including, but not limited to, UST programs, EPA-approved Resource Conservation and Recovery Act (RCRA) programs, and their federal equivalent EPA CERCLA or Superfund programs. Through these programs, voluntary cleanup actions offer an important avenue for states and responsible parties to accelerate cleanup by jump-starting the environmental process, and to quickly protect human health and the environment, and reduce costs. However, voluntary actions can be difficult to negotiate, because agreement between all stakeholders is necessary in order to finalize voluntary actions and develop an agreement and plan for site restoration. Sites that are good candidates for voluntary actions are often sites that are not highly publicized or visible, and where a release or spill has recently occurred. These sites are generally located away from populated areas or sensitive environments, or are sites where site restoration is relatively straightforward. This is not to say voluntary action cannot, or should not, be conducted at sites that are highly visible. Unfortunately, highly visible sites tend to keep project teams "on their toes," leaving less room for flexibility. In these cases, voluntary actions may not always be a priority, especially when the public does not have a trusting relationship with either the responsible party(s) and/or the regulatory agency overseeing cleanup. In these situations, project teams sometimes proceed with great care in order to avoid mistakes, which can be slow and tedious.

EPA strongly encourages voluntary corrective actions and realizes that voluntary cleanup offers advantages such as timeliness, flexibility, and efficient use of funds and resources. However, the EPA is also aware that the representatives of the regulatory community have on occasion complained that procedural barriers have delayed cleanups where parties were willing to undertake voluntarily cleanup (Federal Register 1996). Although there is currently no consistent approach for conducting voluntary cleanup actions at the state or federal level, in most cases a formal agreement or decision document is needed to initiate a voluntary cleanup. At the time this book was written, EPA was planning to issue guidance on the use of state voluntary cleanup programs to address contamination at sites that may be subject to cleanup under CERCLA, including hazardous waste generators, unregulated by Resource Conservation and Recovery Act (RCRA) corrective actions requirements. The Guidance for Development of Memoranda of Agreement (MOA) Language

Concerning State Voluntary Cleanup Programs was in the process of development in partnership with interested states. The guidance will outline general components of MOAs and principals that EPA will use when deciding whether to endorse a state voluntary cleanup program and to assure private parties that subsequent federal action under CERCLA will not be taken except under limited circumstances. The general principles used in CERCLA MOA may apply to state voluntary cleanup programs. Defining the MOA contents and finalizing the memorandum, or other decision document used to document the decision to conduct a voluntary cleanup, is the primary regulatory and administrative step for voluntary cleanup. When this step is completed, then all parties should agree with the language contained in the decision document before the action is undertaken. Prior to finalizing an MOA, or other decision document, understanding the state environmental program from both a cultural and flexibility standpoints is valuable in order to streamline the voluntary cleanup action process. Additional information on voluntary actions is also contained in Chapter 8, Section 8.1.5, under EPA's Brownfields Initiative. Example state voluntary cleanup programs are presented with Brownfields information.

5.2.1 Addressing Voluntary Actions from a Cultural and Flexibility Perspective

Opportunities to accelerate cleanup and reduce costs through voluntary actions under state environmental programs include the technological, regulatory, and administrative approaches described in this book. In addition, state environmental programs offer acceleration opportunities that stem from internal program flexibility. In some respects, this type of acceleration opportunity is dependent on the "culture" of an environmental program, or in other words how the program is administered. Differences in internal flexibility within environmental programs also exist in the inconsistency between state programs. In general, no two state programs are administered in the same way, even though they oversee the same environmental regulations. To maximize acceleration opportunities under voluntary cleanups, an analysis of the freedom or flexibility programs or agencies extend must be considered in order to make informed decisions and proposals from the perspective of both the regulator and environmental industry. An analogy can be drawn comparing highway speed limits from state to state. While all states have limits of some type, they vary from strict adherence to posted highway limits to no speed limit, where "prudent and reasonable" are used as the criteria for judging the speed limit during the daytime. If you are in a hurry, understanding how states manage their speed limits will help reduce the likelihood of getting a ticket. Or, in the case of proposing an acceleration plan under a voluntary cleanup, the likelihood of a proposal being approved depends on how well it is tailored to the state's environmental program. From the regulator's perspective, steering the voluntary cleanup so that it meets program requirements helps streamline the process and gain acceptance more quickly. While challenging the way a program is run, either internally or externally, is also an option to encourage more efficient or effective program implementation, these type of changes generally take a long time to implement and gain acceptance.

Flexibility in state environmental programs also includes the ease in which innovative technologies are approved for demonstration or cleanup actions. One state program may have the tendency to be overly cautious in approving innovative technologies for site investigation or cleanup. Seeking concurrence in order to proceed with an innovative technology under a voluntary cleanup in this situation may be especially difficult. However, by crossing a state border, working with another agency, or even sometimes requesting approval from another staff member in the same program, a more receptive reaction may be realized because that particular state, program, or person may be more accepting of new technologies that have the potential to save time and money, even though there is some inherent risk in using the technology. As discussed in Chapter 7, people, and how they administer their jobs or programs, are a major component of the acceleration equation.

Overall, most state environmental programs encourage, or are open to, consideration of voluntary cleanup actions as an option to quickly implement actions that protect human health and the environment. More infrequently, however, states will also consider innovative technologies or approaches when approving voluntary actions. Determining how a program functions internally and externally and identifying an approach to gain concurrence for a voluntary cleanup is an important step for finalizing an MOA or other decision document. In addition, communicating at the right time, and at the proper decision-making level, is important in order to negotiate voluntary cleanup with both the responsible party and regulatory decision makers. The following steps are recommended to develop and approve a plan for a voluntary cleanup action:

- Provide pertinent project information to key decision makers for review early in the process and prior to having a meeting. The option for a voluntary cleanup should be highlighted in the information package, including the pros and cons of conducting a voluntary cleanup action.
- Determine if there is a general consensus to move forward with a voluntary cleanup.
- Address any data gaps and plan a formal meeting or conference call with the regulatory and responsible party decision makers and develop a preliminary MOA.
- On the state or local side, a bureau chief or department head, for example, may have to be present to make a final decision.
- On the responsible party side, a representative or owner should be present who can make financial commitments for voluntary actions that are agreed upon.
- Follow the process until a final MOA or other decision document is developed, or until another opportunity emerges for site cleanup.

Voluntary Cleanup Case Study

In the north-central plains of Montana, and within a quarter mile of the Canadian boarder, a high-pressure crude oil pipeline burst in an overflow channel below an impoundment located within a small stream valley. The burst resulted in an unknown amount of raw crude oil being released into the shallow sand and gravel vadose zone and underlying sandstone aquifer. On the U.S. side, a water supply well was located downgradient of the release and could be contaminated if the release was not cleaned up. In addition, the Canadian government was concerned that unless immediate action was taken, the crude oil would eventually migrate from the U.S. into Canada. To address these concerns, the pipeline company developed a plan to determine the extent of contamination and collect information needed to design a cleanup plan. To expedite the project, the Montana Department of Environmental Quality (DEQ) quickly reviewed and approved the company's work plan, and requested that the effort include evaluating potential corrective actions. It was at this time that the concept of a voluntary cleanup action was discussed by the state and pipeline company as a possible avenue for site restoration.

In approximately 2 months time, site conditions were characterized and the extent of contamination was determined. The pipeline company's consultant developed a proposed plan to conduct an immediate cleanup action involving excavation of contaminated soil and landfarming techniques to treat the contaminated soil. To formally approve the voluntary cleanup, a state bureau chief and company representative meeting was scheduled where the proposal was discussed and formally agreed upon, and acknowledged in a follow up letter. The state wanted the site cleaned up as soon as possible to ensure protection of human health and the environment, and prevent contaminant migration or impacts to Canadian resources. The company was willing to invest money up front and expedite actions under a voluntary cleanup action as long as the state would be willing to support them in obtaining the necessary permits and quickly approve the effort, saving time and money for the company.

The voluntary cleanup plan was formalized in a combination work plan, plans and specifications, and transmittal letter document. Within about 2 weeks, a contractor was retained by the pipeline company, state permits were obtained to work in the stream valley bottom, and the state gave its approval to use Montana's one-time landfarm rule for petroleum spills. As part of the effort, the pipeline company purchased 20 acres of nearby highlands to operate its landfarm, which met Montana's criteria for one-time landfarm operations rule for depth to groundwater and soil conditions. The work plan included target action levels for petroleum hydrocarbons, a tilling schedule to turn contaminated soil, and a sampling and analysis plan for characterizing post-cleanup and landfarmed soil conditions.

The removal portion of the voluntary action took approximately 3 months to complete, and a total of 25,000 yd^3 of soil and were excavated from the stream bottom. Crude oil was skimmed off the water table surface using a vacuum truck and a series of temporary oil/water separation ponds. The excavation was started at the upgradient end of the release, and moving downgradient, clean fill was placed in remediated areas to systematically close the site. On-site testing and confirmation sampling was used to flag soil that met target action levels and soil that needed to be excavated and hauled to the landfarm area. The haul trucks used a circular path for unloading and loading the contaminated soil, which was modified, as needed, in the field to preclude down time for the excavator and haul trucks. Upon excavating the last of the contaminated material, the overflow channel below the impoundment was armored to ensure slope stability and the surface was seeded with natural grass.

The cost of removal of the contaminated soil and initiation of the landfarm operation was approximately $275,000, or $11/yd^3, which addressed the immediate risk to the downgradient water supply well and the threat to the Canadian boarder. The project was extremely economical considering the volume of contaminated soil that needed to be excavated and costs of similar projects. The landfarming operation lasted several years after the excavation effort was completed before being closed out; however, the speed at which the landfarm effort was completed was not as critical compared to the removal effort, because the pathway to receptors was eliminated when the contaminated soil was hauled to the landfarm area.

5.3 STATE UNDERGROUND STORAGE TANK PROGRAMS

Common UST releases also offer opportunities for environmental professionals to accelerate cleanup actions and decision making. In many cases, regulations governing leaking USTs and cleanup of fuel related contaminants can be addressed relatively easily and inexpensively compared to, for example, CERCLA actions at hazardous waste sites. Soils contaminated with gasoline or less volatile fuels are often amenable to conventional cleanup technologies such as soil vapor extraction (SVE), *ex situ* bioremediation, or soil washing. In addition, cost of disposal and treatment for petroleum-contaminated soil and water is substantially less than that for hazardous wastes. The regulatory drivers for petroleum wastes as compared to hazardous wastes also allow for more opportunities and flexibility to fast-track cleanup through less restrictive disposal and treatment options.

UST acceleration opportunities and techniques are included in this chapter because investigation, corrective actions, and overall compliance of USTs are primarily handled at the state and local levels. Under 40 Code of Federal Regulations (CFR) parts 281 and 282, regulations are in place to allow states to take a lead agency role. Specifically, section 9004 of RCRA enables EPA to approve state UST to operate in the state in lieu of the federal UST program. Program approval is granted by EPA if the Agency finds that the state program (1) is "no less stringent" than the federal program in all seven elements, and includes notification requirements of Section 9004(a)(8), 42 U.S.C. 6991c(a) (8); and (2) provides for adequate enforcement of compliance with UST standards (Section 9004(a), 42 U.S.C. 6991c(a)). Within this regulatory framework, the opportunity to act proactively, save money, and expedite decision-making and cleanup actions resides mostly with individual state UST programs. While acceleration approaches such as SVE as a presumptive remedy for gasoline soil contamination may be appropriate for many UST sites, the regulatory avenue to propose and approve such an action is up to the individual state, which is therefore dependent on which state the UST is in, and how its program is administered. Understanding the UST regulatory framework in the state you work in, or work for, is critical to make informed decisions related to fast-tracking investigation, cleanup, and closure of leaking USTs (LUST). The following sections provide examples and discussion for accelerating actions and reducing costs. For telephone numbers and contacts for each state's UST and LUST programs, see Tom Crosby's web page at the address listed in Appendix A under Recommended Internet Locations on the World Wide Web.

5.3.1 Acceleration through Preventing Future UST Releases

While the focus of this book is accelerating cleanup and decision making at sites that are known or suspected to be contaminated, it is also important to understand that EPA's focus for UST programs is primarily on preventing future UST releases, while still addressing past releases. Past UST design and construction were highly ineffective in precluding product releases as substantiated by Jim Sims, Director of the Pollution Liability Insurance Agency (PLIA), "Sixty to 70% of underground storage tanks will leak or have leaked in the past. Of those, perhaps 20% will be a threat to human health and the environment and require cleanup." Since the mid 1990s, EPA's strategy is to encourage early compliance with the requirement of upgrading UST and piping to new UST installation and monitoring standards before December 22, 1998. The upgraded UST design, regulatory specifications/corrosion protection, and monitoring requirements are much more protective than previously required and are anticipated to substantially limit future UST releases. The longer outdated USTs are in place and active, the more likely they will leak or cause further contamination of soil and groundwater resulting from undetected leaks. Upgrading USTs to 1998 standards is one way of fast-tracking cleanup by identifying potential releases as soon as possible and addressing contamination. In addition, the overall costs of addressing contamination are minimized if contaminants are likely to migrate and spread contamination to adjacent soil and groundwater. Funding can also play an important role in expediting actions at LUST sites; for instance, a lack of funding can hinder progress at LUST sites, making state funding eligibility a major concern for LUST sites. From these perspectives, acting sooner than later can save time and money through cleanup of less impacted soil or groundwater, precluding a future UST release, and ensuring funding eligibility by meeting state reimbursement program requirements, such as EPA's UST upgrade deadline.

UST Reimbursement Programs

Reimbursement programs for owner/operators of USTs are a major source of funding for environmental actions at LUSTs. The incentive driving reimbursement programs is that contamination resulting from USTs is in actuality a broad social issue, not just an issue for the site owner/operator (Gurr and Homann 1996). A significant portion of USTs are operated by small businesses or private individuals, which significantly limits the number of resources available to conduct cleanup actions at UST sites. From a social and community viewpoint, maintaining UST reimbursement programs in order to address releases and mitigate impacts is vital to ensure protection of public water supplies and that regulatory requirements are met. Thus, in order to address the social issues of LUSTs, regulators and owner/operators must understand their state's UST reimbursement program and eligibility requirements for reimbursement in cases where LUSTs are known or suspected. In doing so, those responsible for regulating UST programs can help owner/operators minimize their cost liability through maximum reimbursement eligibility, and allow the owner/operator to more aggressively pursue reasonable and prudent actions because funding is available to help pay for the actions. With most petroleum cleanup costs ranging anywhere from $20,000 to $150,000, reliance on the owner/operators to cover all costs of LUST sites often results in slow or minimal progress, unless pollution liability insurance is available to cover these costs. In cases where partial or maximum reimbursement is not possible and insurance coverage is inadequate, investigation and cleanup may be hindered at LUST sites, which may cause further impacts to soil and groundwater due to limited funding. The following is an example of a UST reimbursement program for North Carolina, which operates several different reimbursement and loan programs for owner/operators of USTs.

Example UST Program — In 1988 the North Carolina General Assembly established a cleanup program for the leaking petroleum USTs in North Carolina. Since its inception, over 80,000 commercial underground petroleum storage tanks have been registered through the State UST Program. There are estimated to be thousands of additional unregistered commercial and residential petroleum USTs in the state. The LUST cleanup program in North Carolina is supported by four funds: a Federal Trust Fund, a Commercial Fund, a Non-Commercial Fund, and a Groundwater Protection Loan Fund.

The Federal Trust Fund is used to reimburse state expenditures for cleaning up contaminated UST sites when the owner cannot be located or is financially unable or unwilling to pay. The Commercial Fund provides for the payment of cleanup costs of environmental damage after the owner of a commercial leaking UST has incurred and paid an initial "deductible" of cleanup costs. The Non-Commercial Fund provides for the payment of cleanup costs of environmental damage when the leaking USTs are noncommercial or where the owner/operator cannot be identified (and federal funds are not available), or the owner fails to proceed with the cleanup because of unwillingness or lack of financial resources. There is no deductible associated with the cleanup costs for leaking USTs qualifying for the Non-Commercial Fund. North Carolina's Groundwater Protection Loan Fund (Loan Fund) was established by the 1991 General Assembly to assist commercial UST owners in obtaining loans to upgrade their tanks to current standards when they cannot obtain funds from a commercial financial institution. The current North Carolina LUST staff consists of 67 personnel with a total of 34 located in the regional offices and 33 in the central office. The certified administrative operating budget for the State Trust Fund for the 1992–93 fiscal year was $1.3 million. Federal funds in the amount of $2.0 million (including $165 thousand for the compliance group) were awarded to the program for the same period.

This typical UST program is vital for supporting and funding investigation and cleanup of LUST sites. Environmental professionals working for the state help owner/operators through the technical and regulatory UST process, usually through an environmental contractor. The funding programs are for reimbursing the contractor and subcontractor fees associated with the UST work after program requirements are met.

5.3.2 Considerations for Pilot Cleanup Technologies and Actions during UST Removals

In many cases, fuel contamination in soil can be cleaned up relatively easily, depending on the site conditions, area contaminated, and the relative level of contamination present. Fuel contamination in groundwater, however, is generally much more expensive and more difficult to remediate or stabilize as compared to soil contamination. In some cases, the tank removal stage can offer construction-oriented solutions that help mitigate or eliminate contamination, such as over excavation of contaminated soil, or installation of cleanup equipment or piping when a UST is being removed. Far too many times, USTs have been removed and the excavation backfilled with clean material, only to have an environmental consultant come back to the former UST site and re-excavate backfill soil for disposal off-site or for installation of *in situ* treatment system.

For soil, over excavation is a relatively simple technique that involves removing not only enough soil around the tank to remove it, but additional soil that is fuel-stained or tested in the field and shown to be contaminated (Figure 5-1). However, depending on infrastructure, buildings, and utilities, the size of a UST excavation may need to be limited. In these cases, it may be less expensive to install piping for *in situ* SVE or bioremediation, or another treatment technology, depending on the

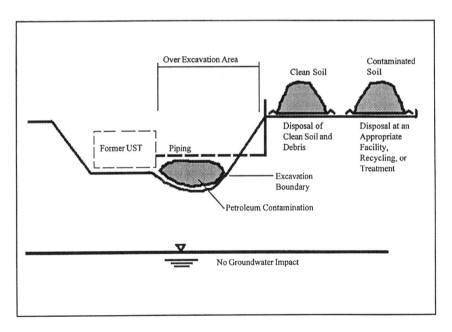

Figure 5-1 A diagrammatic illustration of a UST over excavation action. This approach is often used to remove contaminated soil adjacent to the USTs upon their removal. This approach is essential in order to quickly close UST sites that have limited soil contamination, and also to remove highly contaminated soil (as practicable) that may be a source of groundwater contamination.

amount of soil, level of contamination, and type of contamination present. Alternatively, *ex situ* treatment such as landfarming or soil washing can sometimes be used in situations where the technology is acceptable, and the approximate extent of contamination is known in order to estimate soil volume and handling needs. While preplanning installation of cleanup equipment or piping is an option for fast-tracking cleanup actions, permitting installation of such equipment ahead of time can present an issue, depending upon state regulations and requirements. An example project would be the removal of a gasoline UST in sandy soil. If soil is impacted it may be advantageous to have a variety of blank and slotted PVC piping available on site ahead of time, or readily accessible, to install piping for a simple SVE system when the excavation is open (Figures 5-2 and 5-3). In this example, the cost of purchasing and installing the PVC pipe is relatively insignificant compared to revisiting the site after the excavation is backfilled and installing piping for an SVE system. If installed ahead of time, the piping can be used as constructed for pilot testing, expanded if necessary for full-scale operation, or properly abandoned in cases where SVE is shown not to be effective in remediating soil.

When considering these types of alternatives, the cost of off-site disposal, treatment, or recycling of petroleum-contaminated soil should also be considered. In cases where the release is relatively small, often the UST removal and over excavation approach is the best suited alternative and can be used in order to close a UST site. In fact, many states allow and encourage over excavation during UST removals and collection of confirmation samples to close sites. Larger releases present less opportunity to follow this approach, but are more amenable to *in situ* and *ex situ* treatment alternatives. Prior significant experience related to UST removal is critical in order to make informed field decisions. In the example of the preplanned SVE system, a team member experienced with SVE systems should be on site during the UST removal to advise the team whether placement of piping is needed and to dictate the conceptual design of a simple SVE system. Utilizing an open excavation for installation of SVE piping or other equipment is often less expensive than remobilizing heavy equipment to a former UST excavation that has been backfilled, since few states, if any, would permit a UST excavation to remain open for a period of weeks or months unless there are extenuating circumstances. Once piping is installed, it is available for pilot testing or full-scale operation at relatively little cost.

When UST removals include over excavation or preplanned installation of remedial equipment or piping, documenting decisions to make field determinations must be clearly outlined and agreed upon by all parties ahead of time. In cases where reimbursement funds are available, the proposed actions should be preapproved to ensure that the action will be eligible for federal or state reimbursement. Major issues facing construction-oriented solutions during UST removals include developing a simple and flexible field work plan, ensuring costs are reimbursable, meeting regulatory needs ahead of time, identifying the cost and benefit factors of proceeding with construction oriented actions vs. other alternatives, such as potential for no action, and most importantly being able to make prudent field decisions

STATE AND LOCAL ENVIRONMENTAL PROGRAMS

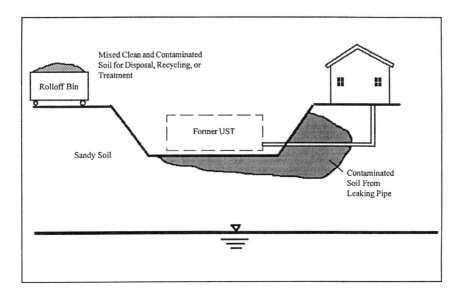

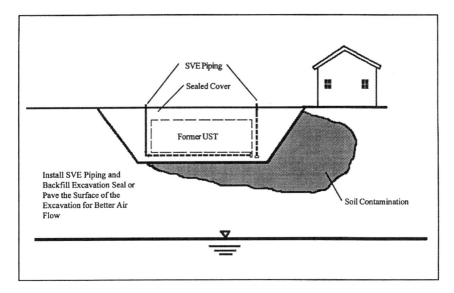

Figure 5-2 A conceptual preplanned corrective action at a UST site. (a) Residual volatile soil contamination is found to extend beneath a structure that cannot be excavated safely because the structure will be undermined. (b) Piping suitable for an SVE pilot test is installed before the UST removal excavation is backfilled. The materials could be on site during UST removal or purchased based on observations gathered during the UST removal. (c) Vapor concentrations from the pilot test are used to support a full-scale SVE operation using the same extraction wells. In some cases, modifications to the SVE system may be needed in order to maximize the performance.

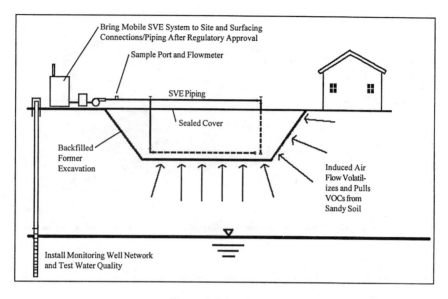

Figure 5-2 (continued)

during UST removals. In general, the more experience the field team has with UST removals, investigation of UST sites, cleanup of petroleum contamination, and estimating cost for disposal or recycling of contaminated soil and groundwater, the more capable the field team will be in making cost-effective decisions for over excavation or installation of cleanup equipment during the UST removal stage. On a cautionary note, blindly installing piping, equipment, or over excavating UST removal excavations without proper experience, guidance or approval should be avoided under most circumstances.

Risk-Based Corrective Actions for UST

Many states are developing framework for total petroleum hydrocarbon (TPH) cleanup based on the American Society for Testing and Materials (ASTM) Risk Based Corrective Action (RBCA) model (designation E 1739-95). The State of Washington, for example, is in the process of adapting the RBCA model and using the general RBCA framework and adding specific components that they considered important, such as pre-Tier 1 assessment and inclusion of ecological and aesthetic considerations. Using the RBCA framework, a consistent approach for investigation and cleanup is possible when concepts of risk and exposure are considered and the use of alternate compliance points (Figure 5-4) is evaluated. In addition, the RBCA model is reported to be relatively cost-effective for dealing with UST releases if compared to traditional approaches. In terms of accelerating cleanup and decision making, the RBCA model offers an approach environmental professionals can use to integrate UST programs and expedite LUST cleanup. A summary of the RBCA model is outlined in Chapter 6; however, for application of the RBCA process, the ASTM standard should be reviewed.

Case Study — Lawrence Livermore National Laboratory 1995 LUST Findings and the State of California's Containment Zone Policy

In October 1995, the State of California and Lawrence Livermore National Laboratory (LLNL) released the report *Recommendations to Improve the Cleanup Process for California's Leaking Underground Fuel Tanks*. This study was conducted at the request of the California State Water Resources Control Board (SWRCB) UST program. The California SWRCB asked the LLNL and the University of California at Berkeley, Davis, and Los Angeles to reevaluate the state's Leaking Underground Fuel Tank (LUFT) Cleanup Procedure program. EPA Region IX UST Program demonstrated its support for the project through partial funding of the study. The following case study outlines the progression of regulatory framework developed by the State of California in response to the LLNL work in order to proactively address LUST sites and petroleum contamination in soil and groundwater based on new findings.

The LLNL study focused on the actual number of public water supply wells that have been impacted by fuel from leaking tank sites. The study found that only 48 of 12,150 public water supply wells tested statewide (less than 0.4%) reported measurable benzene concentrations. Consequently, the study recommended that natural attenuation and biodegradation be used as the preferred remedial alternative for leaking tank sites greater than 250 ft from a public water supply well, once the source of contamination has been removed. These sites were coined "low risk soil case" and "low risk groundwater case" by the State of California. The recommendation to not actively treat contaminated soil and groundwater from leaking tank sites that are more than 250 ft from a public water supply well instigated controversy throughout the UST industry and regulatory agencies. Those who agreed with the LLNL study believed that it was a prudent and cost-effective approach for making UST cleanup decisions. Others felt that the LLNL study was scientifically inadequate and had little or no merit.

Depending on your point of view, the LLNL study can have a significant impact on accelerating UST closures and ongoing cleanup of UST sites. Many sites that were once considered candidates for cleanup actions would be closed or monitored, letting passive remediation take its course to clean up sites. The State of California developed interim guidance (California Environmental Protection Agency (Cal-EPA) 1995a; Cal-EPA 1995b) at the time this book was written in an attempt to apply the LLNL study recommendations to their UST program. The interim guidance stated that the Regional Water Boards and Local Oversight Program agencies should

- Prioritize the review and closure of low impact cases that require little or no action in order to be closed, the most obvious example being the case where the tank has been pulled, there is no free product, and only soil has been impacted.
- Allow passive bioremediation and monitoring, instead of pump-and-treat, SVE, and air sparging for low-risk groundwater-affected cases where shallow groundwater with a maximum depth to water of less than 50 ft exists and no drinking water wells are screened in the shallow groundwater zone within 250 ft of the leak.

Their guidance further outlined the Low Risk Soil Case, recommending closure if the following criteria are met:

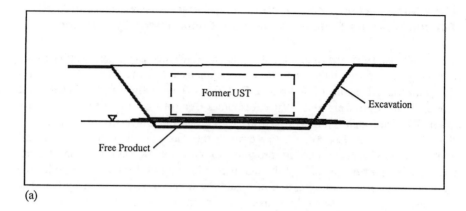
(a)

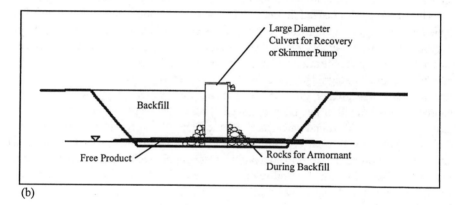
(b)

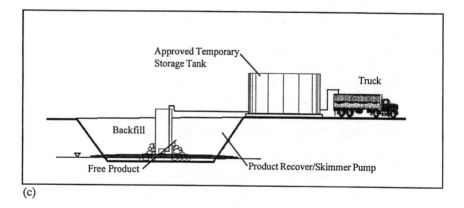
(c)

Figure 5-3

- **The leak has been stopped and ongoing sources, including free product, are removed or remediated to the extent practical per California UST regulations.**
  Free product or soil that contains sufficient mobile constituents (leachate, vapors,

or gravity flow) to degrade groundwater quality above water quality objectives or result in a significant threat to human health or the environment should be considered a source. For old releases, the absence of current groundwater impacts is often a good indication that residual concentrations present in soil are not a source of pollution. In general, if impacted soil is not in contact, or expected to come into contact, with or very close to the groundwater, it is unlikely that it is a significant source of pollution.

- **The site has been adequately characterized to determine if the site poses a threat to human health, the environment, or other sensitive nearby receptors.** The level of detail required at a given site will depend upon the presence or absence of potential receptors and exposure pathways. Delineating plumes to non-detect level is not required at all sites. It is assumed that subsurface conditions are highly variable and there is always some uncertainty associated with evaluating data at sites. However, the cost of obtaining more data must be weighed against the benefit of obtaining that data and the affect the data will have on the certainty of decisions to be made at the site.
- **Little or no groundwater impact exists and no contaminants are found at a level above established maximum contaminant levels (MCL) or other applicable water quality objectives.** By definition according to California law, soil-only contamination cases do not have significant groundwater impacts.
- **No water wells, deeper drinking water aquifers, surface water, or other sensitive receptors are likely to be impacted.**
- **The site presents no significant risk to human health.** The ASTM standard for RBCA (ASTM E-1739-95) provides a methodology to conduct a tier-based risk analysis at petroleum release sites. This methodology incorporates EPA's risk assessment practices to determine specific risk (tier 1), which provides generic risk-based screening levels and more site-specific (tier 2 and tier 3) cleanup levels that are protective of public health and environmental resources (based on site conditions).

The guidance further outlined the Low Risk Groundwater Case for closure if the following criteria are met:

- **The leak has been stopped and ongoing sources, including free product, are removed or remediated to the extent practical per California UST regulations** (see also the Low Risk Soils Case above).
- **The site has been adequately characterized to determine if the site poses a threat to human health, the environment, or other sensitive nearby receptors** (see also the Low Risk Soils Case above). In addition, presence or absence of horizontal and vertical conduits that could act as preferential pathways for the dissolved plume should be evaluated as a part of the site characterization process.

Figure 5-3 A conceptual preplanned corrective action at a UST site. (a) A low flashpoint petroleum product, such as fuel oil, is discovered during a UST removal. (b) A culvert, pipe, or well is installed in the excavation while the UST excavation is backfilled that is suitable for skimmer or product recovery pump to operate in. (c) As a temporary action, free product is recovered and stored in an approved on-site storage system/tank for recycling. In this example there are many issues that should be considered when attempting to preplan a recovery system. These include explosion hazards, well permitting, product storage safety, technical practicality for product removal, and the schedule for long-term response and investigation actions that will likely be employed to clean up and close the site.

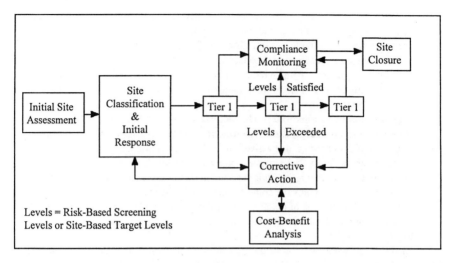

Figure 5-4 An abbreviated adaptation of the ASTM RBCA process.

- **The dissolved hydrocarbon plume is not migrating.** The LLNL concluded that the petroleum plumes in the subsurface tend to stabilize once the source is removed. Natural biodegradation of hydrocarbons is the primary reason why the stabilization occurs. Chemical concentrations of hydrocarbons in groundwater that decrease or do not change with time are the best indicators of a stable plume. Comparison of background and hydrocarbon plume concentrations of inorganic ions such as oxygen, iron, nitrate, sulfate, and other ions can provide evidence of biodegradation at a given site. These data may not be required to determine plume stability, but can supplement other lines of evidence. Stable or decreasing plumes often display short-term variability in groundwater concentrations. These effects are due to changes in groundwater flow, degradation rates, sampling procedures, and other factors that are inherently variable. This behavior should not necessarily be construed as evidence of an unstable plume, but may be the natural variations of a stable plume in the environment.
- **No water wells, deeper drinking water aquifers, surface water, or other sensitive receptors are likely to be impacted.**
- **The site presents no significant risk to human health.** For this analysis, the groundwater ingestion pathway need not be considered if the groundwater is not currently used as a source of drinking water or projected to be used within the life of the plume (see also the Low Risk Soil Case definition above).
- **The site presents no significant risk to ecological receptors.** RBCA has no specific guidance for evaluating environmental risk, although the basic framework is appropriate if site-specific exposure pathways and ecological receptors are included. If the site has a potential to significantly impact surface water, wetlands, or other sensitive receptors, it should be considered low risk.

In October 1996 (Cal-EPA 1996a), the Regional Board requested additional changes and the State Board issued these interim policy modifications for containment language:

- The Regional Boards will determine whether remediation to achieve water quality objectives is economically and technologically feasible.
- The contaminant source (e.g., tank) and any free product must be removed.
- Contaminants must be contained within the containment zone, and containment zone designation may be revoked if migration of contaminants causes water quality standards to be exceeded in other aquifers or surface water.
- The responsible party must develop and implement a management plan to assess and mitigate potentially significant impacts to human health, water quality, and the environment. Containment zone designation may be revoked if the discharger fails to implement the management plan.
- Where a containment zone might impact water supplies, a discharger must provide alternative water supply and reimbursement for increased water treatment and well modification costs. Additional mitigation measures may be proposed by a discharger seeking containment zone designation.
- Containment zones must be no larger than necessary based on site-specific conditions. A containment zone cannot substantially decrease the yield, storage, or transport capacity of a groundwater basin.
- Containment zones cannot be established in a critical recharge area, or where it would be incompatible with a local groundwater management plan.
- The Regional Boards can only designate a containment zone after a 45-day public review period.
- Before designating a containment zone, the Regional Boards must notify federal, state, and local agencies of the proposed designation, and consider any comments provided by these agencies.

Furthermore, under Resolution 1021b, a Low Risk Soil Case can have no soil saturated with petroleum and there can be no detectable petroleum in the soil within 20 ft of waters of the state. For site closure, a Low Risk Groundwater Case can have a maximum concentration of methyl tertiary butyl ether (MTBE) in groundwater less than 35 parts per billion, and if either of the conditions are met (1) benzene concentrations do not exceed 1 part per billion, or (2) there are no surface water bodies or drinking water wells within 750 ft of the petroleum source and benzene concentrations do not exceed 1 part per million. If sites do not satisfy the low-risk criteria, a more detailed conceptual model must be developed. No further regulatory action is required if a conceptual model supports that either (1) benzene and MTBE concentrations within the plume remain constant or decrease over time, and their concentrations at affected drinking water wells do not exceed 1 and 35 parts per billion, respectively, (2) active corrective measures lower the benzene, MTBE, and petroleum vapors to satisfy the criteria for low-risk sites described above, or (3) engineering or institutional controls reasonably abate the risk.

California's interim protocol and Resolutions Number 92-49 and 1021b provides an excellent example of how one state is dealing with low-risk petroleum sites and fast-tracking decision making for LUST closure. Existing California UST sites that become low-risk sites based on the LLNL study recommendation and are operating a remedial system can be shut down as long as the above criteria are met, even though the original cleanup goals were not met. Current trends for UST programs appear to more readily embrace the use and benefits of passive or intrinsic bioremediation on fuel-related spills in soil and groundwater in order to expedite environmental decision making and minimize investigation and cleanup costs, and the State of California is one of the most active in pursuing new policy in this area. More information on intrinsic remediation is presented in Chapter 3, Section 3.2.

5.4 WESTERN GOVERNORS ASSOCIATION

Established in 1984, the Western Governors Association (WGA) is an independent, nonpartisan, primarily federally funded organization of governors from 18 western states, two Pacific-flag territories, and one commonwealth. WGA was formed to provide strong leadership in an era of critical change in the economy and demography in western states. WGA is highly involved with environmental cleanup, demonstrating innovative technologies, and proposing policy to fast-tract environmental action. During autumn 1990, the WGA approached several federal agencies with the idea of developing a project to improve the overall process for cleaning up environmental contamination on federal lands located in western states. The result was a Memorandum of Understanding (MOU) between stakeholders in order to adopt a more cooperative approach for developing improved technical solutions to environmental restoration. To implement the MOU, a federal advisory committee, the Develop On-Site Innovative Technologies (DOIT) Committee was created in 1992. The WGA held a "commercialization roundtable" in August 1993 and a regulatory roundtable in October 1993. The concept of an Interstate Technology and Regulatory Cooperation (ITRC) Working Group under the DOIT Committee was conceived during the roundtables to explore opportunities for interstate cooperation on innovative technologies. Also during the roundtables, it was recognized that, in the view of environmental businesses and investors, the demonstration of innovative technologies is *secondary* to the development of markets for those technologies. The following are excerpts from WGA's two ITRC reports (WGA 1996a; 1996b). These excerpts summarize information related to innovative technologies and improving environmental agreement and communication between states for interstate cooperation. More detailed information and additional topics are provided in several reports, which can be obtained from the WGA (see Appendix A).

5.4.1 Interstate Technology and Regulatory Cooperation Working Group Findings

Innovative environmental technologies can offer significant savings over conventional technologies. Recent Navy research at 320 sites found that at 51 sites where new or innovative technologies were implemented, a net savings of 60% was possible when compared to conventional technologies (WGA 1996a). Total cost savings for the 51 sites was estimated at between $80–90 million. By comparison, a recent University of Tennessee study focusing on regulatory and policy options currently in legislative discussion aimed at reducing cleanup costs found savings in the 20–40% range for affected National Priority List sites. New Jersey reports site-specific findings that are even more compelling. In Carteret, NJ, the state and AMAX corporation selected a low-temperature thermal technology that cost $2 million dollars. Conventional excavation and disposal would have cost $7 million. In Wayne, NJ, an innovative treatment well for groundwater was installed for $630,000 instead of the conventional pump-and-treat remedy, which would have cost $2.5 million. Despite the obvious benefits of innovative technologies, they are often not chosen. This is the result of several factors including risk aversion, lack of comparable cost and performance data, and insufficient signals from the top down encouraging the use of innovative solutions.

As significant as potential cost savings are for utilizing innovative technologies, there are, unfortunately, strong deterrents to their greater use. Some of these deterrents involve regulatory issues; however, the primary deterrent is fear of making a wrong decision. State regulatory personnel often choose conventional remedies, even if they cost more, because of the cost and performance uncertainty they associate with innovative technologies. However, through interstate collaboration, the WGA concludes that if states can increase their knowledge of, and comfort level with, choosing new technologies, most of these deterrents will wane. In fact, ITRC findings showed that, second only to site-specific field demonstration data, states rely on information from one another in order to help guide cleanup decisions. Increasing information transfer at the state-to-state level is therefore essential in order to increase the demonstration and use of new and improved technologies and approaches. The following are ITRC findings related to state and interstate development to improve innovative technology development. A complete summary of the ITRC findings is presented in the WGA's report (1996a), which includes a much more diverse presentation of topics.

Interstate Development

Multi-state participation in technology demonstrations is important for broad regulatory acceptance. From an interstate standpoint, state regulators can collaborate to develop consensus procedures for the demonstration and deployment of innovative technologies. Participation of multiple states encourages broad regulatory acceptance of innovative technologies. To promote interstate relationships, the ITRC developed consensus documents and incentives to encourage participation such as networking, information cross-feed, and program leadership designed to facilitate consensus for employing innovative technologies.

Developing state programs to encourage demonstration and use of innovative technologies helps commercialize new technologies. At the individual state level, regulatory streamlining, policy development that encourages innovation, and targeted regulatory reform can reduce the review time and time needed for selection of innovative technologies and commercialization of technologies. In general, state environmental agencies can facilitate greater use of new technologies and reduce the time it takes to review them if they ensure that other state departments or divisions (i.e., air, water, etc.) of the state issuing permits for the project work cooperatively, communicate and share information, and agree on permitting requirements. If a demonstration is conducted for the purpose of gaining state approval or permitting for a larger project, then the issuing agency should also have an agreement with the applicant on what type and what quantity of data is necessary to evaluate and approve a permit application. Ideally, this agreement should be reached at the beginning of the demonstration, treatability study, or research and development stage for permit application review. The ITRC recommended that discussions should be held as necessary between an agency and applicant as new information becomes available and changes the scope of the project.

States can also stimulate the use of innovative technologies through policies that address specific technology needs. This would include moving toward performance-based (rather than prescriptive) regulations, technology-specific operation and management standards, greater use of RCRA subpart X for permitting certain technologies, and specific policies to encourage the use of innovative technologies. In addition, state regulators can enhance their knowledge of new technologies through information transfer mechanisms. Access to the World Wide Web, for example, is one avenue for information transfer, since there is an increasing amount of environmental information available there. In-house and out-of-house training are also useful in order to introduce new technologies as they are demonstrated and commercialized. Lastly, the ITRC found that most states believe a critical component to greater use of innovative technologies is early and meaningful stakeholder involvement in new technology selection, demonstration, and deployment. Thus, each state should have in place public involvement policies and procedures during demonstration of new technologies.

Case studies are an important source for overcoming regulatory barriers. Case studies are one way to show how other state programs demonstrate, implement, permit, and approve characterization and cleanup technologies. Reporting success stories in technical papers and documents, such as the WGA ITRC's *Case Studies of Regulatory Acceptance In Situ Bio Remediation Technologies* (1996b), act as strong incentives for state programs to pursue and accept new and innovative environmental technologies and approaches.

***In situu* bioremediation is an important remedial technology for site cleanup evaluations.** *In situ* bioremediation (ISB) technologies rely on the capabilities of indigenous or introduced microorganisms to degrade, destroy, or otherwise alter objectionable chemicals in soils or ground water. These technologies can be applied to soils or deep sediments and in arid or wet regions. *In situ* bioremediation is a class of technologies as variable as the subsurface itself. The In Situ Bioremediation Technology Specific Task Group of the ITRC recognized that, given appropriate conditions, *in situ* technologies can remediate contaminants more cost-effectively than conventional technologies. Additional information on *in situ* bioremediation is compiled in the WGA ITRC's report *Case Studies of Regulatory Acceptance In Situ Bioremediation Technologies* (1996b).

Direct push cone penetrometers should be used to fast-track environmental investigations. A series of different sensors are currently being developed for deployment with cone penetrometer systems which have the potential to more rapidly and efficiently characterize hazardous waste sites. One such system is the Site Characterization and Analysis Penetrometer System Laser-Induced Fluorescence (SCAPS-LIF) technology which provides real-time *in-situ* detection of TPH contamination both above and below the water table. Using SCAPS-LIF, a continuous log of TPH concentrations can be measured by extending the cone penetrometer through unconsolidated fine grained materials. A detailed log is also produced for the geologic setting using standard cone penetrometer techniques. Based on the current SCAPS develop projects, future units will likely be able to log volatile petroleum contaminants in addition to TPH.

Other ITRC Findings

Texas — innovative technology program. Established in 1993 by the Board of Commissioners to encourage use of beneficial innovative technologies in Texas. Goals included developing a list of innovative technologies in Texas, reviewing agency rules for barriers to innovative technologies in Texas, supporting agency review of permits related to innovative technologies, and educating agency staff and the public on innovative technology. The program integrates agency response on innovative technology by acting as a clearinghouse for industry and the Agency on New Technology. The program has provided a safe harbor for industry and agencies to work together and review new ideas and technologies. The program has also provided an important filtering function for the state by offering a critical review outside of the permit process to discriminate between valid new approaches and potentially fraudulent ones. This educational, informational role has been effective in opening new avenues and establishing partnerships between regulatory staff reviewing innovative technologies and industry.

Nevada — one-stop permit application. The Nevada Department of Environmental Protection (DEP) is developing procedures for a one-stop permit application. The concept under development is issuing a questionnaire to potential permit applicants that guides the department and the applicant through the various DEP permit programs. The process would establish a single point of contact for the applicant in order to facilitate the permit process and any follow-up activity. This contact would interface with the ombudsman office to coordinate the permit programs for air, water, waste (solid and hazardous), mining, and underground storage tanks. In this manner, the DEP would be able to assist an applicant in determining which permits are needed and the data needs of each permit application. This process would eliminate the submission of the same information to separate permit programs and would streamline the time spent by the applicant and DEP in the permit review process. A coordinated, one-stop process would be beneficial for the greater use of innovative technologies in particular, as often both the innovative technology user and the DEP may not be familiar with which permit programs are required.

Iowa — comprehensive petroleum UST fund. The Iowa UST program contract administrator issues community-based contracts that are based on grouping UST sites in local areas. In turn, the community-based contracts are used to conduct environmental work on multiple sites in the same general vicinity. Grouping sites has resulted in reduced contractor mobilization costs, greater economies of scale, avoidance of issues related to assessing sites with comingled contamination, and improved contractor reports. In several cases, the contractors were able to use data from other sites they worked on, and as a result, streamlined the process because less data had to be collected from the combined number of sites.

5.5 STATE ABANDONED MINE RECLAMATION PROGRAMS

Mining in the U.S. has been popular since early in our nation's history. Thousands of hard rock mines were abandoned after the ore played out or market conditions changed (WGA 1996c). Most of these sites were never reclaimed, and range from small excavation pits to huge underground caverns honeycombed with mine shafts. Some sites have toxic waste rock and tailings piles strewn over large areas. While additional inventory and characterization may be needed to clarify the overall dimension of the abandoned mine land (AML) problem, this effort should not delay action on known problem sites. However, mining waste problems are unique in that the volume of waste is large, the types of problems are diverse, and the geographic dispersion of sites is vast. In addition, many sites are in places that are relatively inaccessible during part of the year because of extreme weather conditions. Also, the ownership and responsibility for cleaning up mining sites, unlike that of many other waste streams, is often uncertain and sometimes features combinations of federal, state, tribal, industrial, and private interests within the same watershed.

The current regulatory approach for abandoned mine reclamation is filled with uncertainty and concerns regarding liability (WGA 1996c). In New Mexico, for example, the state entered into a mine waste cleanup project at the Pecos Mine in coordination with the successor to the original mining company which last produced lead in 1939 at the site. Uncertainty developed over postcleanup liability and the state's roles as both site owner (because the state purchased the mine mill and associated structures) and as regulator. A full summary of the Pecos Mine and other sites is described in an WGA report on AML in the U.S. (1996c). However, some ongoing remediation efforts have taken a new, cooperative approach. Partnerships and coalitions within watersheds, which include stakeholders in the decision-making process, have been developed. One example is an innovative, collaborative process underway in the upper Animas River Basin (southwestern Colorado) for addressing numerous historical mine waste sites using a watershed approach. A group of about 35 diverse interests, including state and federal regulatory agencies, are analyzing environmental impacts from over 400 inactive mine sites located throughout that area. Their focus is to define a set of cleanup actions at selected sites leading to an improvement in water quality at an agreed upon standard. Within the federal government, cooperation among agencies and resource sharing has also taken on new importance considering federal agency budget reductions, which limit restoration potential on AMLs.

Abandoned Mine Reclamation (ARM) is generally a state Superfund or CERCLA regulatory process, because contaminants found at abandoned mine sites are generally characterized as hazardous wastes as defined by CERCLA and RCRA. In addition, most sites were abandoned many years ago and there is often no apparent responsible party to pay for cleanup actions. In these cases, a regulatory process is needed to search for and identify responsible parties, which is generally handled under CERCLA. In cases where AMLs do not reach the National Priorities List (NPL) status and there are no viable responsible parties to pay for actions, limited federal or state funding is generally used to address health or environmental risk concerns at these sites. The following example is the State of Montana's approach for fast-tracking abandoned mine reclamation (AMR) and minimizing restoration costs.

5.5.1 Case Study — Montana's Approach for Abandoned Mine Reclamation

The State of Montana addresses federal- and state-funded AMR work through primarily the Mine Waste Cleanup Bureau (MWCB) under the Montana DEQ. The MWCB program has been able to meet the spirit and intent of CERCLA, while fast-tracking investigation and cleanup of AMR sites. In doing so, they have been able to keep project costs at a minimum while protecting human health and the environment. To accomplish this, the MWCB made relatively simple procedural adjustments in the sequence and name of documents developed under the MWCB program to meet the spirit of CERCLA. Figure 5-5 shows the overall process that the MWCB follows for AMR projects. In many respects, the process is similar to a typical CERCLA program. However, the MWCB changed the name of common CERCLA deliverables into "reclamation" deliverables. By doing so, the perception of bureaucracy associated with Superfund projects was eliminated, the size of documents was reduced, and the level of effort and resources needed to conduct projects were reduced compared to other CERCLA programs. As a result, a more streamlined program emerged that addresses relatively complex investigation and cleanup requirements. For example, the typical remedial investigation/feasibility study (RI/FS) report developed under CERCLA is called a reclamation investigation report under the MWCB program. While similar in use and scope, the reclamation document is less detailed and focused than most of its Superfund counterparts. Some CERCLA documents, such as engineering evaluations/cost analyses (EE/CA), are used in the MWCB program because they offer an avenue to fast-track document preparation and reduce costs. In this case, EE/CAs are used as the primary remedial alternative document under the MWCB program, replacing the often overwhelming feasibility study prepared for remedial actions under Superfund.

The evidence suggesting that minor procedural changes can have a substantial effect on a program's performance is supported by MWCB's success in completing AMR projects quickly and efficiently. From an environmental science perspective, the changes have little meaning. However, from a program implementation perspective where people are influenced by past performance, the changes are significant because they allow the state personnel and their contractors an avenue to streamline the AMR process and meet the spirit and intent of CERCLA. MWCB's success is also demonstrated in the relatively low cost of MWCB projects compared with similar-size Superfund projects. After administering the program for several years, the MWCB established a range of reasonable costs for projects and has been able to keep the costs of investigation, design, and cleanups relatively low. For example, a typical MWCB reclamation investigation costs about $25,000. Similarly an EE/CA costs approximately the same for developing and planning a reclamation construction effort of $500,000 to $1,000,000. Assuming a reclamation construction effort of $750,000, the cost ratio of investigate to design for a typical MWCB reclamation

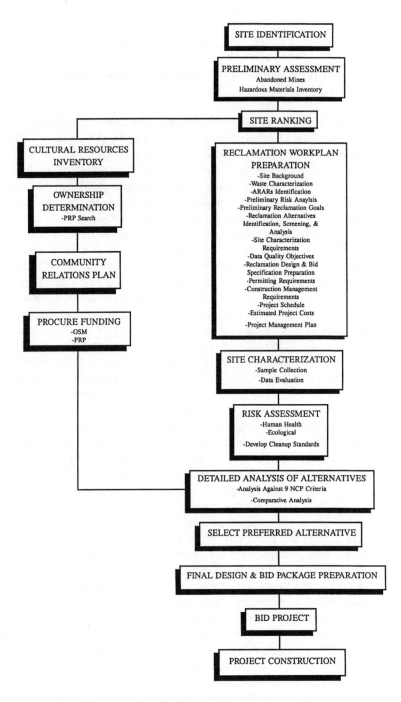

Figure 5-5

job is 0.03, or 6% of the construction cost to complete both deliverables. The MWCB's costs are much lower than typical Superfund efforts, which have a cost to complete investigation ratio to construction of about 0.2 to 1 (20–100%) and a cost to complete design to construction ratio of about 0.1 to 0.2 (10–20%). In summary, the MWCB approach is relatively simple in that it primarily focuses on alleviating a perception that environmental regulators and industry workers have when working on Superfund projects. Through relatively minor procedural changes, and effective communication of the program goals and objective to workers and contractors, the MWCB has been able to streamline their AMR program, fast-track environmental actions, and reduce environmental costs at problem abandoned mine sites.

5.6 DEVELOPING FLEXIBLE RECORD OF DECISIONS UNDER CERCLA

Reaching and finalizing a record of decision (ROD) for an NPL site can take anywhere from 5 to 10 years to accomplish, and sometimes longer. Often, the most difficult aspect of an NPL ROD is reaching consensus on "how clean is clean" for remedial actions. In most cases, there is little room for flexibility within final action levels, because a concise analytical cutoff point is agreed upon for site closure. However, the Streamside Tailings ROD for the Silver Bow Creek NPL site in western Montana contains language to leave in place, remove, and deposit contaminated mine tailings using "flexible" action levels for site cleanup. The ROD represents a substantial amount of time and information collected for the Streamside Tailings project, along with long negotiations between the State of Montana, the EPA, and the responsible party conducting the remedial action.

5.6.1 Case Study — Streamside Tailings Project

The Streamside Tailings ROD was finalized in 1995 by the Montana DEQ Superfund program, which is the lead enforcement agency on this portion of the NPL project. The remedial action plan calls for the removal and treatment of tailings using the Streamback Tailings Revegetation Study (STARS) technology. In general, highly contaminated sediments both within and outside of Silver Bow Creek are found along a 26-mile stretch of the stream. In the Silver Bow Creek case, years of analytical sampling benefited the cleanup phase of the project by determining that a very high level of accuracy is possible using an "order of magnitude definition" to identify soil requiring remediation. According to the ROD, the methodology is expected to provide for an easily defined performance standard for field implementation, while

Figure 5-5 The Montana DEQ AMRB regulatory process used to expedite, streamline, and meet the spirit of CERCLA. Relatively minor changes in the CERCLA terminology, and communicating the goal and objectives of the program to staff and contractors, has resulted in a highly efficient and effective program for abandoned mine site

also yielding a degree of cleanup that will provide adequate protection of receptor species without setting specific chemical action levels (DEQ and EPA 1995). Graphs of data for distinct boreholes showing lithologic, chemical, and physical parameters vs. depth in the soil reveal that often, the point at which the change in each of these parameters is greatest is approximately the same depth for several parameters. At some depth, most metals concentrations decrease at an approximate order of magnitude from concentrations measured in the surface to near surface depth intervals. This order of magnitude decrease in metal concentrations generally coincides with an increase in soil pH and a decrease in electrical conductivity. Although there is no unique base of the tailings with an abrupt, step-like change in chemical and physical parameters, the point that most closely approaches that distinct change can be quantitatively chosen by examination of multiple parameters and depth. While this decrease in metal concentrations is not equal to a specific value for any metal, the observation provides a good "rule of thumb" to semiquantify the location of the base of the tailings. Sampling will be performed during the response action for the Streamside Tailings project to verify that all tailings or impacted soil contaminated above the order of magnitude cleanup criteria are appropriately addressed (DEQ 1996).

In addition to the order of magnitude action level for metals, DEQ developed criteria for determining when specific tailings or impacted soil may be STARS treated *in situ* and within the floodplain. This important aspect of the ROD also provides flexibility in determining which soil can remain in place. Tailings and impacted soil areas that can remain in place meet these criteria:

- Tailings or impacted soil that is not saturated in groundwater during any part of the year.
- Tailings or impacted soil where the STARS treatment will effectively immobilize contaminants in the tailings or impacted soil.
- Tailings or impacted soil that are not located where they may be eroded and re-entrained into the stream system through normal stream processes or major flood events (post-STARS treatment).

The "flexible ROD" for the Streamside Tailings project allows for field determination of "how clean is clean" without using specific action levels and allows for flexibility in which tailings and soil are moved to repositories or kept in place. By allowing flexible criteria to define action levels and utilizing oversight and analytical confirmation sampling, the State of Montana, in association with the EPA and the responsible party, developed an approach that may help others develop flexible criteria for action levels at CERCLA sites. Depending upon the Streamside Tailings project success, other flexible RODs may be developed to allow a range of cleanup action criteria, or determination thereof, while still protecting human health and the environment.

5.7 CALIFORNIA BASE CLOSURE ENVIRONMENTAL COMMITTEE

The California Base Closure Environmental Committee (CBCEC) was formed to help with the timely restoration and reuse of miliary properties under the Base

Realignment and Closure Act (BRAC). The committee includes representatives of the Cal-EPA Department of Toxic Substances Control, SWRCB, Regional Water Quality Control Board (RWQCB), Governor's Office of Planning and Research, EPA Region IX, the departments of Army, Navy, and Air Force, and the Department of Defense (DoD). The CBCEC's mission includes addressing issues affecting timely cleanup and reuse of closing miliary bases and identifying methods and techniques that promote accelerated restoration and expedited property transfer under BRAC. To help expedite completion of the committee's mission, it created several subcommittees, called Process Action Teams (PAT), which investigate specific issues and report back to the CBCEC with conclusions and recommendations. Based on the PAT's findings, guidance documents are prepared to help expedite the safe closure and transfer of military bases. Since California base closures number relatively high compared to other states with closing military operations, quick cleanup and potential reuse are major factors in mitigating economic ramifications resulting from base closures.

To date, manuals released by the CBCEC include the Site Characterization PAT's *Recommended Content and Presentation of Reporting Hydrologic Data During Site Investigations* (CBCEC 1993a) and *Long-Term Ground Water Monitoring Program Guidance* (CBCEC 1994). In addition, the Technology Matching PAT has prepared the *Treatment Technologies Applications Matrix for Base Closure Activities* (CBCEC 1993b). The first manual compiled by the Site Characterization PAT specifies data content and presentation methods needed to support hydrologic investigations. The manual calls for ongoing development and review of hydrologic technical memoranda (Tech Memos) as investigation activities progress. The Tech Memos help provide the development and ongoing amendment of site-specific and installation-wide working hydrologic models to guide current and future site investigations. The second manual by the Site Characterization PAT calls for developing a groundwater sampling plan that provides cost effective groundwater elevation and chemical data needed to support site investigations, feasibility studies, remedial designs, remedial actions, and long-term operation and maintenance activities. The manual emphasizes development of a dynamic sampling program to amend sampling frequencies and the type of analytes tested to reflect changing data needs as a project progresses. The manual recommends quarterly and annual reports as the mechanism for reporting results, which are combined into one submittal to avoid duplication of data and evaluation.

The third manual, compiled by the Technology Matching PAT, is a comprehensive technology guidance matrix. The matrix was developed to assist remedial project managers in identifying and evaluating applicable treatment technologies during the feasibility study or corrective measure study stage. The PAT recommends that the matrix should be used only at the point when all necessary steps have been taken to determine whether further action is warranted to remediate the site, whether institutional controls and engineering controls are ineffective or inappropriate, and whether the application of treatment technologies is the preferred alternative. Overall, the matrix is a valuable source of information for fast-tracking technology evaluations. It contains information on many types of contaminants, lists of contaminants found in common areas of military bases, tabulated information on developed treatment technologies including descriptions, technology advantages and restrictions, reference sites, and technical contacts for matrix user support. Since the matrix is comprehensive and easy to use, it is an excellent resource guide that environmental professionals can use at more than just closing military bases to streamline technology evaluations for corrective measure studies, EE/CAs, and feasibility studies. The three manuals developed by the CBCEC are one of several ways in which the CBCEC is fast-tracking environmental investigation and cleanup under BRAC at military installations. In summary, the CBCEC is an excellent example of how states can integrate federal and state environmental programs to work together and meet not only federal expectations of environmental programs, but also state and local expectations.

5.8 REFERENCES

California Base Closure Environmental Committee (CBCEC). 1993a. Recommended Content and Presentation of Reporting Hydrologic Data During Site Investigations. August.

CBCEC. 1993b. Treatment Technologies Applications Matrix for Base Closure Activities. November.

CBCEC. 1994. Long-Term Ground Water Monitoring Program Guidance. March.

California Environmental Protection Agency (Cal-EPA). 1995a. Lawrence Livermore National Laboratory (LLNL) Report on Leaking Underground Storage Tank (UST) Cleanup. State Water Resources Control Board letter dated December 8.

Cal-EPA. 1995b. Interim Guidance on Petroleum Hydrocarbon Cleanups. Regional Water Quality Control Board letter dated December 8.

Cal-EPA. 1996a. Amended Resolution No. 92-49. State Water Resources Control Board, October 8.

Federal Register. 1996. Proposed Rules. Vol. 61, No. 85, May 1, p 19442.

Gurr, T.M. and Homann, R.L. 1996. Managing Underground. Pollution Engineering. Vol. 28, No. 5, May, pp. 40-44.

Montana Department of Environmental Quality (DEQ) and U.S. Environmental Protection Agency (EPA). 1995. Record of Decision Streamside Tailings Operable Unit of the Silver Bow Creek/Butte Area National Priorities List Site. November, pp. 93-108.

Montana DEQ. 1996. Cleaning Up Montana — Superfund Accomplishments 1983-1996.

Western Governors Association. 1996a. Interstate Technology Regulatory Cooperation Working Group. June.

Western Governors Association. 1996b. Case Studies of Regulatory Acceptance *In Situ* Bioremediation Technologies. February.

Western Governors Association. 1996c. Abandoned Mine Waste Working Group Final Report. June.

CHAPTER **6**

PRIVATE SECTOR, PROFESSIONAL ORGANIZATIONS, AND CONCEPTS OF INNOVATIVE SOLUTIONS

Professional associations, environmental consultants and contractors working for government, and entrepreneurs are a driving force in the environmental industry. They are often the catalysts driving new ways of solving environmental problems, implementing proactive strategies to reduce investigation and cleanup costs, instituting methods of research and development, and the primary source of innovative technologies. In order to help support development of innovative technologies, the U.S. Environmental Protection Agency (EPA), Department of Defense (DoD), and Department of Energy (DOE) offer programs to help private sector businesses and entrepreneurs demonstrate and commercialize new technologies. This chapter provides some of the more widespread strategies, processes, tools, and technologies developed by these entities. Most importantly, this chapter includes examples and case studies that illustrate how these entities have been able to accelerate environmental decision making, cleanup, and cost reduction.

The majority of information in this chapter is derived from the two different perspectives of private businesses and professional organizations. Professional organizations tend to be pragmatic, conservative, and uncompromising with respect to following the scientific method and developing new approaches, strategies, and processes to fast-track environmental actions. Businesses and entrepreneurs, on the other hand, are generally more proactive and cost-minded, implying they are more willing to accept risks related to potential failure in an attempt to achieve greater success, try new approaches, and outcompete their competitors. This observation may be considered, at best, a generalization, and undeniably the philosophies of professional associations and businesses vary in many ways. The contents of this chapter includes information from both perspectives; however, the text does not explicitly identify the subject matter or process as "conservative" or "proactive" in a comparative sense. Nonetheless, it is important to realize there are differences between some of the approaches and methods outlined in this chapter with regard to how they were developed, and the inherent level of risk in the approach or method. Realizing that these differences exist is important in order to evaluate the chances of potential failure associated with conducting accelerated and streamlined actions.

6.1 OBSERVATIONAL APPROACH

Uncertainty is a key technical factor in hazardous waste site remediation (Brown et al. 1990). It can lead to unreasonable data gathering exercises if the point of diminishing information return is not recognized. The observational method has been used to recognize and respond to substantial uncertainty, both within the hazardous waste remediation field and outside of it. Brown et al. (1990) and Corey et al. (1992) published papers in the early 1990s related to application of the observational method at hazardous waste sites, and specifically for groundwater extraction systems. A portion of their work is summarized herein to outline the general concepts of the observational method and its application at contaminated sites requiring investigation and restoration actions. More specifically, the application of the observational method is presented as an approach that can be applied to more than groundwater extraction systems, with the goal of encouraging its greater use in the environmental industry.

The principal advantage of using the observational method is the explicit and up front recognition of uncertainty in conducting remediation projects. Brown et al. submit that recognition of uncertainty contributes positively to the remediation process by recognizing as a "fact of life" that uncertainty is inherent, and that it is not possible to "fully" characterize a site before remediation begins. From this perspective, the observational method offers an approach to manage and plan contingency efforts during remediation construction and operation, thereby reducing the impacts of uncertainties early in a project.

6.1.1 Observational Method

Karl Terzaghi, a soil mechanics engineer, first developed systematic procedures for engineering under conditions of uncertainty and called his method the "observational," "experimental," or "learn-as-you-go" method. While widely used by geotechnical engineers for many years, the observational method has not been widely used in the hazardous waste remediation field, but nonetheless could be. Information provided by Peck (1969) and published in *Geotechnique* included Terzaghi's original work on the observational approach and eight key elements necessary to practice the observational method. The eight key elements include

- Exploration sufficient to establish at least the general nature, pattern, and properties of the deposits (conditions), but not necessarily in detail.
- Assessment of the most probable conditions and the most unfavorable conceivable deviations from these conditions.
- Establishment of a design based on a working hypothesis of behavior anticipated under the most probable conditions.

- Selection of quantities to be observed as construction proceeds and calculation of their anticipated values on the basis of the working hypothesis.
- Calculation of values of the same quantities under the most unfavorable conditions compatible with the available data concerning the subsurface conditions.
- Selection in advance of a course of action or modification of design for every foreseeable significant deviation of the observational findings from those predicted on the basis of the working hypothesis.
- Measurement of quantities to be observed and evaluation of actual conditions.
- Modifications of design to suit actual conditions.

Following Terzaghi's method, the character and complexity of the work being performed will determine the degree to which all of these elements are applied. In addition, as quoted by Peck, Terzaghi said that

> many variables, such as the degree of continuity of important strata or the pressure conditions in the water contained in soils, remain unknown. Therefore, the results of computations are not more than working hypotheses, subject to confirmation or modification during construction. . . . Base the design on whatever information can be secured. Make a detailed inventory of all the possible differences between reality and the assumptions. Then compute, on the basis of the original assumptions, various quantities that can be measured in the field. . . .On the basis of the results of such measurements, gradually close the gaps in knowledge and, if necessary, modify the design during construction.

Most importantly, the observational method is not applicable if the design cannot be altered during construction. In addition, the method should not be applied if the monitoring and response to one of the "potential deviations" (i.e., a contingency action to address problems observed during construction or operation) costs more than a more conservative design augmented by additional field data. While some engineering projects have been initiated using the observational method, there are likely many more projects of which the method has been used as the only avenue left to address unforeseen site conditions after construction has started.

The observational method has limitations if not properly administered and can fail under certain conditions, including

- **Failure to anticipate unfavorable conditions**. This error can lead to no preplanned course of action for a contingency plan, no ability to remedy an unforeseen condition (i.e., there are no identifiable contingency plans that can be developed to reasonably address the problem), or failure to measure a specific uncertainty early in a project that results in a larger than necessary failure later on, as described in the next bullet.
- **Failure to measure the proper observations**. This error can result by not choosing the appropriate observations, not selecting reliable observations, failing to employ personnel with the expertise needed to properly measure observations, or inaccurately interpolating the quantities measured during construction and operation (i.e., the cause of the observation/failure is not properly diagnosed).
- **Failure to consider the influence of progressive failure**. This error can result when there are progressive failures (errors occurring one after another) that go undetected or are underestimated until a massive failure is realized.

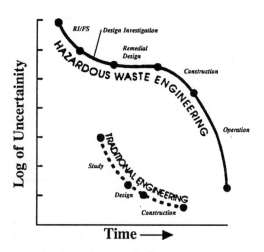

Figure 6-1 Comparison of levels of uncertainty. (Source: Brown et al. 1990. Application of Observational Method to Hazardous Waste Engineering. J. Manage. Eng., 6, 4, 479–501. With permission.)

Through proper management and planning around uncertainty, these potential failures can be mitigated.

Managing and Planning for Uncertainty

From a management and planning perspective, it is often reasoned that funds invested at the investigation and study phases can potentially save large expenses later. Based on experience and standard practices developed around traditional engineering services (e.g., design and construction of roads, buildings, bridges, sewage treatment plants), uncertainty is reduced to manageable levels at the feasibility study stage (Figure 6-1). However, at hazardous waste sites, the high potential risk to public health and the environment, combined with the high cost of remediation, sometimes demand extraordinary accuracy in the work performed. Any errors in the choice or implementation of the remedial alternative can lead to substantial residual health and environmental risk, or financial liability for the client or government entity. The high cost of being wrong determines the degree and intensity of effort necessary to characterize site conditions, which has resulted in the investigation-laden efforts and strategies implemented in the 1970s and 1980s that sometimes went beyond reasonable and prudent.

From a technical perspective, the subsurface environment presents significant uncertainty. In heterogeneous and complex environments, small subsurface features, conditions, or changes in geologic conditions often have substantial impacts on water movement and chemical concentrations. Major uncertainty also plagues source characterization, assessment of chemical fate and transport, assessment of exposure risk and health effects, and remedial action performance, all of which are inherent at most hazardous waste sites. Therefore, the consequences of uncertainty must be managed and planned for early in the life of remediation projects.

Managing uncertainty means embracing the fact that uncertainty is inherent in remediation projects. From this perspective, most uncertainty will be unique, related to site-specific conditions and actions, which must be integrated into the management process for the entire project in order to meet project goals and objectives. Planning for uncertainty means identifying specific measurements of quantities to observe, and identifying, in advance, contingency plans to respond to each situation that are anticipated to potentially develop because of uncertainty during construction and operation. Together, management and planning of uncertainty at hazardous waste sites involves consideration of the following two concepts when applying the observational approach:

- It is generally assumed that more study reduces uncertainty. In fact, however, it must be recognized that the marginal value of further studies at contaminated sites declines rapidly at some point, and more study does not necessarily lead to better information (Figure 6-1).
- The implicit goal of remediation projects has, in the past, been to design the "ultimate remedy" that can be "walked away from" following construction. However, for most complex remediation projects, in fact, this has not been possible no matter what alternative is selected. In addition, continued monitoring is generally required unless all contamination is removed, meaning the site must be revisited.

More information specific to uncertainty during the application of the observational approach is presented by Brown et al. (1990). Overall, their paper is focused on EPA's Superfund process, and Superfund terminology is used throughout their work to illustrate uncertainty and its application for groundwater remediation at Superfund sites. For the purposes of this chapter, however, the observational method is broadened to encompass other remedial programs, such as Resource Conservation and Recovery Act (RCRA) corrective actions program. Terminology used in this text, includes language from other environmental programs, and is not limited to the Comprehensive Environmental Response, Compensation and Liability Act (CERCLA). The following summarizes the application of the observational method for groundwater remediation projects, which is useful to illustrate how the observational method can be applied to other response actions, such as soil remediation projects.

Application of the Observational Method for Groundwater Remediation Projects

Brown et al. (1990) state that the elements of the observational method should be considered at least once, and possibly several times during each phase of groundwater remediation projects. These primary phases include (1) project planning, (2) remedial investigation/feasibility studies (RI/FS) or corrective measures studies (CMS), (3) remedial design (RD) or corrective measures design, and (4) the remedial action (RA) or corrective action phase. The four phases are summarized below:

1. In the project planning phase, goals and objectives of groundwater remediation are established. As part of this effort, the observational method is integrated into the project management and planning scheme, along with anticipation of the necessary response action(s) and consideration of site characteristics. In this phase, the deviations that may be required for the project are considered for planning purposes and are based on the available data and anticipated site conditions.
2. During the RI/FS or CMS phase, information is collected and compiled in order to characterize site conditions and evaluate groundwater cleanup alternatives using the observational method. If this phase is reached, in most cases, the need for a further investigation or a response action has been demonstrated to some degree through cursory sampling and analysis, such as a site inspection. In addition, data related to general site conditions have been collected and compiled and described in planning documents. To implement this phase, three main components must be completed:

   - **Determination of the type of response required**. This component includes an analysis of potential exposure pathways, risk to human health and the environment, and consideration of five common groundwater response actions (as listed by Brown et al.) including, (1) no action, (2) natural attenuation (or intrinsic remediation [see also Chapter 3]), (3) containment, (4) *in situ* treatment, and (5) contaminant removal (i.e., extraction). Information on exposure pathways and contaminant migration is used to evaluate risk, in a qualitative sense, and identify the most likely response action required. In cases where there are no significant risks to receptors, a no action or intrinsic remediation alternative may be the most likely option. However, in cases where there are significant risks, the aquifer properties should be evaluated and divided between "low yield" and "high yield" aquifer conditions. This division is useful to segregate low-yield aquifers that may offer containment or *in situ* treatment alternatives instead of pump-and-treat alternatives that are typically used in high-yield aquifer conditions. High-yield aquifers are historically candidates for pump-and-treat if the contaminants are shown to be an actual or potential threat to human health and the environment, or in cases where there are regulatory compliance issues.

     In this component, the selected response dictates the relative level of uncertainty that will be anticipated, which in turn is used to plan construction and operation actions conducted under the observational method. Brown et al. submit that groundwater pump-and-treat responses tend to have an overall lower level of uncertainty than the other five groundwater response actions listed above. Their reasoning is based on the belief that deviations can be systematically implemented under extraction alternatives that can be used to better predict and control groundwater flow and contaminant migration (i.e., increase pumping rates). In some situations this is true; however, there are many cases where pump-and-treat is shown to be highly inefficient and ineffective. Many editorials and professional papers have since been published on the limited efficiency and effectiveness of pump-and-treat technologies. In addition, the widespread use of pump and treat has been highly disputed. Many now accept that under heterogenous site conditions and in the case of dense non-aqueous phase liquids (DNAPL), the success of pump-and-treat technologies is limited, at best, which directly relates to a high level of uncertainty. In addition, other technologies and approaches have been improved since the early 1990s, which reduces the amount of uncertainty associated with some alternatives. For example, by using the newest approaches for documenting intrinsic remediation, the overall level of uncertainty is likely lower compared to using older methodologies. Furthermore, other alternatives can be considered to address high-risk groundwater contamination scenarios, such as institutional controls.

- **Gather information to establish general site conditions.** Similar to typical RI/FS or CMS, this component includes collecting data related to site contaminant characterization and site conditions, including the hydrogeologic setting. When applying the observational method, the information gathered is not focused on full site characterization, typically required under other approaches; rather, the method is centered on developing a sound conceptual model and inventory of uncertainties. The approximate lateral and vertical extent of contamination must be determined as early as possible when applying the observational method in order to expedite the cleanup action. However, the results do not have to precisely delineate the plume boundaries. General characterization is usually sufficient, because the observational method handles plumes of larger or smaller extent as deviations. Also, because the observational method moves projects more quickly into the design phase, the information gathered should include information that is commonly collected during design phases (i.e., aquifer composition data, contaminant concentrations of water from extraction wells, long-term aquifer yield, etc.), and should not be limited to only characterization data.

- **Identify the most probable conditions and reasonable deviations.** Using the available site information, a conceptual model is developed under the observational method to outline both what is known about the site and what the site "could be." Based on the site data and inventory of uncertainties that exist in the subsurface, the conceptual model identifies explicitly the facts, and assumptions associated with the site. This information generally includes (1) the most probable rate and direction of plume movement, (2) extent of contamination, (3) contaminant concentrations, (4) hydrogeologic and aquifer response conditions, (5) location of receptors, (6) discharge, treatment, disposal options for extracted water, and (7) presence of immiscible-phase liquids.

The key to knowing when to stop an iterative investigation process and model development and testing is determining that the remaining uncertainties can be handled as "reasonable deviations." However, if any of the remaining uncertainties are thought to be insurmountable from a design, construction, or contingency action perspective, additional investigation is necessary to minimize these conditions unless there are other alternatives that can be considered, such as less conservative action levels or institutional controls. Reasonable deviations include, for example, small differences in rate and direction of plume migration, small differences in the extent of contamination, or the presence of another contaminant or higher concentrations of known contaminants that can be handled through existing engineering controls, or by adding an off-the-shelf treatment unit to the treatment system. Large deviations may indicate the conceptual model was not adequately tested and that significant modifications are needed to address the deviation (i.e., a significant increase in the number of extraction wells). Most importantly, deviations are not worst-case conditions; they are specific items that are developed in response to specific types of uncertainties. As a result, and as part of the feasibility study phase, each remedial alternative should integrate the eight observational elements (see the beginning of this section) and address the following items:

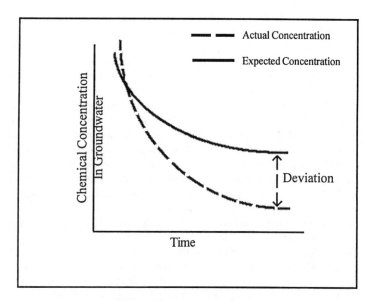

Figure 6-2 Deviation in rate of remediation. (Source: Brown et al. 1990. Application of Observational Method to Hazardous Waste Engineering. J. Manage. Eng., 6, 4, 479–501. With permission.)

- Identify specific expected conditions that form the basis of the cleanup action.
- Describe the cleanup alternative.
- Describe the deviations, monitoring, and contingency plans.
- Describe the potential impacts of the remedial alternatives, deviations, and contingency plans.

Example parameters and information that are generally monitored during application of the observational approach include changes in groundwater gradient, changes in the nature and extent of contamination (Figure 6-2), chemical concentrations downgradient of the extraction wells, influent chemical concentrations into the treatment system, and effluent chemical concentrations (Figure 6-3). In addition, the level of monitoring required under the observational approach may be greater than that of the traditional approaches, depending on the level of site characterization. Lastly, the monitoring system should not only be designed to indicate that a deviation has occurred, but also identify the cause of the deviation because differing circumstances can cause the same deviation.

3. During the design phase of the observational method, a contingency plan is developed that outlines the specific courses of action, or design modifications, proposed to respond to each reasonable deviation. For example, for a groundwater extraction system, specific courses of action would have to be developed for deviations in each of the key parameters being monitored, such as

- Inadequate hydraulic gradient control needed to contain or remove a contaminant plume, or slower than expected cleanup rates. The course of action may include increasing pumping rates or installation of one or more extraction wells.

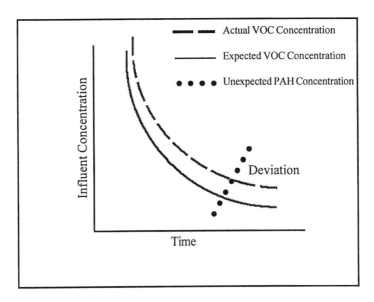

Figure 6-3 Deviation in expected treatment system effluent compensation. (Source: Brown et al. 1990. Application of Observational Method to Hazardous Waste Engineering. J. Manage. Eng., 6, 4, 479–501. With permission.)

- A portion of the plume escapes downgradient of the planned containment or recovery area. The course of action may include no action, monitoring the situation, installation of one or more additional extraction wells within the existing recovery area to limit future plume migration, or expanding the containment or recovery area by adding more extraction wells.
- Influent contaminant concentrations are different than expected or new chemicals or contaminants are detected. A response may include modifying pumping rates to change influent composition, diverting flow from other wells to equalize composition, adding or deleting treatment units and processes, modifying the treatment efficiency, or diverting the influent water to a basin for subsequent treatment off site.
- Posttreatment effluent concentrations are found to be higher or lower than expected. A response may include diverting the effluent to holding basin for subsequent on-site or off-site treatment, or stopping the treatment system temporarily for upgrading.

In order to propose a specific response for each reasonable deviation, the design needs to be flexible and incorporate contingency responses. In addition, it is critical to consider the sensitivity of the remedial action in relation to responding to a deviation. It is relatively easy to envision installation of another extraction well; however, it is much more difficult to evaluate the potential impact it will have on another part of the remedial action, or whether it is actually possible to implement.

A remedial design using the observational method must identify the key parameters of remedial performance, and include, or reference, a monitoring plan with these parameters for use during the construction and operation periods. In addition, the design must incorporate the contingency actions into the scope of work for the remedial action contractor. Therefore, it is important to calculate, estimate, and

identify methods that provide a basis for detecting deviations, such as computer modeling results. Finally, clear criteria for what constitutes a deviation must be defined, which may include statistical approaches to evaluate performance, especially in cases where the difference between expected and observed conditions are relatively small.
4. During the remedial action or correction actions phase, it is likely that it will be necessary to actually implement contingency plans and respond to deviations. The actions that are taken may not be straightforward, depending on site conditions and the type of remedial action underway. The party responsible for overseeing the remedial action will have to use judgment, and have the required experience necessary to determine if a deviation has occurred and to select an appropriate response should action be taken. It is possible, and likely, that observations measured during remedial operations will indicate a deviation has occurred; however, the quantity measured, or observed conditions, will not always reveal the cause of the deviation. In cases where more than one cause is possible for a deviation, considerable judgment may be required to select the most appropriate response action. A fair amount of operational experience may be required before this level of judgment is acquired. Figure 6-4 summarizes the observational approach for groundwater extraction.

Greater Use of the Observational Method

Greater use of the observational method is possible in the environmental industry, and should be considered in cases were expediting a project schedule is desired, or when fast-tracking protection of human health and the environment. While the papers by Brown et al. (1990) and Corey et al. (1992) on the observational approach are focused on groundwater remediation, the observational method can also be applied to other methods or types of environmental cleanup and media. In addition, the method could potentially be applied to investigation programs in order to improve performance of investigation efforts by "observing" real-time analytical results where uncertainty principles and data are used to adjust and focus investigation efforts. Most importantly, however, the observational approach must fit within the regulatory framework and be the preferred approach by all members of a project team, including separate entities such as the responsible party, regulatory personnel, and the contractor. Also of great importance is the development of a contracting strategy that matches the observational method. A fixed price contract, for obvious reasons, would not allow implementation of contingency plans. Instead, a more flexible contracting strategy must be used in order to allow for deviations, yet keep the overall costs reasonable (see also Navy section [Section 3.3] in Chapter 3 for more information on contracting).

6.1.2 Sequential Risk Mitigation

Sequential risk mitigation (SRM) combines several acceleration approaches described in this book, and quantitatively integrates the concepts of relative risk to human health and the environment posed by contaminated sites. By using a step-by-step strategy, the initial response under the SRM approach is to first implement

PRIVATE SECTOR, PROFESSIONAL ORGANIZATIONS, INNOVATIVE SOLUTIONS 215

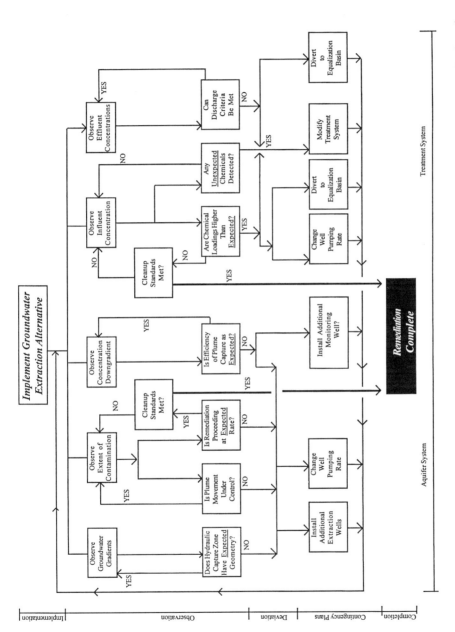

Figure 6-4 Application of the observational approach for groundwater extraction. (Source: Brown et al. 1990. Application of Observational Method to Hazardous Waste Engineering. J. Manage. Eng., 6, 4, 479–501. With permission.)

Table 6-1 Sequential Risk Mitigation Levels

Level	Numerical Risk Range	Example Description
1	Extreme risk to 10^{-4}	A serious situation threatening public health and the environment.
3	10^{-2} to 10^{-4}	Hot spots, exposed free phase chemicals, dense nonaqueous phase liquids or light nonaqueous phase liquids in drinking water aquifers with clearly definable pathways and/or rapid movement of chemicals.
4	10^{-4} to 10^{-5}	Subsurface buried chemicals of medium concentrations, dissolved constituents in groundwater with slow-moving or ill-defined pathways.
4	10^{-5} to 10^{-6}	Chemicals of low concentration in soil and groundwater at significant depths below ground surface with poorly defined pathways.
5	Less than 10^{-6}	Minute concentrations in soil and groundwater with no realistic or remotely possible pathway of exposure.

Source: Smith Technology Corporation (undated).

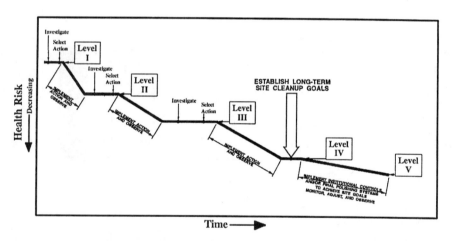

Figure 6-5 Site remediation by SRM. Risk is sequentially reduced and quantified over time through actions and observations (see Table 6-1 for risk levels 1 through 5). (Source: Smith Technology Corporation, Portland, OR.)

an action to address the most obvious source of high-risk contamination, and then to continue following the process to sequentially reduce risk until it reaches an acceptable level. Specifically, sites are reclassified from level 1 to level 5 over time, depending on actions completed and the overall risk to humans and the environment. In the hierarchy, level 5 represents the lowest level of risk (Table 6-1 and Figure 6-5).

The benefit of using this approach over traditional ones is that it focuses funding primarily on cleanup and protection, which is similar to the technical process inherent in EPA's Superfund Accelerated Cleanup Model (SACM) (Chapter 2). In many respects, the SRM approach integrates a similar technical process for conducting early actions up front, and later, long-term actions. However, the SRM approach also incorporates the need to consider the inherent uncertainty associated with remediation of hazardous waste sites by allowing the integration of the observation method into the process (Figure 6-5). As described in the preceding section, uncertainty can be dealt with by using the observational method to develop deviations and contingency plans to expedite the overall cleanup process. Thus, the SRM approach is an example of how the concepts of early actions and the observational approach can be integrated in order to streamline protection of human heath and the environment, and accelerate critical response actions.

6.2 RISK-BASED CLEANUP APPROACH

The American Society for Testing and Materials' (ASTM) *Standard Guide for Risk-Based Corrective Action Applied to Petroleum Release Sites* (Designation E1739) details a process developed to outline a consistent decision-making process for the study and cleanup of petroleum release sites. Benefits of the Risk-Based Corrective Action (RBCA and pronounced "Rebecca") process include the use of alternate compliance points, a consistent framework for the investigation and cleanup of petroleum releases, and integration of risk and exposure concepts into the evaluation and cleanup process. In addition, the process, if properly administered, is designed to efficiently and effectively use resources. Similar to most environmental methods and approaches, the RBCA process is based on protecting human health and the environment; however, it differs in that a "tier" process is used to characterize site conditions and risk, and evaluate restoration alternatives. Conceptually, the RBCA process integrates the variance of site complexity, petroleum contamination characteristics, and the risk posed to human health and the environment in order to develop an action plan or to close sites. In addition, the tier process utilizes a consistent strategy between sites to propose closure or plan corrective actions as soon as possible. On a site-specific level, RBCA response actions are tailored to address site-specific needs and potential exposure to contaminants that present a risk. Many environmental professionals support the underlying principles of RBCA, as a means to bring consistency into the investigation and cleanup of petroleum. However, they are concerned that the overall RBCA process may oversimplify restoration analysis to the point where, at some sites, specific conditions may not be considered in order to maintain the process rather than deviating based on professional judgment and experience (see Trotti 1996 for an example).

In the RBCA standard, the ASTM did not limit its approach to a particular class of compounds; however, examples related to petroleum releases are emphasized. The overall RBCA process and strategy can be applied to other types of releases beyond petroleum; however, there must be a regulatory avenue and agreement to use the RBCA approach. In addition, from an ecological risk standpoint, the RBCA process does not provide a quantitative evaluation process. Rather, the RBCA process includes a qualitative assessment of the actual or potential impacts to ecological receptors, which may not be sufficient for all situations, especially under programs that regulate releases other than petroleum contamination. Nonetheless, the RBCA decision-making process integrates risk and exposure assessment methods with site assessment activities and remedial measure selection in order to ensure that the chosen action is protective of human health and the environment (ASTM 1995a). Figure 6-6 shows the RBCA process flow chart, which outlines the three-tier approach and 10 overall steps of the RBCA process. As each tier is completed in the RBCA process, the process includes evaluation of whether additional site-specific analysis is required or if response actions, or closure, are warranted. The general sequence of tiers and steps for the RBCA process is summarized below.

- **Site assessment** (step 1). In this first step, a baseline site assessment is required to compile background information, chemicals of concern, migration pathways, exposure potential for human and ecological receptors, and other site-specific information that may affect decision making or prompt the need to conduct a response action.
- **Site classification** (step 2). Initial corrective actions are considered in this step based on the assignment of site classifications used in the RBCA process. The site classifications are based on the importance of a response and include (1) immediate threat to humans, safety, or sensitive environmental receptors, (2) a short-term (0 to 2 years) threat to humans, safety, or sensitive environmental receptors, (3) a long-term (>2 years) threat to humans, safety, or sensitive environmental receptors, and (4) no demonstratable long-term threat to humans, safety, or sensitive environmental receptors (see ASTM guide for specific criteria for each classification and limitations). Actions may be implemented while the site classification evaluation is completed, and subsequent reclassification is required after early or interim actions are completed.
- **Tier 1 evaluation** (steps 3 and 4). In the third step, a RBCA Look-Up table is developed (or used from a previous RBCA project) to compare Tier 1 Risk-Based Screening Level (RBSL) chemical concentrations with the criteria used to support site closure or prompt requirements for additional site-specific characterization. The Look-Up table RBSLs may include soil, groundwater, and vapor concentrations, depending on site conditions and contaminant type, and they are derived using U.S. Environmental Protection Agency (EPA) reasonable maximum exposure (RME) scenarios and toxicological parameters for specific compounds in petroleum contamination, such as benzene. The ASTM guide describes the procedures required to prepare a Look-Up table. As part of the forth step, the need for further tier evaluation is completed. This step includes consideration of site closure and monitoring if chemicals of concern are less than RBSLs, need for interim cleanup actions if RBSLs are exceeded (refer to Tier 2 evaluation below), use of the RBSLs as remediation target levels (bypassing Tier 2 evaluation), and evaluation of the need for additional site-specific classification information.
- **Tier 2 evaluation** (steps 5 and 6). Under the fifth step of the RBCA process, site-specific target levels (SSTL) and point(s) of compliance are determined and used to compare with site chemical data and develop a plan of action for cleanup or site closure. As part of this step, the need for further site characterization is also evaluated. Overall, the Tier 2 SSTLs and Tier 1 RBSLs are developed based on meeting similar numerical risk criteria; however, for Tier 2 SSTLs, site-specific exposure information replaces the non-site-specific exposure information used in Tier 1. As part of the sixth step, the need for further tier evaluation is considered. In addition, potential implementation of interim remedial action and use of Tier 2 SSTLs or Tier 1 RBSLs as remediation target levels are evaluated.

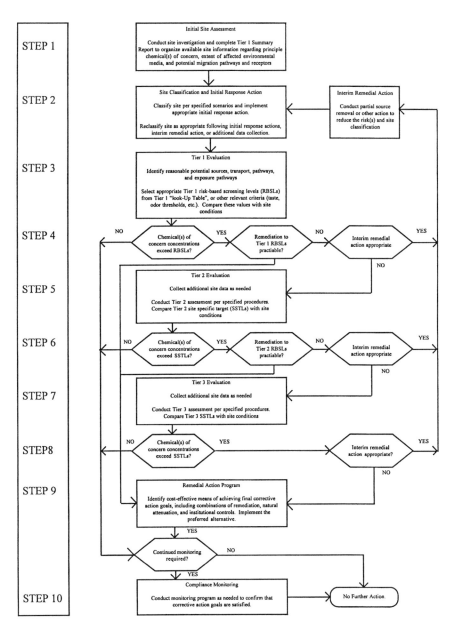

Figure 6-6 ASTM's RBCA process flow chart and 10-step process. (Source: ASTM. 1995a. Standard Guide for Risk-Based Corrective Action Applied to Petroleum Release Sites. Designation E 1739-95. With permission.)

- **Tier 3 evaluation** (steps 7 and 8). In cases where the Tier 3 evaluation process is warranted, step 7 involves a much more in-depth evaluation of SSTLs. At this tier level, the process may require additional site assessment information, integration of probability evaluations in the process, or application of solute fate and transport models in order to refine SSTLs. These actions are used to refine the SSTLs for direct and indirect pathways, and points of exposure and compliance. Upon reaching this tier level, either monitoring, site closure, or additional data collection is warranted in order to continue the process. In step eight, comparison of the site chemicals of concern at the points of compliance (or source areas) with the Tier 3 SSTLs is completed to make this determination. In this step, the process either moves the project back into the site characterization stage or progresses the site toward the final decision making stage and/or closure.

- **Remedial action** (step 9). The ninth step involves development of a remedial action plan to meet SSTLs (or RBSLs if applicable). The remedial action stage includes consideration of contaminant reducing alternatives or application of institutional controls, or engineering control (such as capping), as a way of achieving the level of protection to human health and the environment and regulatory compliance required to close the site.
- **No further action** (step 10). Pending completion of any compliance monitoring required, a decision for no further action is made in the tenth step. This step also includes ensuring that any institutional controls used to close the site remain in place.

Included in the RBCA guide are five appendices, including (1) information on physical/chemical and toxicological characteristics of petroleum products, (2) derivations and an example of a Tier 1 RBSL Look-Up Table, (3) uses of predictive modeling relative to the RBCA process, (4) considerations for institutional controls, and (5) examples of RBCA applications. Together, these appendices provide a comprehensive package needed to implement the RCBA process and develop cleanup and closure strategies. From a regulatory perspective, the ASTM's RBCA process is intended to compliment federal, state, and local regulations, and use of the RBCA process would likely require regulatory approval ahead of time before implementing the process. Several states, such as Washington and California, have adopted the RCBA process for their underground storage tank (UST) release sites and regulatory framework (see Chapter 5).

6.2.1 Other ASTM Standards

ASTM has several other standards, in addition to the above RBCA standard, that focus on environmental restoration activities and improved investigation or response performance, all of which are available from ASTM (see Appendix A for an online information source). One example is ASTM's Provisional Standard Guide *Accelerated Site Characterization for Confirmed or Suspected Petroleum Releases* (Designation PS3-95). This provisional guide provides ASTM's approach to rapidly and accurately characterize a confirmed or suspected petroleum release. However, because the standard is provisional and only partial consensus was realized in the ASTM approval process by the sponsoring subcommittee members, the standard does not carry the same weight as the RBCA standard. The following is a brief outline of the general concepts presented in this provisional standard. A more in-depth description of their approach is available from ASTM (1995b).

Overall, the provisional guide is a framework for private and government entities to fast-track the site characterization process, reduce investigation costs, and quickly make informed corrective-action decisions. The guide outlines the approach for one-time site characterization studies, use of rapid sampling techniques and on-site analytical methods, and application of on-site interpretation of field data to continually refine site characterization needs and the site conceptual model. Information in their approach is used to determine or consider (1) the need for interim remedial actions (IRA), (2) the overall site classification and relative prioritization of projects, (3) need for further actions, and (4) alternatives for remediation. In general, their approach can be integrated into existing cleanup programs and is designed to be used in conjunction with other ASTM guides, such as the RBCA standard. The standard contains general information, references, terminology, identifies the significance and use of the approach, describes the accelerated site characterization process, provides an example of a data quality classification system, provides a list of physical and chemical properties and geologic/hydrogeologic characteristics applicable to site characterizations, and provides a case study. The provisional standard is a good example of how professional organizations are developing approaches to expedite the investigation process.

6.3 COMPUTER MODELING

This section contains a relatively brief and rudimentary technical discussion on computer modeling. The purpose of presenting this information is to introduce computer modeling as an important tool in the environmental industry and illustrate how computer modeling can be used to streamline decision making, planning, and reduce costs.

Vadose zone, groundwater, and solute fate and transport modeling can be applied fully, or in part, to help resolve regulatory compliance issues, predict future conditions, and fast-track development of restoration plans. Many proprietary and nonproprietary computer models are available to environmental professionals that work under a wide variety of conditions. Some of the models are relatively inexpensive, ranging from free to several hundred dollars, while others are graphically sophisticated and cost thousands of dollars. Cost of computer models generally correlates well with how sophisticated, complex, and versatile programs are. In addition, higher-cost programs often offer "user friendly" operation when running the software. In many instances, the computer code itself may be inexpensive, extremely powerful and validated, yet difficult to use, or require that it be linked to another software program in order to view scaled modeling results. However, a preprocessor may be purchased, often for more money than the modeling code, to more easily simulate complex sets of data in two and three dimensions.

In relation to accelerating environmental decision making and fast-tracking projects, computer models offer environmental professionals a powerful and proven tool that historically has been used to support and plan environmental investigations and cleanup. Proper use of computer models at the correct stage of a project, such as the remedial investigation or remedial design phase, can limit the amount of data needed to be collected in the field, streamline the decision-making process by predicting or anticipating future conditions, and reduce costs. The purpose of most groundwater flow and solute fate and transport modeling efforts is focused on predicting the consequences of a proposed action, or simulating unusual conditions to predict future remedial performance scenarios (Anderson and Woessner 1992). In addition, models are used in an interpretive sense to gain insight into controlling parameters in site-specific settings, or as a framework for assembling and organizing field data and formulating ideas about system dynamics. In addition, models are also used to help establish location and characteristics of aquifer boundaries and assessment of the quantity of water within the system and recharge characteristics. Lastly, models are used in a generic sense to study hypothetical systems. Similar applications also apply to most vadose zone and chemical modeling efforts.

6.3.1 Modeling Techniques and Decision Making

Modeling natural systems is often relatively complex and scientifically in depth, especially three-dimensional numerical applications. In order to use modeling approaches for accelerating cleanup and decision making, a relatively strong modeling background is useful, and should include a working knowledge of practical and research modeling applications. The content of this section is based on the assumption of having a overall basic understanding of computer modeling, knowledge about the types of computer applications available, and a general understanding of the specific parameters and conditions related to soil, geology, hydrology, and chemistry that are input into models. This level of technical background is useful, because it allows modelers and planners working together to more easily identify opportunities that fast-track investigation and site characterization, expedite cleanup, and streamline decision making.

Most importantly, the purpose of this section is not to propose modeling for the sake of modeling. Rather, the establishment of clear benefits, cost reduction, or combined time and cost savings obtained through simulating site or hypothetical conditions. For more information on modeling, Anderson and Woessner (1992), Wang and Anderson (1982), and Wilson et al. (1995) are all useful resources. The following basic and advanced modeling concepts should be familiar in order to evaluate and identify opportunities to fast-track and streamline projects through modeling efforts:

- Basic vadose zone, geology, hydrogeology, hydrology, and chemistry concepts.
- Water movement within consolidated and unconsolidated media under saturated and unsaturated conditions.
- Recharge and discharge relationships and concepts of mass balance in natural systems.
- A general familiarity with the types, reliability, and usability (EPA 1992a) of common models available and selection criteria thereof used to determine which model may be well suited for a project.
- Familiarity with the governing equations and computer codes in selected models.

- Differences between analytical and numerical modeling approaches.
- Application of one-, two-, and three-dimensional and profile models.
- Development of conceptual models and model grids for two- and three-dimensional applications.
- Differences between finite element and finite difference methods.
- Establishment of model boundary conditions and simulation of sources and sinks.
- Simulation of steady-state and transient conditions.
- Use of sensitivity analyses in applications and model calibration approaches and application.
- Simulation of solute fate and transport processes and use of particle tracking techniques.
- Use of postaudits to check predictions.

Evaluation of Modeling Needs

Prior to deciding to use a model and simulate site conditions to fast-track decision making or reduce costs, an evaluation must be completed in order determine whether a modeling effort is warranted, and to evaluate whether modeling will actually accomplish the intended modeling and project goals. Many environmental sites do not require a modeling effort, such as most UST sites with limited soil contamination. However, there are environmental sites where specific conditions related to specific media or movement of contaminants exist that justify the use of computer models.

The complexity and size of environmental sites must be considered when evaluating the need for modeling and ability to model a site. In the case of relatively large or overly complex sites, the level of site characterization can sometimes be significant. However, through scheduling and comprehensive data acquisition planning ahead of time, the level of characterization can be minimized by using modeling techniques to help identify data gaps, or in some cases actually help fill data gaps, and minimize the amount of field data that needs to be collected. However, the latter is far less common, and only an extremely well-calibrated model can be used for this purpose. Most importantly, when considering modeling applications, the user must recognize that simulating natural systems is only as reliable as the data used in the model, and that the overall model design and parameters input into the model must be representative of the natural system in order to yield useful results. Thus, the available data must be able to satisfy this basic criteria before a modeling effort is proposed, unless there are specific assumptions, conditions, or applications that preempt this requirement. Contaminated sites that are relatively "straightforward" in understanding and assessment, or are small in size and exhibit relatively homogeneous and isotropic conditions, also can be modeled to fast-track site evaluations, predict future conditions, and realize cost savings. However, the level of effort for this type of modeling effort must be proportional to the overall project size and effort. For large or complex sites, the modeling effort may require an in-depth, three-dimensional numerical modeling approach in order to accurately predict future site conditions. A small site with relatively homogeneous and isotropic site conditions generally requires only a one- or two-dimensional analytical modeling approach or simplistic numerical approach in order to accurately predict future site conditions.

In addition, the cost of the modeling effort should be proportional to the overall cost of a project, unless the effort is a requirement, or if the effort will add substantial benefit to the project. In most cases, the benefits of modeling site conditions should at least pay for themselves by expending fewer resources in other areas of effort (such as design or evaluation), saving time by streamlining the project schedule, or minimizing the need to use traditional field methods to fill data gaps and address regulatory issues. From these perspectives, the elements that justify a modeling effort include evaluation of whether the effort will be representative and the results are anticipated to be accurate and precise, whether the model is economical and timely, and whether the effort will have an overall net benefit to the project. The following are several example situations where modeling local and/or regional conditions are beneficial.

- Supporting *in situ* biodegradation under an intrinsic remediation alternative.
- Evaluation and screening of various pump-and-treat alternatives to streamline remedial design.
- Simulation of a landfill cover performance or engineered caps for contaminated sites in order to quickly quantify leachate generation potential and predict potential groundwater impacts under various wet and dry climate conditions.
- Numerically predicting future impacts, exceedances in regulatory limits or risk, or exposure to contaminants migrating in groundwater.
- Simulation of chemical load and attenuation processes in groundwater and the impact of continued contaminant leaching into the groundwater system.
- Simulation of future site-specific conditions to help determine if the current investigation or cleanup proposals are on track, such as simulating high recharge conditions or impacts of the construction for a new impoundment near a site.
- Simulation of the commingled contaminant plumes or separate contaminant sources in order to estimate contaminant source contributions.
- Simulation of site conditions needed to satisfy regulatory requirements or address regulatory concerns that otherwise would be difficult or costly to complete using other methods.
- Minimization of the need to use traditional data collection methods to fill data gaps or characterize site conditions.

Model Selection and Assessment Framework

For the purposes of this book, model selection and assessment framework are limited to the preference between analytical and numerical modeling approaches. For more specific information on model applications, the EPA's Robert S. Kerr Environmental Research Center, in association with the International Groundwater Modeling Center (IGWMC), has compiled an extremely comprehensive on-line modeling information resource database containing specific model background information for most analytical and numerical computer codes. More importantly, the on-line service provides the opportunity to download modeling software, codes, user guides, and reports. The online modeling information available from the research center, and other associated information, is extremely useful for selecting specific model codes, and completing a framework assessment of model applications (see Appendix A for EPA's Robert S. Kerr center).

Benefits and limitations of analytical and numerical codes are described below in order to outline differences between the two types of codes for improving, fast-tracking, and streamlining environmental actions and decision making. The elected model and approach must be able to simulate site-specific conditions required to address the site-specific questions and regulatory concerns raised. In addition, the parameters used in models must be verified to ensure that they are realistic values of site conditions, e.g., identification of a conspicuously low specific yield of 0.0001 incorrectly calculated for an unconfined sand and gravel aquifer. For additional information on model selection and framework assessment, EPA has compiled a 74-item checklist for assessing groundwater and contaminant modeling applications entitled *Ground-Water Modeling Compendium* (EPA 1992a) and also prepared *Fundamentals of Ground-Water Modeling* (EPA 1992b).

Analytical Approaches

The advantages of analytical models are that they work well for modeling relatively "simplistic natural systems" (described below), the codes are relatively easy to run, and they can be used to simulate conditions that otherwise would be difficult to estimate or determine by hand. These primary benefits, uses, and limitations of analytical models are described for streamlining environmental decision making, planning, and reducing costs.

In this text, simplistic natural systems refer to a variety of soil, geologic, and hydrologic settings where the lithology, units, etc., are relatively homogenous and isotropic, water flow and other natural processes are uniform across the site, steady-state conditions can be assumed to a reasonable degree, and one or two dimensions are sufficient to simulate site conditions, particle movement or contaminant migration, and the associated processes. In addition to being relatively easy to use, analytical models can also be entered into a computer in a relatively short period of time compared to more complex numerical applications. Most analytical models require the following in order to develop and run them: defining the goals and objectives of the effort, developing a site conceptual model of which the computer model will work, assigning model assumptions, and imputing modeling parameters such as hydraulic conductivity, aquifer thickness, soil moisture content, porosity, etc. Analytical models do not often require a detailed evaluation of boundary conditions, grid design, and other factors required in order to model site conditions, although they should certainly be qualitatively or conceptually considered when assigning the modeling assumptions and imputing parameters. Analytical models can also be run on standard personal computers; more expensive or sophisticated computer hardware is not generally required.

However, because analytical models are relatively easy to use, they are sometimes misused when applied to overly complex settings, or in cases where the modeler does not understand the governing equations and approaches used in the model code, or because the effort may not provide any tangible benefit (i.e., modeling is not necessary to address the project goals and objectives). To avoid misuse, a modeling background is helpful, and also a clear understanding of not only what the goals and objectives of the modeling work may include, but also how they would fit into the overall project goals and objectives. Modeling is only one part of environmental projects, and like any action it must be carefully planned, scheduled, and designed to streamline efforts and reduce costs. In addition, the modeler should have a general understanding of analytic approaches useful to simulate physical processes, approaches used to develop site conceptual models that fit within the site conditions and help assist with the modeling effort, proper use and quality of input data, defining modeling assumptions, and evaluation of modeling limitations of the modeling effort. By fulfilling these requirements, the potential for misuse is decreased.

Less frequently, analytical models are used to simulate relatively complex natural systems, which may require using several analytical models in tandem in order to meet the modeling goals and objectives. In complex systems, analytical modeling may not address comprehensive site issues, or the entire site. However, the modeling results may be focused on evaluating one or more specific unknowns, such as the biodegradation of petroleum constituents or movement of contaminants in the vadose zone. In this example, the modeling results are used as one of many pieces of the site "puzzle," where the different pieces are put together in order to make recommendations. In the latter example, a leachate simulation model could be used to simulate wetting fronts moving down under gravity flow through the vadose zone. In turn, the modeling results could be tied into a computer spreadsheet application and used to calculate the resultant groundwater quality using the leachate chemistry, groundwater chemistry, and volumes of each. In most cases, however, where site conditions are not uniform, such as sites with nonuniform groundwater flow direction, numerical modeling approaches are almost always more appropriate.

More involved use of analytical models include, for example, applying a predictive sensitivity analysis approach to outline a range of possible solutions that exist within a particular site conceptual model in order to improve project decision making. This could include determining the optimal placement and configuration of recovery wells for the purpose of containing a plume. In this example, a model could be used to simulate the most efficient and effective extraction well configuration using particle tracking techniques, which is in turn used to evaluate whether the plume will be captured and estimate how much clean groundwater will be collected along with the contaminated groundwater by the extraction wells. Simulating a range of possible pumping rates using an analytical particle tracking model can be useful to refine the range of optimal pumping rates and extraction well configuration for engineering design development. Lastly, more advanced modeling techniques that are used in numerical modeling efforts can be applied to analytical models to broaden their use and scope. One example is Marquis and Dineen's (1994) approach for developing performance tables to rank multiple remediation scenarios and fast-track decision making. Using their "scoring" approach with analytical models, comparison of remedial alternatives can be ranked more quickly for relating simplistic natural settings requiring groundwater remediation or stabilization. Other more advanced modeling techniques can also be applied to analytical modeling efforts in cases where site conditions are relatively homogenous and isotropic, and groundwater flow direction is uniform.

Case Studies

In EPA's *Ground-Water Modeling Compendium* (1992a), two case studies are presented to illustrate how analytical models can be used to help solve surprisingly complex problems quickly and reduce costs. In one of the case studies, the EPA's contractor proposed additional field investigation to obtain more site information related to contaminant fate and transport in order to complete a groundwater extraction and collection system design. However, as a cost reduction measure, EPA elected to address these data gaps by supplementing the existing data with results of analytical groundwater modeling applications. The modeling approach included addressing data gaps through four modeling objectives:

- Delineate the maximum extent of the contamination plume when extraction is scheduled to begin operation.
- Estimate the mass of contamination per unit volume of the aquifer.
- Conceptualize extraction well alignment design.
- Evaluate design alternatives.

During the development of the modeling objectives, the modeling team attempted to use both a simplistic analytic approach and a finite difference contaminant transport model before abandoning the former for its lack of specificity, and the latter because of limited available site data (EPA 1992a). Instead, a three-phase modeling approach utilizing four separate models (Random Walk, PLUME, Wellhead Protection Area Model (WHPA), and WELFLO), were selected by EPA. This approach resulted in reducing the magnitude and cost associated with additional site sampling.

The three phases of the modeling effort included (1) determining the maximum extent of contamination using Random Walk and cross-checking results with PLUME, (2) conceptualizing design alternatives using a graphical type curve analysis and WHPA for contaminant time-related capture analysis, and (3) using WELFLO to simulate aquifer drawdown and evaluate the remedial design alternatives. The approach selected represented a compromise between the level of detail and accuracy desired by the design engineers, the available site data, and the modeling budget. One of the more interesting aspects of the case study is that the modeling effort resulted in reducing the design cost by $450,000, mostly from the model results simulating the plume boundary in lieu of an extensive drilling program. It was also noted that if a relatively simple analytical model would have been used earlier in the project, the results could have been used to improve the sampling plan and monitoring well placement. This in turn would have increased the possibility that a more accurate numerical model could have been used during the design phase, possibly resulting in a more efficient design.

Numerical Approaches

The advantages of numerical models are that they work well in complex settings, model codes have the ability to simulate heterogenous and anisotropic conditions in three dimensions, they can be used to simulate transient conditions, and they can be used to simulate conditions that otherwise would be extremely difficult or impossible to determine by hand or with analytical models. These primary benefits, uses, and limitations of numerical models are described below for utilizing numerical models to streamline environmental decision making, planning, and reducing costs.

Numerical modeling codes for two- and three-dimension models employ an iterative approach in order to simulate steady-state and transient site conditions. Most importantly, numerical applications are able to simulate site-specific heterogeneity and anisotropy common at complex environmental sites. These conditions include, for example, multiple aquifer systems, areas of low and high hydraulic conductivity, variations in groundwater flow gradient and direction, transient recharge conditions, and many other factors that affect contaminant migration and remediation performance. Analytical models are generally not sophisticated enough or sufficiently adaptable for simulating these types of site conditions and accurately predicting outcomes. However, because of site complexity, and the need to simulate variability in natural settings, the level and representativeness of data needed for numerical models is relatively extensive and expensive to obtain compared to analytical model data. Understanding numerical model data needs, having a thorough understanding of the model code itself, and characterizing the site to a relatively high degree (such as a remedial investigation) are basic requirements in order to effectively apply numerical modeling approaches.

In order to maximize the use of numerical models, the timing and overall goals of the modeling effort must be considered in the project planning stage. This is required to ensure that the proposed effort is initiated and completed within a reasonable time frame and budget. In addition, the site data must be evaluated to determine whether they will satisfy the overall modeling requirements. For example, hydraulic conductivity calculated using aquifer testing data generally provides much more reliable input values than laboratory permeameter values. In this example, before investing in a large numerical modeling project, it may be beneficial to conduct several aquifer tests in cases where only permeameter results are available, and compare aquifer testing values with the permeameter values. Relying on poor hydraulic conductivity data can significantly over- or underestimate impacts on predicted pumping rates and recovery time, for example. Therefore, all input data must be evaluated to be sure it is representative of the site in order to simulate accurate results. In addition, sensitivity analyses need to be performed in order to evaluate the effect varying input values have in the model and sensitivity of input parameters.

The time needed to develop a numerical modeling approach, conduct the modeling effort, and prepare a report can take anywhere from several weeks to several years, depending on the level of complexity of the model data quality and coverage, and reporting requirements. For efforts that take several years to complete, a decision to simplify the modeling approach may be needed to expedite the process, unless such a large amount of time is truly necessary to complete the effort. Alternatively, the effort may need to be postponed, or discontinued, if another approach will economically address the project goals, or if the modeling effort slows a project to the point where a response to serious or potential health and environmental risk is delayed. By integrating the modeling effort into data collection and site characterization phase, the project and overall modeling schedule can be accelerated. In this case, the modeling effort may be conducted simultaneously as part of an RI/FS, thereby using the schedule of the RI/FS to report the results at the same time, instead of conducting the modeling study after the RI/FS report. In this scenario, most importantly, close communication with all project team counterparts is needed to ensure site data and new developments are communicated to the individual(s) conducting the modeling effort.

When considering project needs and goals and objectives, numerical modeling approaches can be integrated into site plans in order to predict and evaluate site conditions, identify ways to increase remedial alternative efficiently, streamline decision making, and reduce costs. For example, prior to completing a hypothetical RI/FS, the project team determines that a nearby gaining river is a significant hydrologic sink for dissolved groundwater contaminants, yet the overall impact to the river quality cannot be measured quantitatively because of dilution. The project team is informed that a water supply well will be installed on the opposite side of the river about 1 mi away in order to pump large amounts of groundwater to serve a nearby community with potable water (Figure 6-7). A numerical model could be developed to predict future site conditions before the well is operational, assuming there is a relatively thorough understanding of local and regional hydrogeological conditions. This could include developing a three-dimensional numerical modeling approach and simulating groundwater flow and contaminant fate and transport needed to evaluate potential contaminant migration underneath the river toward the proposed water supply well, using several different pumping rates. In addition, the results of the modeling effort could be used to calculate contaminant influx, and the resultant concentration of contaminants in the river, which cannot be documented using standard sampling and analysis method detection limits. Actual case studies for numerical model applications are compiled in EPA's *Ground-Water Modeling Compendium* (1992a). More detailed case studies are compiled by Anderson and Woessner (1992).

Presentation of Modeling Results

Modeling reports generally contain an introduction, description of the hydrogeologic setting and conceptual model, model design and results, model limitations, summary and conclusions, and appendices (Anderson and Woessner 1992). Most importantly, however, modeling reports must contain graphical displays and tables that are easy to understand, yet retain sufficient detail to outline modeling results for planning purposes. Color graphical presentations are especially useful to enhance modeling points of interest, such as plume boundary location or the points of compliance. Presenting the modeling results in this manner is important because most modeling reports contain highly technical information that can be overwhelming for nonmodelers, and are used by others who develop the site-wide plans and recommendations. In order to streamline the regulatory review time, help make decisions, and more efficiently convey modeling results, the modeling results and conclusions must be easily understood by all parties, which may also include the public.

230   STRATEGIES FOR ACCLERATING CLEANUP AT TOXIC WASTE SITES

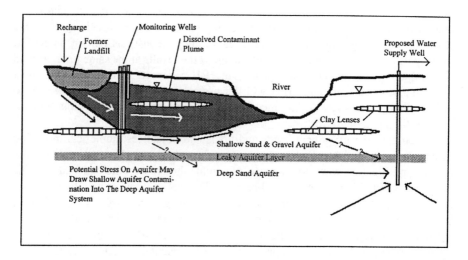

Figure 6-7   Hypothetical physical and chemical conditions. Using groundwater flow and solute transport computer modeling techniques, it is possible to quantify water quality impacts, if any, on a proposed water supply well from nearby groundwater contamination. In addition, the chemical load into the river can be estimated using modeling data. However, in order to reasonably predict future potential conditions, sufficient local and regional physical and chemical characterization is essential. In addition, well construction specifications and a range of pumping rates are needed for the modeling effort.

6.4 GEOPHYSICAL TOOLS AND APPLICATIONS

This section contains a relatively brief and rudimentary technical discussion on geophysical tools and applications. The purpose of this section is to introduce geophysics as an important tool available in the environmental industry and demonstrate that geophysics can be used to streamline investigation work, quickly survey shallow and deep site conditions, and reduce costs.

The underlying premise for using geophysics is that the contaminant (or objects) being investigated or sought has properties substantially different from the properties of the host soil, rock, or groundwater (Blaricom 1980). For environmental sites, geophysics are technology-based tools that can be used to accelerate site characterization and reduce investigation costs. These benefits can be realized under specific, and sometimes limited, site conditions assuming the appropriate geophysical tools are used to investigate a known or suspected contaminated site. In general, there are three primary uses of geophysical applications at environmental sites:

1. Detection of contaminants or anomalies (such as free product).
2. Measurement or identification of site conditions that could influence movement, attenuation, or migration of contaminants (such as buried river channels).
3. Detection of buried objects that are associated with contaminants sources (such as USTs).

The advantages of geophysics are that (1) the need for intrusive activities, such as installation of numerous monitoring wells, can be minimized, (2) geophysical data can be used to focus the characterization efforts into more localized areas streamlining investigations, (3) the amount of time required to survey large areas using many geophysical tools and methods is often considerably smaller compared to intrusive or traditional investigation methods, such as defining the edge of free product using boreholes, and (4) successful use of geophysics can reduce overall project costs. In addition, in the last 10 years computer software and hardware have been improved, allowing geophysicists to improve filtering capabilities and automatically log large quantities of electronic signals, which in turn has allowed geophysicists to differentiate smaller and smaller differences between contaminants, buried objects, and host material. In order to maximize acceleration and cost reduction using geophysics, the end user must not only have the proper background, but also keep up to date with the new developments and advantages and limitations of new approaches and processes. In turn, environmental professionals can make informed decisions for or against geophysics and specific geophysical tools. As with most technologies, the proper educational background and experience includes understanding the governing physics and equations, field plan development and implementation, equipment use and limitation, and data interpretation and reporting. General and practical geophysical information is described by Vogelsang (1995), Ward (1990), and Telford et al. (1991). The following section provides a brief overview of the selection and planning of geophysical surveys in relation to streamlining investigations and site characterization, and reducing costs.

6.4.1 Selecting and Planning Geophysical Applications

Table 6-2 contains information compiled by Northeast Geophysical Services of Maine (undated) and outlines the most common types of geophysical applications used in the environmental industry today. Table 6-2 is useful for helping differentiate geophysical applications applicable to specific conditions and uses, and aids in evaluating which type of general geophysical applications will address site-specific needs. Additional resources should be consulted to help select a specific geophysical tool and design a survey, such as the references listed above.

The processes associated with the general types of applications, and the governing physics behind most of the technologies have not substantially changed in the last 20 years; however, data collection, data processing, data presentation, and data evaluation have substantially improved, increasing the accuracy and precision of geophysics. Selecting the proper geophysical method, tool, and data compilation and evaluation approach is critical in order to ensure the effort will be beneficial. This is generally conducted by completing a thorough review of related scientific literature and vendor information, and by working with internal/external geophysicists and consultants. In addition, as part of the selection process, it may be beneficial to propose multiple geophysical tools, different arrays or station configurations, or backup geophysical tools in order to compare results or augment the preferred tool or approach result if the primary approach is ineffective. Multiple tools are especially useful if there is some doubt in the capability of the preferred geophysical tool to

Table 6-2 Common Environmental Geophysical Methods and Applications

Method	Process	Limitations	Comments
Seismic refraction	The seismic refraction method utilizes sound waves. Sound travels at different velocities through different materials and is refracted at layer interfaces.	Layer velocity (density) must increase with depth. Layers must be of sufficient thickness to be detectable. Data collected directly over loose fill (landfills) or in the presence of excessive cultural noise will result in substandard results. Single narrow fractures are generally too small to be detected.	Used to evaluate depth to watertable, depth to top of indurated till, depth to bedrock. Also used to evaluate fractures zones in bedrock, aid bedrock contour mapping and location of bedrock lithologic contacts, rippability, and bedrock structure and topography control related contaminant migration. Seismic refraction is a valuable tool for mapping bedrock troughs and fractures. It is usually more cost-effective and gives better coverage than drilling alone. A seismic wave is usually generated by a small explosive charge, a shotgun shell, or a sledge hammer. The wave's travel time from the sound source to refracting layers, along those layers, and back to geophones is precisely measured. From the time–distance relationships, subsurface layer velocities and thicknesses can be calculated. Fracture zones can often be detected because they usually have a lower seismic velocity than solid bedrock. The velocity of sound through water-saturated material is about 5000 ft/s while the velocity through crystalline bedrock generally ranges from 12,000 to 18,000 ft/s. By calculating the velocity of sound along the bedrock surface, low velocity zones, which may represent fractures, can be delineated. The geophones can be linearly spaced any distance apart, but most often are spaced 10 to 50 ft apart. In general, the greater the expected bedrock depth, the greater the geophone spacing. Shorter spacings and sometimes radial patterns are used in fracture zone detection studies.
Seismic reflection	Seismic reflection is a geophysical technique in which acoustic waves, reflected directly from underground surfaces with density contrasts, are used to map soil and bedrock stratigraphy.	Reflection surveys are highly site-specific. A shallow watertable is generally required. Reflection surveys are useful for exploration depths of 50 ft to several hundred feet.	Seismic reflection is useful for graphically profiling subsurface stratigraphy. It is used to map clay and sand lenses and bedrock troughs. Applications of this technique include determination of depth to bedrock, aquifer studies, and mapping of overburden stratigraphy. Successful application depends upon the ability of the ground to transmit high-frequency seismic energy (saturated clays, for example, transmit high-frequency energy quite well). The method overcomes some of the potential problems encountered in refraction surveys (such as the assumption that subsurface layer velocities increase with depth and that layers are thick enough to be detectable). The equipment used is nearly identical to that used in a seismic refraction survey. The field technique, however, differs and ground coverage is usually slower than with a refraction survey. Graphic profiles of subsurface structures are developed with data.

| Ground penetrating radar (GPR) | GPR utilizes high-frequency radio waves to probe the subsurface without disturbing the ground surface. GPR data is collected continuously as the instrument is towed over the ground surface. Radar pulses are transmitted downward from an antenna and are reflected from underground surfaces. The reflected signals return to a receiver creating a continuous graphic profile of the subsurface. Reflecting surfaces appear as bands on the profile. | Exploration depth can be limited by soil or water with high conductivity. Detectability depends upon a dielectric contrast between the subsurface feature and the surrounding material. Closely spaced survey lines are required to locate small objects. A relatively smooth surface is also necessary. |

GPR provides a graphic image of the subsurface and has applications such as locating USTs, mapping utilities and re-bar, determining fill thickness, mapping geologic strata, aquifers, aquicludes, voids, shallow bedrock units and fractures, and archaeology site clearance for drilling. GPR is commonly used as part of Phase II environmental site assessments and other environmental studies to locate underground features. GPR is extremely useful for determining the location, depth of burial, and orientation of tanks, pipes, utilities, and other objects.

Reflection of radar waves occurs at interfaces having contrasting electrical properties (which are controlled largely by composition and moisture content of the material). Examples of reflecting surfaces are soil horizons, soil–rock or air–rock interfaces, watertables, and solid metallic or nonmetallic objects. Radar penetration and resolution vary with the antenna used, as indicated below. They also vary with soil and rock conditions.

Antenna	Penetration	Resolution
900-MHz	~0–5 ft	high
500-MHz	~0–15 ft	high
300-MHz	~0–30 ft	average
80-MHz	~0–80 ft	low

Continuous graphic profiles are generated with data. When travel time/depth relations are known, a depth scale can be substituted for the travel-time scale on the profile, allowing estimation of absolute depths.

| Magnetics | When a ferrous material is placed within a magnetic field (such as the earth's), it develops an induced magnetic field. The induced field is superimposed on the earth's field at that location creating a magnetic anomaly. | Utilities, power lines, buildings, and metallic debris can cause interference. Solar magnetic storms may cause fluctuations in readings. The size and depth of objects affect detectability. |

Magnetometer surveys are rapid and efficient, and can be used to detect buried ferrous metal objects (tanks or drums) or bedrock features with contrasting magnetite content, and archaeology. Detection depends on the amount of magnetic material present and its distance from the sensor. A single steel drum can be detected at burial depths up to 15 or 20 ft. Burial depth can be estimated from magnetometer data collected using the gradient method.

A magnetometer survey for hydrogeologic and engineering applications is conducted on foot, by one operator. The survey can be along single lines or along a series of parallel traverses with readings taken every 5 to 50 ft. Spacing of traverses and readings depends on the width of the expected anomaly. For instance, tank searches may be conducted at a 5-ft spacing while geologic mapping may be conducted at a 50-ft spacing. In the gradient method, the total field is measured simultaneously at two elevations by using

Table 6-2 (continued) Common Environmental Geophysical Methods and Applications

Method	Process	Limitations	Comments
			two sensors on a staff separated by a fixed distance. The difference in magnetic intensity between the two sensors divided by the distance between them is the vertical gradient. This technique reduces interference from solar magnetic storms and regional magnetic changes. This technique is particularly useful for locating small, shallow objects and is also useful for estimating burial depth of objects.
Electromagnetics	Electromagnetic induction surveys (terrain conductivity and metal detector) work by inducing current into the ground from a transmitter coil. The resulting secondary electromagnetic field set up by any ground conductors is then measured at a receiver coil. The presence of metals, ions, or clays increases the ground conductivity. Conductivity readings are reported in millisiemens per meter (equivalent to millimhos per meter). Metal detector readings are generally reported in parts per thousand of the total field.	Measurements are affected by power lines, metal fences, metal debris, and utilities. Fracture detection is affected by overburden thickness, soil conductivity, and orientation and dip of the fractures.	There are a variety of inductive electromagnetic (EM) methods that measure subsurface electrical properties. Three common applications of these methods are terrain conductivity measurements, metal detection, and bedrock fracture detection. EM is also used in borehole geophysical logging. Terrain conductivity surveys are ideal for locating contaminant plumes from salt piles, landfills, and other sources. They can also provide baseline conductivity data around landfill sites prior to construction, and periodically after landfill establishment, to detect and monitor any contaminant plumes. Metal detector surveys can locate buried or hidden metal objects. They can also complement terrain conductivity or GPR surveys by mapping buried waste. Bedrock fracture detection can be accomplished using very low frequency EM (VLF-EM). Locating fractures is often useful in groundwater supply or contamination studies. EM surveys are noninvasive, rapid, and economical, making them well-suited for hazardous waste and hydrogeologic studies.

Detection depth of EM instruments is a function of the transmitter-to-receiver coil separation and the coil orientation (horizontal or vertical). Small coil separations, as in metal detectors and pipe locators, may "see" 2 to 6 ft into the ground. Larger coil separations can be used to detect conductive materials up to several hundred feet deep. Very low frequency EM (VLF-EM) is an inductive technique which measures very low frequency horizontal EM signals from remote military transmitters. Localized conductors, such as water-filled fractures, cause angular disturbances in this signal which are measured with the VLF-EM instrument (in degrees from the horizontal). VLF-EM can best detect linear, steeply dipping conductors oriented in the direction of the transmitter. Detection depth depends largely upon overall ground conductivity, but is commonly over 100 ft. |

PRIVATE SECTOR, PROFESSIONAL ORGANIZATIONS, INNOVATIVE SOLUTIONS   235

Method	Description	Application	Limitations
Resistivity	An electrical current is introduced directly (as opposed to inductively as with electromagnetic surveys) into the ground through a pair of electrodes. The resulting voltage difference is measured between another pair of electrodes. The subsurface apparent resistivity is then calculated. Resistivity is the reciprocal of conductivity. Thus, measuring resistivity provides information on ground conductivity.	The electrical resistivity method is used to characterize vertical and lateral changes in subsurface electrical properties. Vertical changes are measured using the VES technique (in conjunction with borehole data to characterize electrically distinct layers). Lateral changes are measured using the resistivity profiling technique. Resistivity profiling is used to map spatial changes in subsurface electrical properties. Applications of a resistivity survey are similar to those of electromagnetic (terrain conductivity) surveys. Resistivity profiling is most commonly used to map contaminated groundwater plumes.

For the VES method, a series of resistivity measurements are made at various electrode spacings centered on a common point. Sampling depth is increased by increasing electrode spacing. Data is generally interpreted using an appropriate computer program. Data are used to develop modeled depths, thicknesses, and resistivity of subsurface layers. In the resistivity profiling method, four electrodes are positioned at a fixed distance from each other. A current is introduced between two of the electrodes and a voltage potential is measured between the other two electrodes. The electrode pairs are moved along a surveyed line and the electrical measurements result in a horizontal profile of apparent resistivity. Different electrode spacings can be used to yield a cross section of resistivity changes with depth. Data are used to map the apparent subsurface resistivity values from profile stations. Results may be plotted as profile lines or contour maps (isopleth resistivity map), or in other presentation formats. | Vertical electrical soundings (VES) are affected by changes in surface relief and lateral changes in resistivity. The electrode array length is about 10 times the depth of investigation. Profiling: resistivity profiling is slower and more expensive than EM surveying. |
| Borehole logging | Borehole geophysical logging includes the measurement of various chemical and physical characteristics of materials and fluids in and around a borehole. | Borehole geophysical logging techniques used together with surficial geophysics, well data, and knowledge of the local geology can be essential in solving problems in groundwater hydrology. Boreholes are often logged to map vertical overburden and bedrock stratigraphy, location of bedrock fractures, borehole conductivity, location of well screens cross-hole studies, zones for setting packer test intervals, and setting well screens. Another common application is mapping the thickness and continuity of aquifers. Log interpretation includes identification of potential fracture locations and distinctive hydrogeophysical units which can be compared with fractures and rock or sediment layers identified in the descriptive log. Data are collected and stored digitally, and can be presented at any scale requested. The following information gives brief descriptions of some measurements that are commonly logged in boreholes, and their applications. | Limitations vary with the method used. The type of casing and borehole diameter are factors to consider. |

Table 6-2 (continued) Common Environmental Geophysical Methods and Applications

Method	Process	Limitations	Comments	Parameters Measured/Applications
			Log	
			Caliper	Borehole and casing diameter. Fracture identification, lithologic changes, and well construction.
			Gamma	Natural Gamma radioactivity. Lithology and estimation of clay content in overburden.
			Temperature	Temperature of borehole fluid. Indicates geothermal gradient, and water flow in borehole or between borehole and fractures.
			Resistivity	Resistivity of borehole fluid. Indicates water flow within borehole, or between borehole and fractures, and water quality.
			Single point resistance	Resistance of materials between probe and ground surface electrode.
			Normal resistivity	Apparent resistivity of material. Lithology and water quality.
			Spontaneous	Electrical potentials between probe and surface electrodes. Lithology, water quality, and in some cases, fractures in potential resistive crystalline rock.
			EM conductivity (induction)	Electrical conductivity in medium surrounding borehole. Location of contaminant plumes, conductive clay units, or bedrock fractures.
			Flowmeter:	Monitor water quality changes over time. Continuous or point measurements of water flow in borehole. Identification of permeable zones and apparent vertical hydraulic conductivity and flow direction.
			Borehole video	Provides visual record of lithology, fractures, well construction.
Gravity	Gravity is the attraction between masses. The strength of this force is a result of the mass and distance separating the objects.	Gravity surveys are relatively slow and expensive. Detectability varies with target size, depth, and density contrast. Interpretation		Gravity surveys can provide useful information where other methods do not work. For example, gravity may be used to map bedrock topography under a landfill, where seismic refraction is limited. Gravity can also be used to map lateral lithologic changes, faults, large metallic mineral deposits, locating subsurface caverns, and locating contacts between geologic units of differing mass and density.

PRIVATE SECTOR, PROFESSIONAL ORGANIZATIONS, INNOVATIVE SOLUTIONS   237

Method	Description	Limitations	Details
		of data often requires control data from drilling, outcrops, or other sources. Detailed surface topographic survey data is also required.	A gravimeter is used to measure the earth's gravitational attraction at various points over the area of interest. Gravity anomalies are due to differences in density of underlying materials. Gravity anomalies are extremely small relative to the total field and are usually measured in micro-Gals (one micro-Gal is about 1 billionth of the earth's total gravitational field). The equipment used in a gravity survey is extremely delicate and precise. Data interpretation is time consuming even with the use of sophisticated computer programs. Results are used to generate contour map and gravity data tables.
Induced polarization (IP)	Induced polarization is the capacitance effect, or chargeability, exhibited by electrically conductive materials.	IP cannot be done over frozen ground or asphalt because good contact with the ground is required. IP is affected by changes in surface relief and lateral changes in resistivity. The electrode array length is about 10 times greater than investigation depth.	In nature, the induced polarization (IP) effect is seen primarily with metallic sulfides, graphite, and clays. For this reason, IP surveys have been used extensively in mineral exploration. Recently, IP has been applied to hazardous waste landfill and groundwater investigations to identify clay zones. As with electrical resistivity surveys, vertical or horizontal profiles can be generated using IP. IP can also be used in borehole logging. Measurement of IP is done by pulsing an electric current into the earth at 1- or 2-second intervals through metal electrodes. Disseminated conductive minerals in the ground will discharge the stored electrical energy during the pause cycle. The decay rate of the discharge is measured by the IP receiver. The decay voltage will be 0 if there are no polarizable materials present. Generally, both IP and resistivity measurements are taken simultaneously during the survey. Survey depth is determined by electrode spacing. Results are reported similar to those of resistivity surveys.

Source: Northeast Geophysical Services (undated). Methods and Applications Reference Manual. Yarmouth, ME.

yield the desired results. Similar to other acceleration approaches and cost reduction methods, geophysics should only be used when site conditions favor their use, and limitations (Table 6-2) and costs are not anticipated to impact the overall project success. A relatively thorough evaluation of the applicability of the proposed geophysical methods should be completed ahead of time to help ensure the proposed approach will be efficient and effective, and the anticipated results will address site-specific goals and objectives.

Finally, it is necessary to plan the field activity itself, plan the data compilation and evaluation, and organize the field effort so the field data are properly collected and compiled for analysis. More importantly, however, it is important to realize that geophysical tools are used to supplement site characterization, and that the results usually must be compared, or strengthened, with existing or future site information or data (such as monitoring well logs) in order to finalize conclusions and make recommendations. Basic considerations for selecting and planning geophysical projects include

- Determine whether the media, contaminant, and/or objects in question are compatible with the proposed geophysical method(s).
- Evaluate whether site-specific conditions favor a successful outcome using the proposed geophysical method(s) or if there may be background noise, etc., that may interfere with the project (such as overhead power lines, underground utilities, or unique geologic features).
- Determine whether existing or future site data are/will be available to confirm, augment, and/or compare with geophysical data.
- Evaluate whether the overall anticipated benefits of the geophysical survey will be cost-effective.
- Integrate the geophysical effort into the entire project schedule to ensure that the geophysical survey will not impede the overall project schedule and that data from all efforts can be used to complement each other.
- Ensure that the field team has sufficient background and experience to operate, interpolate, and report the geophysical results. Inexperienced personnel should not plan or operate geophysics unless supervised by someone with the appropriate background and experience.
- Develop a conceptual report and presentation format ahead of time that is useful for planning purposes, including preparation of concise graphical and tabular displays highlighting findings, which may need to be suitable for the public to understand.

For actual use geophysics, more advanced information should be consulted, such as scientific literature and tool-specific advantages and disadvantages criteria cited in vendor information. However, not all vendor claims may be totally accurate, and a closer evaluation of each specific tool may be required before going into the field with the equipment, including demonstration information if it is new technology. The following case study outlines how geophysical applications can be used to streamline investigation tasks and reduce costs. The case study is one example of many innovative tools available in the environmental industry that can be used to increase project performance and aid site characterization.

6.4.2 Case Study — Electromagnetic Offset Logging

Under the Navy Environmental Leadership Program (NELP), electromagnetic offset logging (EOL) technology was demonstrated at the Buildings 379 and 397 study area at the Naval Aviation Depot (NADEP) located at Naval Air Station (NAS) North Island, California. The Gehm Corporation (Gehm) is the sole provider of the proprietary EOL technology. The EOL technology was demonstrated at NAS North Island to identify new and better ways to conduct environmental investigations of petroleum products and compare costs of the EOL technology to those associated with a more traditional drilling and sampling approach. (Additional information on NELP is presented in Chapter 3.) Petroleum-contaminated sites are typically characterized using conventional methods that include drilling boreholes and collecting soil and groundwater samples. The boreholes and wells provide contaminant analysis information for only specific locations, and multiple borings, wells, and samples are typically required across sites. In turn, interpolations are made among sampling points to map the magnitude and extent of subsurface contamination. To increase resolution of the resulting isopleth maps, the number of sampling points must be increased. In most cases, there are substantial costs associated with collecting the increased quantity of data.

To address light non-aqueous phase liquid (LNAPL) petroleum product in the study area, the Navy tasked one of its contractors with characterizing the study area and the installation of a free product recovery system. The contractor installed 62 product extraction and monitoring wells at the site and determined that the LNAPL petroleum product on the watertable in the study area was up to 5 ft thick at some locations (Figure 6-8). In order to further characterize the extent of the LNAPL plume at the study area, a demonstration was conducted using the EOL technology to streamline the LNAPL characterization. The Navy opted to demonstrate EOL instead of using more traditional methods in order to determine if the technology would expedite the characterization process and reduce costs for future Navy projects.

Electromagnetic Offset Logging Technology Description

The EOL technology uses surface to borehole geophysics to identify subsurface petroleum contamination. Using an induction electrical resitivity survey, a magnetic field is induced into the ground which causes electrical currents to be generated in the subsurface. Petroleum products are highly resistive to electrical currents and the EOL technology identifies these areas within survey areas as highly resistive anomalies. The data are then processed in order to develop three-dimensional representations of underground LNAPL contamination.

As part of the initial step of data collection, the study area is surveyed into a grid. A transmitter coil is then placed at each grid point, and one at a time an electric signal is transmitted. To receive the induced signal from the transmitter, an offset induction receiver is placed in a nearby drill hole or well. The receiver is mechanically pulled up the drill hole, measuring the primary and secondary electromagnetic fields produced at the transmitter coil location. A large, long wavelength signal is

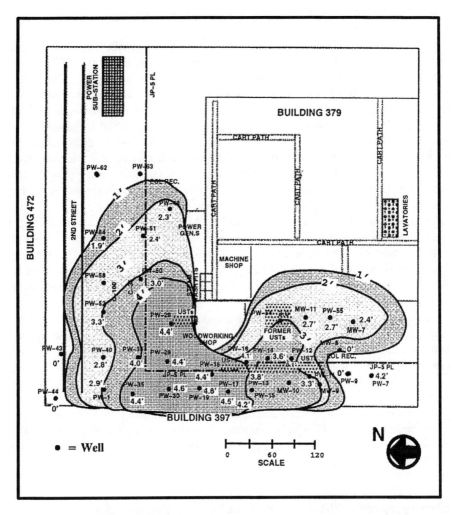

Figure 6-8 Well location and free product plume map based on data collected during the same period the EOL survey was completed. (Source: Navy Environmental Leadership Program.)

created representing the primary electromagnetic field. Superimposed on this signal are responses related to the secondary electromagnetic fields caused by eddy currents moving around the boundaries of resistivity contrasts (or anomalies) in the earth. These raw data are computer processed to remove the primary field and calculate and verify the secondary fields. The secondary fields are in turn converted to apparent resistivity, measured in ohm-meters, and compared directly to the physical properties in the earth.

To accommodate the problems related to making an electromagnetic measurement in and around man-made electrical noise, the transmitter coil and receiver are tuned to around 270 Hz with a narrow bandwidth. This tuning procedure, along with choosing receiver wells in low noise areas, helps filter out most of the excess and unwanted electrical noise and allows the EOL technology to be used in and around most man-made structures and other sources of subsurface electrical noise. The computer-processed data can be presented either in three dimensions or as depth-specific slice and cross section images. Contours of relative resistivity in either of these formats can be developed and used to track the resistivity patterns of the soils or other near surface materials.

In general, the presence of LNAPL is represented by higher resistivity values. However, this relationship may not always be linear due to lateral or spatial changes in geology and fluid chemistry from point to point. This must be considered in the interpretive analysis of the EOL results. Most importantly, the EOL technology cannot be used by itself to document and characterize soil contamination. The EOL technology merely images highly resistive fluids and materials.

Demonstration Design

A 4-acre study area was surveyed and defined on a 30-ft grid outside the buildings and in the buildings, where possible. Other accessible transmitter locations in the buildings were also used. These points were the transmitter points for the resistivity measurements (Figure 6-9). Since the Navy contractor had already installed 62 drill holes at the study area, these points could be used to help conduct the survey. Two of these drill holes were selected for receiver points. Receiver points were selected to provide the best possible subsurface information both under Building 379 and in the area north and east of Building 379, where contamination was suspected.

Two drill holes were selected as down hole receiver wells (PW-63 and MW-8, Figure 6-8). A down hole offset induction receiver was placed in the first selected well. The transmitter coil was placed on the ground surface at the first grid point. The receiver was pulled up the well mechanically, and resistivity measurements were logged until it reached the top. The process was repeated until "reachable" grid points, with adequate signal-to-noise quality, were logged using the first receiver point (Figure 6-9). The process was performed from the second receiver, as well as for the rest of the grid points and overlapping about 20% of the first receiver well grid points. The EOL technology demonstration took approximately 3 days to record data from grid points over the 4-acre area.

The superimposed primary and secondary electromagnetic field data between the grid points and the receiver wells were recorded from the offset induction receiver directly onto a data logger. These raw data were downloaded to an on-site personal computer and processed into initial resistivity logs. The resistivity logs were processed into final three-dimensional models using a Silicon Graphics UNIX work station with Dynamic Graphics' Earthvision Reservoir software at the Gehm facility.

Results and Discussion

The digital data were stored on floppy disk for further processing at a later date, at which time Gehm conducted a proprietary analysis of field data (Figures 6-10

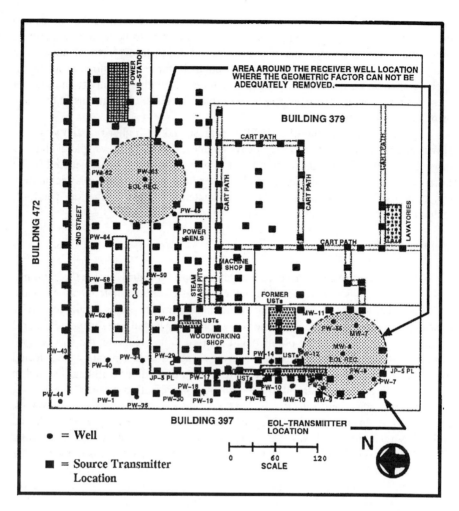

Figure 6-9 EOL source transmitter location base map. Approximately 200 locations, with about 20% overlap locations. (Source: Navy Environmental Leadership Program.)

and 6-11). Gehm's analysis of the data collected during the effort resulted in several observations based on analysis of area resistivity anomalies:

- No LNAPL had migrated under Building 379.
- The LNAPL had not migrated under and north of Second Street (on the northern side of Building 379).
- The possible migratory pathway for the released contamination is the buried abandoned JP-5 pipeline trench.
- The resistive anomalies correlate with the western portion of the Navy's contractor conventional plume.
- The resistivity data suggest that some DNAPLs are present at 30 ft below ground surface (bgs) or deeper.

PRIVATE SECTOR, PROFESSIONAL ORGANIZATIONS, INNOVATIVE SOLUTIONS   243

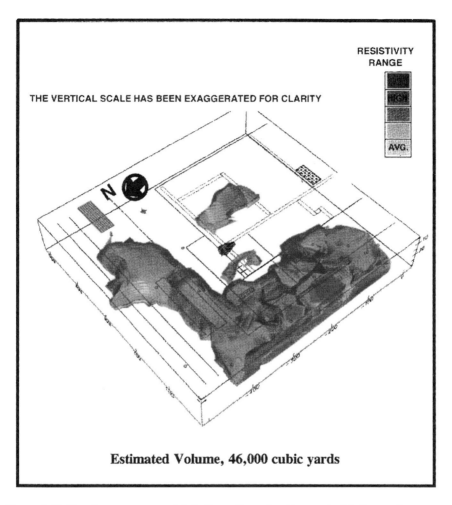

Figure 6-10 The above-average resistivity three-dimensional image depicts the maximum area or volume that may be impacted by free phase hydrocarbons. (Source: Navy Environmental Leadership Program.)

The final primary objective of this technology demonstration was to compare estimated costs of a more traditional drilling and sampling program to a combined approach utilizing the EOL process together with limited drilling (Table 6-3). In order to compare costs for these two approaches, it was necessary to make numerous assumptions. The major assumptions used for this cost comparative analysis are

- This cost comparison is most applicable for this project.
- For comparison, it was assumed that 2 acres have known hydrocarbon contamination.
- The unconfined watertable was at approximately 20 ft bgs.

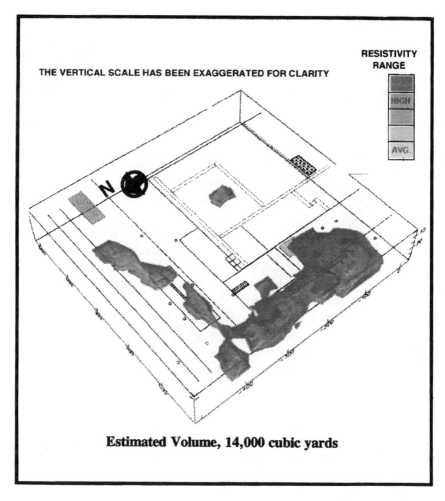

Figure 6-11 The high-value resistivity three-dimensional image depicts the probable area or volume that has been impacted by free phase hydrocarbons. (Source: Navy Environmental Leadership Program.)

As previously mentioned, for nearly all sites, the EOL technology should not be used alone. Drilling and sampling information is required to validate and correlate the resistivity measurements to empirical subsurface hydrocarbon contaminant data.

Conclusions and Recommendations

The EOL technology is a useful technology that can result in an overall cost savings compared to using conventional drilling approaches for investigations at sites with extensive LNAPL contamination. In terms of cost savings (Table 6-3), the use of the EOL technology — assuming limited drilling — resulted in a savings of

Table 6-3 Cost Comparison of Traditional Drilling Approach and EOL with Drilling Approach

Comparison Category	Traditional Approach	EOL With Drilling Approach
1. Background information known prior to comparison	2 acres with known hydrocarbon contamination; 2 acres unknown	2 acres with known hydrocarbon contamination; 2 acres unknown
2. Typical monitoring well construction	40-foot, 2-inch diameter PVC/Schedule 40	40-foot, 2-inch diameter PVC/Schedule 40
3. Monitoring wells needed outside buildings	20 wells (10 wells/acre)	8 wells (3 wells — EOL receiver wells; 3 wells — data correlation model validation; and 2 wells for contingency)
4. Monitoring wells needed inside buildings	2 wells	0 wells
5. Mobilization/demobilization (drill rig)	$4,000 ($2,000/rig) (1 rig outside; 1 low-profile rig in buildings)	$2,000 (1 rig outside)
6. Drilling costs (includes rig, drill crew, materials, and incidentals)	$1,000/well (outside) $1,500/well (inside)	$1,000/well (outside)
7. Well logging, development, and investigative derived waste disposal	$370/well (2 hours labor at $60/hr + $50/drum + $200/drum disposal cost)	$370/well (2 hours labor at $60/hr + $50/drum + $200/drum disposal cost)
8. Sampling and analytical costs (4 samples collected per well; 4 TPH-diesel and 1 SVOC analysis)	$600/well	$600/well
9. EOL costs (2 days field work, mobilization/demobilization, prepare report)	$0	$17,800 ($7,000/day for 2 days + $3,800 mobilization/demobilization)
TOTAL ESTIMATED COSTS	**$48,340**	**$35,560**

Note: Table based on the following assumptions: The number of monitoring wells required to characterize the 2-acre site would be 10 wells per acre (20 wells total). The combined EOL and limited drilling approach would require 8 wells (3 receiver wells without hydrocarbons, 3 wells for data correlation and model validation, and 2 wells for contingency). A building is located on the site requiring 2 wells to characterize potential subsurface hydrocarbon contamination under the building using the traditional approach. TPH = total petroleum hydrocarbons; SVOC = semivolatile organic compounds; EOL = electromagnetic offset logging.
Source: Navy Environmental Leadership Program.

36% compared to conventional drilling and sampling. The cost difference between using conventional drilling and sampling vs. using the EOL technology and limited drilling could be greater if a relatively large number of wells or boreholes are needed to delineate the boundaries of a plume. The EOL technology is applicable to other sites with LNAPL plumes as long as limited background information is available. Characteristics of sites where the EOL technology would be useful include

- Sites with limited access due to buildings or other access limitations.
- Sites with utilities that prevent traditional drilling and sampling.
- Sites with relatively low electrical noise.
- Sites with large LNAPL plumes.

6.5 INNOVATIVE TECHNOLOGIES AND APPROACHES

The term "innovative" yields different definitions depending on who you ask in the environmental industry. As defined by Webster's Collegiate Dictionary, 10th edition, "innovation" refers to (1) introduction of something new, or (2) a new idea, method, or device. In a broad sense, this definition holds true for innovative technologies in the environmental industry. However, for the purpose of this section, and as viewed by many environmental professionals, the term innovation is not limited to a new idea, method, or device. Rather, innovation refers to technologies or approaches that are more efficient or effective when compared to established technologies or approaches used by their counterparts (if any), which in most cases but not all cases are new to the environmental industry. In other words, innovation refers to a new solution or an improvement in existing approaches or technologies that saves time and money. This may be in the form of expediting actions, more efficient processes, and/or reducing costs. In addition, this definition includes the perception that there is an increased inherent potential for failure when using an innovative technology or approach, depending on the application, supporting demonstration data, and site-specific conditions. This definition of innovation held by many environmental professionals is a more all-encompassing definition compared to the definition found in Webster's dictionary.

The purpose of this section in not to list all of the innovative technologies and approaches or somehow summarize, in a vague sense, the type of innovations currently being promoted. Rather, the purpose of this section is to outline the underlying concepts that help identify, understand, develop, and implement innovative technologies and approaches, and outline the differences between "innovative technologies" and "innovative approaches." The contents of this entire book provide a relatively lengthy compilation of innovative technologies and approaches available from vendors, many of which accelerate environmental investigation, cleanup, or reduce costs.

In addition, sections is this chapter focus on the innovative approaches and technologies developed predominately by private businesses and professional organizations. However, in terms of development, it is important to keep in mind that innovative technologies and approaches are not solely thought of and developed by only private sector businesses and organizations. Many discoveries and innovative approaches originate from federal, state, and local government agencies, such as DOE's innovative technology programs (Chapter 4).

6.5.1 General Concepts of Environmental Innovation

The following is a list and explanation of general concepts associated with innovation in the environmental industry.

1. **The number of innovative technologies and approaches is rapidly expanding.** The number and promotion of innovative technologies and approaches continues to grow rapidly as new methods and ideas emerge through technological breakthroughs and changes in regulatory requirements. Keeping track of new developments

is, in and of itself, a difficult task due to the sheer volume and variety of new developments every year. Efficient and effective information transfer is of particular importance in order to keep up to date with new and innovative technologies and approaches. The electronic information highway is currently a prime example of where information on the newest and most effective environmental technologies and approaches are available. A compilation of environmental resources is given in Appendix A, and many of these resources are useful to track new and innovative developments.

2. **Innovative technologies and approaches are developed and implemented from different levels.** Different levels include program levels (i.e, CERCLA), project and contract levels, site levels, or bench-scale levels. For example, SCAM and presumptive remedies, which are innovate approaches, originated from the CERCLA program level. Program-level innovation receives the most attention in the environmental industry because the level of communication and awareness reaches many people through many forms, such as the federal register and EPA guidance. Developing an innovative and accelerated project schedule, however, originates from the project or site levels. Surprisingly, it is primarily at the project and site levels where innovative opportunities present themselves in developing and applying innovative solutions. These opportunities arise because of site-specific conditions, as well as ideas as well as solutions the project team formulates. In many ways, innovation is in the "eye of the beholder" (i.e., the project team), and in the desire to accomplish a task a little bit faster and efficiently with less money.

Examination of projects during each phase, from the planning through the closure process, is required in order to identify opportunities for innovation and formulate new ideas and approaches. This type of approach differentiates projects that follow a more proactive, or streamlined approach, from those projects that follow a more conventional, or traditional approach. Unfortunately, many innovative opportunities are overlooked at the project and site levels, or are purposely not acted upon, because of potential for failure. Establishing the need or requirement to complete tasks more quickly and efficiently is critical in order to create, build, plan, implement, and realize the benefit of innovations. The following summarizes the primary concepts associated with use and development of innovative technologies and approaches:

- Environmental innovation refers to the application of a technology, approach, method, or idea that will expedite, streamline, and/or reduce costs associated with environmental investigation, actions, and decision making.
- Innovative technologies and approaches are not necessarily new, and some technologies and approaches are gradually developed, or have been overlooked, such as the observational approach, and are later reevaluated and integrated into the industry. In addition, innovation includes improvements in established technologies and approaches.
- Application of an innovative technology or approach may involve taking on increased risk of potential for failure, especially in situations where the innovative technologies or approaches are not fully developed, demonstrated, or commercialized. However, proper planning and honest appraisals of potential limitations can mitigate the most serious potential failures through modification or selection of another alternative.

- Identification, use, and development of innovative technologies or approaches are dependent on the desire or necessity for improved performance, or meeting specific program or project goals. Without catalysts such as these, traditional or more established approaches may be selected as the preferred choice over innovative technologies and approaches, even if they are less effective or more expensive when compared to innovative technologies or approaches.

6.5.2 Concepts of Innovative Technologies

Innovative technologies include (1) laboratory and chemical analysis methods and technologies, (2) computer software and hardware technology, (3) investigation tools and technologies, and (4) remediation technologies and processes. The following concepts relate to specifically innovative technologies.

1. **Investigation and remediation technologies**. The most common innovative technologies are those associated with investigation and remediation, which encompass many of the newest developments and emerging technologies. Appendices B and C contain a relatively lengthy compilation of investigation and remediation technologies. Before using such technologies, it is necessary to research the development, demonstration, and commercialization of the tool or process in order to determine the overall applicability and limitations of the technology. In most cases, it is safer to avoid *emerging* technologies or *untested* technologies unless there are plans for involvement with technology development or actually developing a new technology needed to address site-specific conditions.
2. **Computer technologies**. The area of innovative computer technologies is less conspicuous. However, this area continues to expand as hardware and software improvements are realized, allowing larger and more complex problems to be modeled by computers. One example of an innovative computer modeling approach is the graphically oriented Visual MODFLOW computer model, which is a significant improvement over the original personal computer-based U.S. Geological Survey MODFLOW model. The visual version of MODFLOW employs a user-friendly graphic interface for entering input parameters and quickly designing model grids, which is much easier, accurate, and simple to use compared to the original U.S. Geological Survey version. This improvement saves time and money when setting up complex three-dimensional flow problems. In all cases, consideration of new software should, at a minimum, be validated before selection in order to ensure that the new software will deliver the reported accuracy, precision, and desired results.
3. **Laboratory and field analysis technologies**. Innovative technologies also include laboratory methods, procedures, and field screening tools. Analytical data are the foundation on which project decisions and recommendations are developed. When using these types of innovative technologies, the limitations, data validation process, and overall quality control and representativeness of the data collected must be established. The useability and representativeness of chemical data is critical in order to plan future actions or closure and in order to make recommendations. In cases where the representativeness is suspect or unknown, both the accuracy and precision of the results should be evaluated using standard validation protocol or comparative methods to quantify data representativeness. For innovative technologies, this could include comparing a statistically valid number of immunoassay

results with standard laboratory results. In cases where there are no comparisons available, such as analyzing extremely low metal detection limits with a new technology, there may be no standard chemical analysis methods that can accurately and precisely meet the new technology detection limit needed in order to compare results. Unless the data quality and useability can be compared or properly validated, innovative laboratory analytical results must be considered questionable or suspect. There are very few exceptions for using suspect or unvalidated chemical data, and if such data are improperly used in a project for making decisions or recommendations, the result can be having to go back into the field and collect new data and/or revalidate existing data. If data quality is not planned ahead of time, costs increase, projects are impeded, and concerns are raised regarding the overall quality of work, which slows the review process. Unless there are established validation procedures, methods, tools, or protocol to validate or compare the innovative technology with industry standards, the use of these types of innovative technologies should be discouraged.

4. **Comparing innovative and established technologies.** Comparing established technologies and newly discovered technologies must be completed systematically in order to select the preferred technology that best fits the site-specific requirements. The actual comparison of technologies should be initiated by contacting the vendors, or agencies, that offer or developed the technology (such as those listed in Appendices B and C) and by collecting sufficient information to evaluate the pros and cons of each technology from cost, implementability, and effectiveness considerations. Innovative technologies also require evaluation of the development and demonstration the technology has undergone in order to ensure the technology can do what the vendor claims. Based on these screening criteria, the technology of choice, which may or may not be innovative, can be selected when the limitations are evaluated along with the advantages.

5. **Developing innovative technologies.** As described above, innovative technologies require development, demonstration, and commercialization prior to being widely employed. In addition, technology development often requires sufficient funds in order to carry the idea through the commercialization phase (or public domain status) and final promotion of the technology. Also, development of innovative technologies can be difficult due to the time required for proper development, demonstration, and commercialization. Most importantly, there has to be a market for the new technology in order for it to be widely used and accepted. A leading example of an innovative technology that has undergone this process is the Site Characterization and Analysis Penetrometer System Laser-Induced Fluorescence (SCAPS-LIF), which provides real-time *in situ* detection of total petroleum hydrocarbon (TPH) contamination above and below the watertable (see Chapter 5 for more information on SCAPS-LIF). After rigorous development and demonstration, SCAPS-LIF is being more widely transferred to industry and has become more widely acknowledged as a bona fide technology for fast-tracking characterization and reducing investigation costs. In order to facilitate this process, numerous federal programs operate to encourage and partially fund the development process necessary for new and improved technologies. Leading examples of this type of program are NELP, Superfund Innovative Technology Evaluation (SITE) program, and the Air Force Center for Environmental Excellence (AFCEE) technology program.

6. **Selecting innovative technologies over established technologies.** When selecting innovative technologies during project planning, only those technologies that

should function effectively based on the site conceptual model and are demonstrated in similar environments and conditions should be seriously considered. Important site-specific conditions that may affect performance may include climate, geology, mixed waste streams, shallow bedrock or coarse materials, anaerobic conditions, low pH water, etc. If the technology does not fit within the site conditions, then the proposed technology(s) will not yield the level of performance desired, may fail, or may have to be modified during use in order to achieve the desired results, which requires additional resources. In situations where an innovative technology is selected as the preferred technology, process, or tool, gaining approval from project decision makers to go forward with the technology may be a difficult task, because some of the project decision makers may be skeptical of the proposed technology or feel the potential for failure is too high. Therefore, the proponent of the technology must prepare a clear and concise presentation in order to promote and equitably compare the use of innovative technologies with established technologies. The presentation should include not only technical information, but also case studies where the innovative technology has been successfully used. Using this approach it is possible to reach consensus more quickly and proceed with the preferred alternative.

6.5.3 Concepts of Innovative Approaches

Innovative approaches involve the use of established and new technologies; however, they also involve the use and application of regulations, strategies, ideas, and methods that expedite, streamline, and reduce environmental costs. Technology integrated with innovative approaches does not necessarily drive acceleration and cost reduction. Rather, acceleration and cost reduction are the result of how actions and innovative technologies are implemented, planned, and communicated, and how they are scheduled within regulatory frameworks and guidance. Innovative approaches can occur in a variety of forms, including the application of the observational method, or using operable units (OU) to group media, contaminants, or sites in order to fast-track environmental decision making. Most importantly, innovative approaches are primarily site-specific actions, approaches, and processes that are planned and scheduled by project teams and that accelerate, streamline, and reduce costs. For example, while SACM is not by any means a new approach, its application may be considered innovative if the technical processes used in SACM are used to fast-track actions under another environmental program, such as a state Superfund program. Project teams in this example would be the group responsible for implementing the innovative approach at sites. In another example, one of the more recent highlighted innovative approaches is the Brownfields Initiative (see Chapter 8). As states begin to take advantage of Brownfields, many success stories will likely be published regarding its application as an innovative approach to efficiently and effectively operate environmental programs and restore contaminated lands for public use. The following concepts relate to specifically innovative approaches.

1. **Innovative approaches are often difficult to visualize and must be customized for each site**. Compared to innovative technologies, which are often "off-the-shelf

units," innovative approaches exist on paper, presentations, flow charts, schedules, decision documents, and in the minds of the project team. Furthermore, they are evident in the internal and external communication between project decision makers and stakeholders as well. Instead of relying on technology to fast-track projects and save money, innovative approaches require the project team, decision makers, and community to be creative and use existing concepts or approaches in new and different ways to drive performance, or develop totally new concepts, approaches, and methods based on site-specific conditions to fast-track actions and reduce costs. To a limited degree, some innovative approaches are available as "semi-off-the-shelf units;" however, they still often require customizing in order to accommodate site-specific conditions and regulatory requirements. For example, EPA outlines the acceleration approaches in SACM guidance and how they can be applied within the CERCLA regulatory process, such as removal actions, interim remedial actions, or presumptive remedies. Project teams use the SACM framework and guidelines as the map for developing innovative approaches in order to achieve project goals and objectives more quickly and efficiently. Other innovative approaches do not have a program framework to follow and must be customized from the beginning to ensure proper implementation and regulatory compliance. For example, the observational method relies on an in depth consideration of site-specific needs and site assumptions in order to implement the method efficiently and effectively. The process used in the observational approach does not include specific types of response actions, such as removal actions under SACM, which have to be identified, proposed, scheduled, and implemented following the applicable regulatory process. Thus there are varying degrees of identification, implementation, and difficulty in developing and using innovative approaches, and, in most cases, the approach must be customized to some degree in order to address site-specific conditions and drive performance, as described below.
2. **Innovative approaches are flexible and must be custom fit to site conditions.** Innovative approaches are generally flexible and accommodate project-specific needs, such as project goals, regulatory requirements, site conditions, action levels, land use, etc., by considering site-specific conditions and requirements. For example, real-time data collection is now an established technology used to characterize sites. In order to maximize the benefits of real-time data collection, an innovative approach can be used to improve its overall efficiency and effectiveness. In another example, investigation by excavation through proper planning and approval can be used to quickly characterize site conditions and streamline environmental cleanup. To illustrate innovative approaches, these two hypothetical examples are further described below to outline (1) the process for selecting innovative approaches and (2) developing innovative approaches to meet site-specific investigation and restoration needs. Many other innovative approaches are available, but require planning, identification, and customization ahead of time in order to maximize their benefit.

Using an Innovative Approach to Maximize Real-Time Data Collection

Real-time data collection involves developing an on-site approach for collecting and analyzing chemical data. The quality of data may vary from qualitative screening data to quantitative on-site chemical analysis following EPA protocols. More importantly, the analytical results (or field measurements) are available to the field team in a relatively short period of time, from a matter of seconds to several hours, depending on the type of testing equipment used for making real-time decisions. In this case, the technology itself fast-tracks the project by providing data more quickly than an off-site laboratory following traditional procedures. When applying an innovative approach for investigation, it is possible to maximize the useability of data, minimize the time required to collect data, minimize the amount of data needed, and reduce costs.

Most soil sampling and analysis plans locate each sample collection point on a grid, the depth from which samples will be collected, and the analyses to be tested. However, by developing a sampling and analysis plan that provides only the framework for making in-field decisions using real-time for determining where to sample next, at what depths, and what to analyze for, the investigation process can be streamlined. In this example, the sampling and analysis plan may show the general area of concern, list the known contaminants of concern and at what depth they are anticipated to be present, list the procedures and methods for sampling and analysis, and describe the level of effort and budget available, the project schedule, and the goals and objectives of the field effort. However, only a limited number of sample locations would be identified, and the target analyte list would be provided, along with optional analytes (and methods) that can be analyzed depending on what is detected in the initial round of samples. In this scenario, the field team is empowered to make decisions to determine where to sample next and what to analyze for in order to characterize the site more efficiently and effectively compared to following a standard grid sampling approach. Most importantly, the field team would need to be highly experienced, and be able to anticipate uncertainties ahead of time in order to deviate as necessary and make real-time decisions based on the real-time data. This approach can focus the field effort on defining the extent of contamination and characterizing the site conditions in the areas posing the most risk. The benefits of this type of approach are that only one investigation, conducted in short phases (in order to preliminarily plot, validate, statistically evaluate, and review data), may be required to adequately characterize the site for restoration purposes. In addition, unnecessary sampling and analysis is reduced when compared to grid sampling techniques, as long as prudent field decisions are made to meet the project goals and objectives. Clearly, this is an innovative and streamlining approach that should only be proposed in cases where all stakeholders agree ahead of time that the field team is capable and empowered to make real-time sampling and analysis decisions with the objective of streamlining the investigation process.

Investigation by Excavation

During the investigation stage, boreholes are usually drilled to collect soil samples. Under heterogenous conditions, evaluation of borehole data does not provide a clear understanding of site conditions, even if the boreholes are placed relatively close to one another. Investigation by excavation is an investigative approach to expedite the site characterization process, accelerate risk reduction, and potentially reduce costs. Instead of utilizing a drill rig to collect soil samples, a backhoe (or tracked-hoe) may be used to collect samples and observe soil conditions. Most UST programs use this approach to remove USTs, over excavate limited soil contamination, and collect confirmation samples to close UST sites or propose further actions. Investigation by excavation is similar to this type of UST investigation approach, as long as sample integrity and representativeness can be maintained if collected from an open excavation, such as soil analyzed for metals and not volatile organic compounds, and as long as disturbing the surface material is acceptable and safe.

The field team is able to collect soil samples underneath objects, enabling them to characterize site conditions that otherwise would have to be sampled using angle drilling technologies. Soil and objects removed during the excavation activities can be disposed of as investigation-derived waste. However, cost differences of disposal between petroleum contamination and hazardous waste and the anticipated waste volume must be considered ahead of time, along with the relative size of objects. For example, angle drilling may be more appropriate instead of investigation by excavation for large objects, such as buildings. These cost differences must be considered ahead of time to ensure that the overall cost of disposal, etc., will not negatively impact the project if a large amount of hazardous soil is excavated. In cases where the waste is not contaminated, or contains only residual amounts of contamination, it may be possible to return the soil to the excavation upon regulatory approval and close the site or area.

At some sites this investigation technique is useful to reduce risk if there are isolated sources or "pockets/hot spots" of contamination. In addition, the time and cost required to collect samples and observe site conditions is much less compared to conventional borehole characterization, where layers of soil and product are more easily identified in excavations. Nonetheless, investigation by excavation should only be implemented if site-specific requirements indicate it provides a superior alternative over conventional methods and that it is safe and acceptable to have an open excavation(s). In addition, going too far with investigation by excavation is a concern, in that it is an investigative technique and not a cleanup or removal technique. While contamination may be removed, including the source of contamination, these benefits are not the reason for using investigation by excavation. Primarily it should be used as a site characterization approach, and should not be proposed as an cleanup action. This distinction is critical, because under most cleanup actions, public participation is required, and if not properly addressed, this deficiency will result in a serious compliance issue. Therefore, when designing and implementing investigation by excavation actions as part of site characterization, it is critical to differentiate it from a cleanup action.

6.6 REFERENCES

American Society for Testing and Materials (ASTM). 1995a. Standard Guide for Risk-Based Corrective Action Applied to Petroleum Release Sites, Designation E 1739-95.

ASTM. 1995b. Accelerated Site Characterization for Confirmed or Suspected Petroleum Releases. Designation PS3-95.

Anderson, M.P. and Woessner, W.W. 1992. Applied Groundwater Modeling. Academic Press, New York.

Blairicom, V.B. 1980. Practical Geophysics for the Exploration Geologist. Northwest Mining Association. Spokane, WA.

Brown, S.M., Lincoln, D.R., and Wallance, W.A. 1990. Application of Observational Method to Hazardous Waste Engineering. Journal of Management Engineering, Vol. 6, No. 4, October, pp. 479-501.

Corey, M.M., Lorey, R.L. and Farmer, M.A. 1992. The Use of an Observational Approach for Implementing a Groundwater Recovery System. 85th Annual Meeting and Exhibition for the Air and Waste Management Association, Proceeding No. 92-3-06, June 21-26, pp. 1-17.

Marquis, S.A. and Dineen, D. 1994. Comparison between Pump and Treat, Biorestoration, and Biorestoration/Pump and Treat Combined: Lessons From Computer Modeling. Ground Water Monitoring Review, Vol. 14, No. 3, Spring, pp. 105-119.

Northeast Geophysical Services. Undated. Methods and Applications Reference Manual. Yarmouth, ME.

Peck, R.B. 1969. Advantages and Limitations of the Observational Method in Applied Soil Mechanics. Geotechnique, No. 19, p. 171.

Telford, W.M., Geldart, L.P., and Sheriff, R.E. 1991. Applied Geophysics. 2nd ed. Cambridge University Press, NY, NY.

Trotti, R. 1996. Getting Comfortable with Rebecca. Remediation Management, Vol. 2, No. 5, p 6.

Smith Technology Corporation. Undated. Personal communication. Unpublished table and figure. Portland, OR.

U.S. Environmental Protection Agency (EPA). 1992a. Ground-Water Modeling Compendium. OSWER, EPA-500- B-92-006.

EPA. 1992b. Fundamentals of Ground-Water Modeling. EPA/S-92/005.

Vogelsang, D. 1995. Environmental Geophysics: A Practical Guide. Springer Verlag, NY, NY.

Wang, H.F. and Anderson M.P. 1982. Introduction to Groundwater Modeling. W.H. Freeman and Company, San Diego, CA.

Ward, S.H. 1990. Geotechnical and Environmental Geophysics. Society of Exploration Geophysics, Vol. 1-3, No. 5.

Wilson, L.G., Everett, L.G., and Cullen, S.J. 1995. Handbook of Vadose Zone Characterization and Monitoring. Lewis Publishers, New York.

CHAPTER 7

Communication, Teamwork, Leadership, and Trust

Communication and teamwork are, without question, essential cornerstones required for meeting project goals and objectives and catalysts for accelerating cleanup and streamlining decision making. As scientists and managers, we depend on technology for environmental investigation tools and laboratory methods, data evaluation and report generation, remediation processes, and specialty technologies such as geophysics. This reliance depends greatly on site conditions and the contaminants encountered, if any. In addition, we also depend on our ability to plan, implement, communicate project needs and decisions, and work together as teams in order to keep projects moving forward and to meet goals and objectives. As a result of our dependence on coworkers and counterparts to produce results and make decisions, communication, teamwork, leadership, and trust have importance significance in the environmental industry. Specifically, effective communication is essential for transferring information and exchanging ideas. Effective communication is also the foundation of meaningful presentations, productive dialogue, and quality documents and reports. Effective teamwork is essential in order to work efficiently with internal and external counterparts, including coworkers, clients, agency personnel, and community members. Together, effective teamwork and communication, along with strong leadership and high trust factors, are as important as the technologies we depend on for fast-tracking projects and reducing costs. Following this line of reasoning, the concept of cleaning up contamination could be similar to the adage "guns do not kill people, people kill people," where the adage "technology does not clean up sites, people clean up sites" applies.

7.1 GENERAL CONCEPTS

Do not underestimate the importance of this chapter. The scientist in us, and to some degree the manager too, desires to follow the logical pathway from discovery through site closure at environmental sites. Unfortunately, this pathway may not meet all of the *needs* and *expectations* of others, whether the individuals are internal or external to the organization that we work for. Some scientists and managers tend to ignore, or perhaps do not comprehend, the importance of the expectations and needs of others, and how they directly influence our ability to meet goals and objectives. Some environmental professionals use less effective communication styles or older management models, which they feel are logical and adequate in order to do their job and complete projects. However, ignoring the importance of these four fundamental elements (communication, teamwork, leadership, and trust) can affect how quickly environmental decisions are made, and may impede projects due to a variety of pitfalls including misunderstandings, inadequate technical explanations, poor planning, failure to identify win–win alternatives, long review periods, or resampling because other alternatives need to be considered late in the project. The purpose of this chapter is to further outline the importance of communication, teamwork, leadership, and trust, along with other related topics, as they pertain to the environmental industry.

7.1.1 Evidence Supporting the Need to Value People in the Environmental Profession

This chapter alone may not convince all readers of the importance of communication, teamwork, leadership, and trust. However, by observing and researching what other businesses, industries, and government agencies are currently and have been experiencing in the mid- to late 1990s, there is substantial credible evidence to show that the environmental industry is experiencing a transition and changing dynamics similar to many other industries, and it is not immune to change. The 1990s "age of information" and world economy, and emphasis on improving environmental performance with less money, has changed the environmental industry forever. In order to understand the current dynamics affecting each of us in the environmental industry, the question of who we are must be asked. Are we scientists, managers, or regulators? Or, like others in various professions, are we a combination of many titles, who have to work in teams and be flexible in order to successfully deal with change in our workplaces and lives. Increasing work performance and the ability to accelerate and streamline are directly related to how well we communicate and work in team settings, lead people, and encourage trusting relationships.

The environmental industry is changing, not only from a regulatory perspective, but also from a work perspective — there will never again be the same Superfund program that employed tens of thousands of environmental professionals as in the 1980s and early 1990s (Figure 7-1). The purpose of this chapter is to offer approaches to help embrace these changes in order to thrive in a dynamic marketplace and regulatory arena. Thrive, in this case, refers to both private and government personnel, where the industry as a whole must adjust to meet the needs of society while still complying with environmental regulations. In today's environmental marketplace, that means (1) working together, (2) effectively communicating project information, decisions, and requirements, (3) accelerating the cleanup and investigation process, (4) reducing costs, and (5) meeting the goals and objectives of an effective industry that is directly and indirectly subsidized by the public.

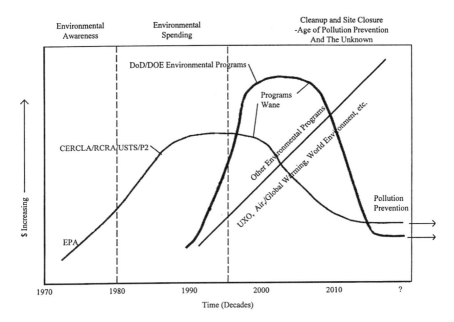

Figure 7-1 Spending trends in major environmental programs for past, present, and future.

The importance of this chapter is also evident in relation to the way in which agencies and businesses are being reorganized, restructured, closed, or acquired. It is evident that environmental performance must improve in order to meet the needs and expectations of society. The environmental industry is far from being alone when it comes to having to change and survive in chaotic times. As scientists and technicians, however, we tend to ignore the key changes experienced by others in different industries, and focus almost entirely upon the technical aspects of our jobs, utilizing the old communication and teamwork approaches that have been in place for years but that do not necessarily emphasize leadership or encourage trust. These outdated models are quickly being replaced. Consider the business models touted by Tom Peters, where building great organizations is based on "Embracing Chaos" and using innovative approaches to improve communication, teamwork, leadership, and trust. In addition, successful businesses like Tom's of Maine, McDonalds, the Xerox Corporation, and the Walt Disney Corporation, and agencies such as the Air Force Center for Environmental Excellence (AFCEE), are all examples of why we should consider investing in and learning new ways of improving communication and teamwork, developing new leadership skills, and building trusting relationships. In doing so, environmental professionals can better ensure themselves and the industry a long and productive future, and meet the needs and expectations of society through acceleration and cost-reduction initiatives.

In many cases, such as when implementing the observational method, working under chaotic conditions is part of the technical approach in order to address uncertainties when they arise and implement contingency plans. In doing so, it is possible to expedite and streamline projects. In fact, many environmental professionals enjoy the technical challenge associated with dealing with uncertainties in real time under observational conditions. However, dealing with chaos also becomes necessary in order to meet most client, public, and government expectations because they, like a groundwater plume, are often moving targets. To temper the effects of moving targets and chaos while fast-tracking projects, environmental professionals must consider the importance of communication, teamwork, leadership, and trust, and improve them in order to increase performance. For example, consider how much effort is currently devoted strictly to technical work in the environmental industry. Experience suggests that 80% of environmental professionals typically spend only about 20% of their time doing technical work they learned in school. The other 80% of their time is spent reviewing budgets, going to meetings, developing project schedules, negotiating contracts, ordering equipment, addressing interoffice or interagency issues, resolving conflicts, and marketing, to name a few examples. Certainly, there are exceptions to this rule — there are some individuals that are strictly field-based personnel, researchers, designers, or computer modelers who spend all their time on the technical side of their profession. However, most decision makers and leaders are generally too involved with day-to-day project needs to spend more time doing the technical work they enjoy most. These individuals are responsible for communicating technical work done by others, spending their time working with other people, and simply trying to keep their heads above water in order to breathe! This chapter covers what a large portion of environmental professionals primarily do for a living: communicate and work in team settings in order to achieve project goals and keep supervisors, clients, or the public, satisfied. In terms of accelerating cleanup, streamlining decision making, and reducing costs, the importance of communication, teamwork, leadership, and trust become critical in order to propose ideas and implement actions.

As expected, some leaders and project decision makers are not as gifted as others, and perhaps would do better by limiting their time to technical subjects only. Training, however, is one avenue to help individuals improve their leadership and communication skills, learn new teamwork approaches, build trust in organizations or teams, and learn how businesses and agencies thrive. Historically, most training in the environmental industry is technically based, designed to introduce new technical approaches and continuing environmental education programs. Certainly, this type of training is important. Yet a significant portion of work environmental professionals do is nontechnical, learned on the job, and does not always relate to the education they received in college or graduate school. This chapter outlines some of the more obvious nontechnical areas that can help improve performance. However, reviewing the references in this chapter is advised in order to gain the full benefit of the topics in this chapter.

The Cost Benefit of Valuing People

Improving communication, teamwork, and leadership can result in increasing performance and help accelerate response actions, however it is important to recognize that additional up-front costs and level of effort may be required in order to improve those skills, along with determination. For example, leadership training requires additional funds and resources that take money away from budgets and/or profits. Yet, improving leadership skills can reduce long-term costs of projects, accelerate investigation and cleanup, minimize the level of effort needed to characterize sites, and help integrate values, principles, and common sense into environmental programs and projects. Most importantly, leadership skills can help keep clients and the public satisfied. Investing in people and teams is wise because we are a service-oriented industry; however, it may require additional up-front time and effort in order to be effective.

7.1.2 Science-Based Information

For some individuals who read this chapter, the topics may seem nebulous, unimportant, or perhaps even unfounded. However, the references contained in this chapter are by no means nonscientific. In fact, several of the references are current standards of excellence for communication and teamwork, developed following the scientific method. In order to understand the usefulness of this chapter, the reader must consider other sciences and observations in order to understand how and why these topics are central to the environmental industry. To illustrate this point, the following analogy is provided: most environmental businesses, such as consulting firms, are run by environmental professionals that have environmental, earth science, or engineering backgrounds. However, few environmental business owners, or high-level managers, have ever taken education courses or training on "how to run a business." Consequently, it is reasonable to believe that there is much to be learned from business professionals who have the education and experience useful to improve environmental business success and profitability. Similarly, concepts of high-level communication and teamwork are seldom included as part of environmental or engineering educations. However, bona fide improvement theory and methods are continually being published that can be used to improve performance and success in the environmental and engineering fields. Acknowledgement of this type of scientific research and development in these areas is important in order to increase the environmental industry performance, and accelerate cleanup and streamline decision making.

7.2 COMMUNICATION

Effective communication is essential for increasing performance and knowledge between coworkers, promoting accelerated response action, and streamlining decision making. Communication includes two primary types of information transfer: (1) verbal communication, such as dialogue in meetings, discussions, and presentations, and (2) written and electronic formats, such as reports, memorandums, professional papers, letters, and electronic mail. Written and electronic communications are relatively standardized in that grammar, punctuation, spelling, and general organization of thoughts and ideas are taught in our primary, secondary, and higher education school systems. The ability to write well generally improves with practice and mentoring, and after environmental professionals have the opportunity to write many reports, work plans, and/or papers. Verbal communication is also relatively standardized, but to a lesser degree when compared to written forms of communication, and only a minority of environmental professionals have been taught more than basic verbal communication skills during their education or training. More often in the environmental profession, verbal communication skills are based on styles or models acquired from cultural backgrounds, personality type, and through mimicking styles of communication used by mentors. For this reason, there is a wide range of verbal communication styles that have varying degrees of effectiveness. Information in verbal communication is emphasized in this section in order to stress its importance, and as a means to illustrate how environmental professionals can improve verbal communication in everyday work environments and in meetings between stakeholders. Topics covered include

- The learning process
- Understanding basic dialogue, storytelling, and the importance of case studies
- Communicating success and failure
- I and Thou concepts
- Behavior styles and gender differences.

7.2.1 The Learning Process

Directed Business Concepts (DBC) developed a business improvement system called *ah ha!*™. Using their system, participants learn basic and advanced skills on various topics including the learning process, communication, quality, service, listening, and improving business and organizational skills. As part of improving communication skills, an understanding of the learning process is crucial in order to balance the level of communication with the level of knowledge. Balancing the level of communication with the level of knowledge can save time and avoid unnecessary problems associated with ineffective communication. The steps of the learning process a baby goes through learning to walk are used as an example to illustrate how the level of knowledge must balance with the level of communication. The steps of the learning process include four progressive levels (DBC 1997):

	Skill Level	Level of Knowledge
1)	**Unconsciously unskilled:**	I'll just lay here and look around.
2)	**Consciously unskilled:**	Big people seem to be able to move around the room ... maybe if I trying rolling?
3)	**Consciously skilled:**	If I keep one hand on the chair, then move one foot at a time, I can walk around the chair.
4)	**Unconsciously skilled:**	Let me run and get that for you!

On a daily basis, environmental professionals work with a variety of people, both internally and externally. These professionals have varying degrees of knowledge in environmental science, regulations, engineering, and technologies. Experienced environmental professionals understand the science, regulations, engineering, and technology, or portions thereof, needed to do their job, and are in general consciously skilled or unconsciously skilled within most, if not all of these areas. In addition, environmental professionals discuss and write about these areas in order to communicate with administrative help, clients, contract specialists, auditors, etc., who need to have some inkling of what the environmental professional is talking or writing about in order to provide them with what they need to do their job. Administrative help and clients, for example, may have a level of knowledge of the environmental industry at the unconsciously unskilled and consciously unskilled because their job is not specifically about the technology or regulations. Under these circumstances, it is easy for the environmental professional to communicate at a level way above the level of knowledge of these individuals whether they are reading or listening, unless they are an experienced environmental professional aware of the differences existing in knowledge levels. The result can be disastrous to projects if the level of communication does not match the level of knowledge. For example, an administrative assistant may provide the wrong service, a potential client may find another business to do the work because they do not understand or trust your plan, the agency contract specialist may procure a fixed price contractor instead of a cost contractor, or a federal auditor may disallow certain costs because they feel they are not "technically" justified based on the information provided. If communication is balanced with the level of knowledge of the person(s) involved, these types of problems can be avoided. Therefore, considering the level of knowledge individuals possess on a specific topic is important in order to adjust the level of communication and balance it with the person's level of knowledge, in order to facilitate successful outcomes.

Finally, environmental professionals may have a level of knowledge in the unconsciously unskilled or consciously unskilled for specific areas. For example, a geographical information system (GIS) is determined to be the best avenue to organize and report data for a large and complex project. The project leaders, clients, and regulatory staff may not have a thorough understanding of GIS applications or their use. Therefore, when working with the GIS specialist who will do the work, the relative level of knowledge must be balanced in order to produce useful results, etc. Other examples of where ineffective communication can occur often relate to computer aided design (CAD) drawings, discussions between different technical disciplines (i.e., engineering and hydrogeology), utilization of software programs, and discussions about regulations.

7.2.2 Understanding Basic Dialogue

Meetings and negotiations are critical in the environmental industry in order to move projects forward and finalize decisions. The basic dialogue used in these settings is far more complicated than simply using words to transfer information, intent, or agreement. Most environmental professionals have limited training or education related to verbal presentations and negotiation skills and they often use a style or approach they think makes sense, or may follow a mentor's style that has been effective in the past. Unfortunately, this approach may not be successful in transferring the desired information or intent, or in reaching consensus in a timely or graceful manner if the communication style used is ineffective. In order to streamline decision making and the information transfer process, it must be realized that verbal skills are as important as written skills in the environmental industry, if not more important, because decisions and proposals are not agreed upon, nor are issues resolved, on paper. Only after there is verbal agreement on decisions, proposals, and issues are they finalized in written formats, such as letters or decision documents. Knowing how to communicate in meetings, negotiations, and presentations is essential in order to move projects forward. Training and mentoring to improve dialogue skills should both be considered for enhancement of communication skills.

Words by themselves are not the most important elements in dialogue. As Doug Sharp of DBC says, "Don't try to hide your feelings with words, it just will not work because words alone will not hide feelings" (DBC 1997). Good communication involves

- 7% words
- 23% voice tone
- 35% facial expression
- 35% body language.

Often in meetings and discussions the intended information is not conveyed to the listeners, even though the words are carefully chosen in order to address a specific topic. The image in the speaker's mind does not match the image in the listener's mind. One example of this is the classic dialogue in arguments between two lovers when one person says "I love you!" and the other person says "No you don't!" Is the first lover really lying? Whether or not the first lover is lying is unimportant because the second lover is going with a "gut" feeling that tells them that their lover does not love them. After the emotions cool down, they realize that really they do love each other, a happy ending for these two individuals. However, in the workplace, the ending may not be as happy. Paying attention to more than words when speaking to others is important if you want to convey the true intention, image, emotion, and information to them. In addition, effective communication involves attentive listening, where 80% of good communication is listening (DBC 1997). A good listener

- Is attentive
- Is interested
- Asks questions to get the full story
- Does not interrupt
- Responds to facts, not emotion.

In order for others to quickly consider and understand proposals and new ideas, dialogue must be effective and move beyond words. Effective communication involves listening, paying attention to how we communicate with others, and learning effective communication styles in order to convey the true meaning of what we are saying and writing. If the true meaning is not conveyed, misunderstandings will occur, impeding progress. In cases when communication is ineffective, what people think they are agreeing to may in fact be something else, resulting in having to revisit the issue or proposal and further expending resources and time. To help avoid these type of misunderstandings, use paraphrasing to confirm what you think the person speaking is saying. Once they agree with your interpretation, you have successfully understood what they are saying. Furthermore, focus on asking "I" statements vs. "you" statements. The "you" statements are often interpreted as confrontational by the person listening. Where, "I" statements speak about yourself, which helps explain what you mean vs. what you *think* about someone else. Using "I" statement effectively takes time and practice, however, communicating in this style is highly effective in avoiding confrontations, which helps promote win-win decisions.

Storytelling

Storytelling may seem a rather weak reed with which to build an organization, but that is to misunderstand the power of stories (Owen 1990). The use of storytelling in the environmental industry is significant in that we all use stories in one form or another to communicate. Some professionals may prefer to call storytelling "case studies" or "examples;" however, the fact is that when discussing complex issues, it is often far easier to explain the meaning or intent through storytelling based on experience and knowledge than by simply discussing the topic itself. Stories help others understand how actions or decisions were made in the past, whether the consequences or the "moral of the story" were from a success or failure. In addition, storytelling that is based on experience adds credibility to the spokesperson, and allows others to see what has been done before and that the person telling the story has valuable input into the process because of their previous experience.

Another indication that storytelling is valuable becomes evident when listening to how effective communicators deliver their ideas and experiences to us. For example, Tom Peters, Stephen Covey, and many others use examples and stories to help convey meaning and concepts. Using these same storytelling methods, it is easier for environmental professionals to convey meaning, illustrate ideas or concepts, and gain credibility. Most importantly, storytelling is extremely valuable for proposing innovative technologies and approaches, or streamlining concepts, because it illustrates new ideas and approaches more effectively than, for example, theory alone. In turn, the idea or approach may be more quickly accepted by decision makers.

Publishing Case Studies

Communication includes publishing case studies in order to transfer information on how and why an approach or method is successful, and also publish lessons learned to help others understand what does not work in the environmental industry. Published case studies are the written equivalent of verbal storytelling outlined above. Case studies are extremely important for promoting accelerated and innovative response actions or streamlined decision making, especially if they are new or unfamiliar to industry professionals. In addition, it is important to publish cost-reducing initiatives that not only meet the goals of protecting human health and the environment, but also save money. By communicating these findings, the findings and theory can be used at other sites to help determine how to, or how not to, complete actions and make decisions. An ancillary benefit of publishing case studies is helping society as a whole understand what we do. In turn, society will be more inclined to continue funding environmental programs and maintaining the current environmental awareness because there is a perception that progress at environmental sites is important and we have to hasten it. From this perspective, more mainstream publications should be used to promote environmental accomplishments.

7.2.3 Communicating Success and Failure

In order to make informed decisions, people need to learn from the past, which involves communicating successes and failures. Communicating success is common in many agencies and businesses, although not universal. Past successes are often used to promote new approaches, improve methods, or innovative technologies that can be used to complete projects more quickly, save money, or simply illustrate how a difficult job was successfully completed. Communicating successes also brings recognition to the people and teams behind successful efforts, helping to build loyalty and honor people in the workplace. Communicating failures is also necessary because people often learn by their mistakes, which is especially important in cases where trial and error is used in order to remedy a problem. Limiting future trial and error is possible by communicating not only what works, but what does not work, so the same mistakes are avoided.

Communicating *future* potential success and failure is also important. This is especially true when pursuing proactive environmental approaches, developing acceleration strategies, and using innovative technologies and approaches, because the potential for failure may be higher than when following a more conservative or traditional approach. Limiting communication to only a potentially successful future is risky because if there is a failure, the skeptics and cynics will be sure to let everyone know it did not work. This may result in a poor reception, meaning a minor mishap can essentially stop future use of the new approach or method that in fact has possibilities. However, if potential failure is properly communicated up front and there is consensus to move forward with the idea or approach, the impact of failure is lessened, and opportunities to try again or improve the idea or method can emerge. While potential failure should not be dwelled upon, it should be communicated so that the entire project team and stakeholders understand any added risk. In cases of excessive risk, the project team and stakeholders may choose a more conservative or more costly approach, which may be advisable if there truly is an excessive level of potential for failure. In cases where the potential for failure is relatively low, focusing more on the benefits and less on the risks may be justified depending on the precedent-setting impacts of the action. Most importantly, an equitable balance must be struck between potential success and failure when communicating new ideas, methods, and approaches.

7.2.4 I and Thou Concepts

In Tom Chapple's *The Soul of a Business* (1993), the concepts of *I and Thou* as written by philosopher Martin Buber, are discussed in how they were integrated into his business, Tom's of Maine. The following quotations explain the fundamental aspects of "I and Thou" and "I and It" in Tom Chapple's book:

> In one attitude toward the world, we expect something back from all the relationships we have. We organize, use, abuse, control, and dominate everything — and everybody — in our lives because we want something from these relationships. Buber described this mind set as an "I–It" relation when we treat even other people as objects. The other attitude we can have, however, is our relation to another human *not* for anything in return but in simple respect, love, friendship, and honor for their sake. Buber called this an I–Thou relationship (Chapple 1993).

Chapple explains that by treating people as worthy of respect (i.e., I–Thou attitude), his company was able to increase its success and maintain the overriding values of producing natural products without, for example, preservatives or sensationalized packaging that his competitors use. This topic has relevance in the environmental industry because of the perceived attitude of "us and them" when working on environmental projects. There are untold numbers of examples where an "us and them" attitude substantially impedes progress at environmental sites. Multiple stakeholders, responsible parties, community members, and government agencies have often had this mind set which pits one or more entity against another. Communication, in any form, becomes ineffective and difficult under these circumstances. Therefore, the concepts of "I and Thou" do have relevance as a way to mitigate the impacts of "us and them" in the environmental industry.

The difficult task becomes modifying the attitude of project team members and stakeholders. In order to begin moving away from the "us and them" mind set and into the "I and Thou" mind set, the attitude of the team members and involved entities must embody the concepts that people are worthy and respectable. In turn, communication improves, fostering open and honest discussions and negotiations related to environmental actions and decisions. In addition, using the "I–Thou" attitude, the values and benefits of accelerated response actions, innovative technologies, and cost reductions, for example, become much easier to communicate to others. Chapple's work presents a case study in how the "I–Thou" attitude can be integrated into a business or team. This same integration of the "I and Thou" concepts, to a degree, is possible within environmental project teams. However, these concepts must first be popular, and accepted by the entire project team, in order to be effective. More information related to team relationships, and honoring and respecting people, is presented in the sections on partnership models and the power of cynics in this chapter.

7.2.5 Behavior Styles and Gender Differences

Effective communication involves understanding the four basic human behavior styles as described in William Marston's *The Emotions of Normal People* (1928). Marston believed that people tend to learn a self-concept that is basically in accord with one of four factors (listed below). It is possible, using Marston's theory, to apply the powers of scientific observation to behavior and to be objective and descriptive, rather than subjective and judgmental. In Marston's work, all people exhibit four behavioral factors in varying degrees of intensity. These four factors include

- **Dominance–Challenge**: How you respond to problems and challenges.
- **Influence–Contacts**: How you influence others to your point of view.
- **Steadiness–Consistency**: How you respond to the pace of the environment.
- **Compliance–Constraints**: How you respond to rules and procedures set by others.

Since Marston's work was published, various companies have developed processes on paper and on computer to analyze human behavior styles. These types of tools can be used for self-evaluation and to characterize another person's behavior style. As part of understanding behavior styles, these tools also suggest ways to improve communication, identify ideal work environments, explain differences between styles, and identify ways to motivate and manage certain behavior styles. In addition, understanding how to improve communication with coworkers and project team members is also connected to how well you know yourself, and the behavior style of others you work with. It is possible to adjust your communication, work environments, management styles, etc., in order to reduce the likelihood of miscommunication or conflict, and improve overall communication with others.

In Tony Alessandra's book *The Platinum Rule* (1996), he compares the concepts of the golden rule, "Do unto others as you would have them do unto you," and his "platinum rule," which is "treating others like they want to be treated." The golden rule implies that other people would like to be treated the way you want to be treated, which often may not be the case based on the variety of personalities and behavioral styles humans have. Alsessandra's work relates to "how to be people smart" in order to improve communication. His work focuses on first understanding what people want and then providing it to them. Understanding what people want relates to understanding what drives people, how to better relate to them, and recognizing options for improving relationships with them. His work is important in relationship to the environmental industry because it connects how we as scientists, regulators, staff workers, and individuals can improve communication and streamline decision making by treating one another, and others within the industry, like they want to be treated instead of treating them like we want to be treated. Alessandra's work also follows the theory that there are four basic types of behavioral preferences people have (see factors above) and that understanding these preferences is useful in considering communication *options* while maintaining your own values and belief system.

By using tools such as TTI Performance System's *Managing for Success*™ *Behavioral Style Analysis* software or Carlson Learning's *DISC Dimensions of Behavior,* the difficult task of quantitatively characterizing human behavior and identifying ways to improve communication is relatively easy and inexpensive. As a means of improving internal and external communication in the environmental industry, more widespread use of these types of tools, or similar products and profiles, would likely be beneficial. This is especially true in cases where communicating advantages and disadvantages of an acceleration program, new method, or different strategy is required, and the proponent wants to use the most effective communication style to convey the idea or image needed so that others can easily understand the project.

Different communication styles not only relate to behavioral differences, but also to location and gender differences. When speakers from different parts of the country or of different ethnic or class backgrounds, talk to each other, it is likely that their words will not be understood exactly as they were meant (Tannen 1986). Furthermore, male–female conversation is also cross-cultural communication. If women speak and hear a language of connection and intimacy, while men speak and hear a language of status and independence, then communication between men and women can be like cross-cultural communication, prey to a clash of conversational styles (Tannen 1990). As environmental professionals, we are faced with cross-cultural and cross-gender communication. Realizing that what we want to communicate as leaders, managers, engineers, technicians, and field staff is not always understood by clients or the public, for example, helps us realize that we need to evaluate and modify communication styles in order to achieve our goals. In Tannen's work (1986; 1990), she outlines how cross-cultural and cross-gender communication affects our lives and our work, and presents ways to improve these lines of communication. Reviewing this literature, or seeking guidance from communication experts, is the best avenue to fully grasp the limitations of cross-cultural and cross-gender communication, and ways to improve it.

7.3 TEAMWORK

Communication is a process by which information is transferred between people. Teamwork is a process by which the information is acted on by a group to achieve some desired result. Environmentally speaking, teamwork is a catalyst for accelerating cleanup and streamlining decision making, both internally and externally. In group settings, teamwork defines the ability to efficiently and effectively work together, understand the group and individual strengths and weaknesses, and accelerate the process needed to meet project goals and objectives. In this section, basic teamwork models and principles are discussed in relation to increasing performance and efficiency, and accelerating response actions and decision making. These topics include

- Mission, vision, and identification of common goals and objectives
- Partnership models
- Empowerment and accountability.

7.3.1 Mission and Vision

Mission identifies the reason for being, and vision clarifies how you will operate in the future (Robinson and Robinson 1995). Together, mission and vision describe the destination headed by a group, organization, or project team. Goals and objectives are components of the mission and vision that relate to the site-specific or project-specific tasks needed to arrive at the desired destination. Together, mission, vision, and goals and objectives are important because they embody the common philosophy, operation, and direction for a project team working at the program or project level. Vision statements are of particular importance because they can captivate the project team, help improve efficiency, and focus resources in order to arrive at a common destination. Powerful, effective visions have three qualities: they are big, attractive, and doable (Owen 1990). They are big in a comfortable way, but not grandiose. They are attractive in that the team is comfortable with the concepts and direction. And they are doable in that they are technically and historically feasible.

While an actual mission and vision may not be needed for relatively small environmental projects, such as an underground storage tank (UST) removal, the larger, multi-million dollar projects can benefit from developing mission and vision statements as long as the project team believes in them. Mission and vision statements for projects goes beyond the individual agency, responsible party, or business mission and vision statements. Collective statements, developed by the entire project team (i.e., agency, responsible party, contractor, public, etc.), establish and integrate concepts of the project team, such as accelerating actions, streamlining decision making, expediting response actions, integrating community concerns, and developing goals and objectives in order to realize the vision. Most importantly, these statements are developed by the entire project team. They incorporate the collective needs of the group, vs. those of the individual, which gives meaning and purpose to the team, and a desire to want to work together as a team and arrive at a common destination. After the mission and vision are developed, it is possible to identify common goals and objectives that support that mission and vision.

Common Goals and Objectives

The second order of business, after developing mission and vision statements, is agreement on common or "shared" goals and objectives, which are the primary goals and objectives upon which the project team can find common ground in the beginning of a project. Identification of common goals and objectives is important for starting a project off on the "right foot" and include concepts, values, and direction that can be agreed upon and prioritized so work can begin as soon as possible and meet regulatory requirements. Goals and objectives that cannot be agreed upon in a timely manner must be resolved as progress continues, and new information is gathered on the project for making long-term decisions. Agreement on common goals and objectives involves team members working together and finding common ground. However, there are important distinctions between internal and external common goals and objectives as outlined below.

- **External goals and objectives.** These goals and objectives are determined by the entire project team (i.e., the responsible party, lead regulatory agency, other agencies, etc.). They include concepts and values that the entire project can agree upon and how the decision-making process will function, and identify common ground for all stakeholders (i.e., the destination).
- **Internal goals and objectives.** These goals and objectives area determined by the client or agency conducting specific work for the project that are subtasks of the external goals and objectives. These are designed to help meet the external goals and objectives. Communication in this team setting should be limited to within the business or agency; however, the identified goals and objectives would be shared with the *entire* project team in order for them to understand the direction and actions the business/agency is taking.

Common goals and objectives include strategic and tactical elements, where strategic elements are associated with planning, and tactical elements are associated with action. In addition, common goals and objectives can focus on the technical aspects of investigation, data evaluation, and remediation, and they also focus on the administrative, regulatory, and cost aspects. Lastly, common goals and objectives can include such diverse subjects as maintaining values and local community culture. In all cases, common goals and objectives should be relatively stable and adjusted only if there is new information, regulations, or data that cause a deviation (such as identification of a new responsible party). An example of a common goal would be to "adequately characterize site conditions in order to make prudent and cost-effective decisions related to protection of human health and the environment." In this, the goal is not to characterize every contaminant down to nondetection; rather, as essentially all regulations and guidance prescribe, professional experience may be applied to determine when to stop the iterative investigation process. A common goal and objective of this nature is sometimes needed because some environmental professionals believe that we should define all contaminants down to nondetection, regardless of the cost and extenuating site-specific circumstances. Working out differences between team members ahead of time, such as how much investigation is required to make decisions, is more effective than addressing the issue midway through a project, which can substantially impede progress. Another example of a common goal would be to "consider future land use of the site and focus investigation and cleanup efforts on the most likely future land use scenario." More specific common goals and objectives could include implementation of a product recovery system in a specified time period, or preparation of a draft proposed plan and preliminary Record of Decision (ROD) in a specified time period. Alternatively, goals and objectives can be associated with how the project is run. For example, goals could be set forth to "discuss major technical information ahead of time, and before they are submitted in major draft reports to help streamline the review process," or "encourage local involvement in the environmental review process." There are no limits on what common goals and objectives can include, with the exception that they must work within the regulatory framework and that all stakeholders or entities agree with them. By agreeing with them ahead of time, task-oriented common ground can be found to initiate progress, maintain progress, avoid conflict, and aid decision making.

For the teams lacking a high level of trust or ability to work together, agreement on common goals and objectives is difficult to realize, if not impossible. Consideration of trust between entities is therefore important, and is further evaluated later in this chapter (see Section 7.5, Trust).

7.3.2 Partnership Models

In Chapter 3, the Department of Defense (DoD), and in particular the Navy, are discussed in relationship to partnership models. The two main partnership models outlined were formal partnering and informal partnering, as models to efficiently and effectively work as teams on environmental projects. The formal partnership model involves organizing a 1- or 2-day gathering of military personnel, regulatory personnel, and contracting personnel, etc., that are working on large projects and encouraging those in attendance to learn more about each other as individuals and environmental professionals, and to understand the mission, vision, goals, and objectives both individuals and entities have for the project. At the end of the gathering, which may or may not be facilitated by a third party, each group member signs a contract, agreement, or a memorandum of understanding that acts as the formal understanding that individuals will work together as team, avoid conflict, and streamline the overall environmental process.

The informal partnering model is, as it implies, up to the team to develop and administer. In general, the same overall goals are applicable to an informal partnership; however, a formal agreement may not be signed by the team members. In addition, the partnering discussions may be held over several lunches, dinners, or during other times set up instead of an official 1- or 2-day arranged gathering designed specifically to accomplish similar goals. A third type of partnership model is an anonymous partnership, which are in many ways informal partnerships, where the individuals and entities happen to get along well with each other and have a proactive relationship in place without officially trying to establish a partnership. For all three partnership models, understanding the needs and expectations of everyone is important in order to maintain the partnership and nourish it. In these partnership models, the public can be part of the process; however, a clear distinction between "decision makers" and the "public" must be established in the beginning of the partnering process. Otherwise, the public may incorrectly assume that they are an equal *decision maker* in the environmental process, which under most regulatory situations they are not, following current environmental processes for community involvement. The concept of not making the public an equal decision maker, however, is not in agreement with some partnership models as described below.

In Peter Block's (1993) book *Stewardship*, the concepts of partnerships are discussed from a more philosophical perspective, where he compares partnerships with patriarchies. Patriarchies are designed around control, consistency, and predictability. These three elements strike important cords for most environmental professionals, because control, consistency, and predictability are often considered marks of quality for technical work completed in the environmental industry. However, in terms of working with people, the concepts of patriarchies may not provide the most efficient and effective model to follow. According to Block, partnerships offer an alternative model for working in a patriarchal system by distributing ownership and responsibility of the actions and decisions to all team members. Furthermore, as Block points out, if you truly want to improve performance, partnerships are the model to follow instead of patriarchies for the following reason: *When patriarchy asks its own organization to be more entrepreneurial and empowered, it is asking people to break the rules that patriarchy itself created and enforces* (Block 1993).

In his model, partnerships thrive when control shifts from manager to the core worker, supplier to the customer, doctor to the patient, teacher to the student, caseworker to the citizen. In an external partnership model (one that goes beyond the organization and/or business), the public would be in control instead of the lead regulatory agency. Environmental professionals could argue that the public cannot make sound *technical decisions*; however, citizens certainly have the right, and even the obligation, to comment on environmental actions and decisions and be involved in the process. A similar situation occurs when medical doctors may contend that their patients are incapable of making the proper heath decisions in situations involving a complicated diagnosis, or in determining which disease treatment alternative is best. Thus, there is a dilemma between how far partnership models should go, and what environmental regulations prescribe. Please do not take these statements out of context. No argument is presented for or against unprecedented changes of making the public an equal or primary decision maker for environmental sites. The current laws are relatively explicit in how environmental decisions are made, which includes solicitation of written comments and conducting public meetings, depending on the program and type of environmental actions. However, no significant decision-making responsibility has been given to the public. Shifting environmental *control* to the people has yet to be attempted and would likely result in substantial changes in what happens at contaminated sites, which may or may not be perceived as improvements by many. More importantly, and in accordance with the current environmental regulatory framework, it is possible to sincerely consider the public's needs, expectations, and ideas, and involve them in environmental processes. The environmental regulations of the future will dictate how far partnerships can go for environmental restoration actions.

Block's partnership model is presented as a broad business or organizational strategy where there are specific requirements that have to be met in order to have a partnership. The fundamental requirements of his model are listed below. Detailed explanation of Block's partnership model is contained in his book Stewardship (Block 1993).

- Exchange of vision, values, and purpose
- The right to voice your opinion and say no
- Joint accountability
- Absolute honesty
- No abdication.

Summary of Partnerships

The underlying purpose of environmental partnerships is not only to help people get along, but also to streamline and hasten protection of human health and the environment. In addition, partnerships can help reduce environmental costs. From a fundamental viewpoint, partnerships are one way to encourage teamwork and integrate local community members into the environmental process or program. In addition, partnerships may go beyond the general principles that teamwork encompasses. For example, partnerships can be designed to distribute control and accountability throughout the entire team. Teamwork, on the other hand, may or may not distribute control and accountability, and may more closely follow the patriarchy model where the main goal is simply working well together vs. distributing control and accountability. In addition, as with many other types of acceleration approaches and models, there are associated risks that may hinder progress or increase costs if not properly planned or handled. Partnerships are no different, where careful planning, and integration of the site-specific requirements and local concerns are needed. Furthermore, careful planning will increase the likelihood that the partnership will be beneficial and long-lasting, help meet the project goals and objectives, and help meet the needs and expectations of the public.

Do Partnerships Work?

The DoD, U.S. Environmental Protection Agency (EPA), and the Department of Energy (DOE) promote partnering as a way to encourage teamwork and streamline decision making on environmental projects. However, in the environmental industry, partnering may not always be effective because of substantial differences of opinion, low trust factors, inadequate leadership, or simply because one or more entities or individuals have no desire to make it work. As an analogy, soil vapor extraction is a presumptive remedy for volatile organic compounds (VOC) in soil. Soil vapor extraction (SVE) is highly effective in many environments, as long as it is properly designed and operated. SVE is highly ineffective, however, in extremely clayey soil and, in general, should not be automatically assumed to be the preferred cleanup technology in these environments. Partnerships are similar in that they must be properly planned and nourished and fit the cultural and/or professional setting, otherwise they may be ineffective and a waste of time and money. Therefore, it is important not to partner because this book, DoD, or others promote partnerships. Rather, partnerships must be custom fit, and used to enhance teamwork in order to work in a particular cultural and/or professional setting. In other words, they cannot be forced into being.

7.3.3 Empowerment and Accountability

Empowerment is a relatively new term, used extensively in the 1990s, that describes the concept of entrusting individuals with making decisions, and encouraging them to follow through with tasks and projects under limited supervision. In addition, empowerment means shifting accountability to an individual or group responsible for the quality of work performed, keeping the client satisfied, and meeting project goals and objectives. Generally, empowerment is an intrabusiness or intraagency concept, where supervisors choose not to follow traditional supervisor roles in favor of empowering staff to make decisions, accept responsibility, and become accountable for work, projects, or an entire business. This type of empowerment is *internal* when it is limited to a specific agency, project, or organization.

External empowerment is less common, and involves different entities such as government agencies, contractors, businesses, or community members working together. As a means of improving teamwork, empowering project teams to take on responsibilities can be beneficial in order to encourage innovative solutions and fast-track actions. Spreading responsibility between entities can maximize the use of resources, minimize duplication of effort, and encourage meeting common goals and objectives. In addition, it advocates increased communication with the rest of the team and thus spreads accountability, to some degree as limited by regulations, among the project team, depending upon how individual entities or groups are empowered and how the project is funded.

In cases where an accelerated investigation or cleanup action is underway, potential for failure due to uncertainty may be greater compared to following more traditional environmental approaches. However, through empowering staff, problems can be mitigated as they arise. For example, uncertainty encountered in the field can be addressed by empowering the individual(s) responsible for the field effort to address problems as they arise and quickly remedy the problem. This approach is more effective than going through several layers of decision makers that would most likely arrive at the same decision sometime later. As described by Block (1993), empowerment embodies the belief that the answer to the latest crisis lies within each of us. Many crises can be addressed relatively easily, and may not be best addressed most effectively by a single decision maker who is not directly involved with the actions underway. This is not to say communication between team players should be abandoned for the sake of letting people make their own decisions, or that uninformed decisions or actions should be undertaken by empowered individuals. In cases where relatively significant ramifications are possible, such as those that will impact protection of human health and the environment or substantially increase costs, joint decision making is essential. Open lines of communication are important to ensure that informed decisions are made, and that decisions that impact other individuals, agencies, or responsible parties are considered carefully during the decision making process. Empowerment includes the understanding that individuals or entities are responsible for seeking help and guidance in order to address major crises, and the recognition that the values and importance of others on the project team must be integrated into the decision-making process in order to streamline project closure.

External Empowerment Limitations

In cases where environmental professionals are working for a particular client, the client may choose to be involved in all decision making, which may limit the flexibility of the decision-making process and increase costs. In this case, the potential for external empowerment is low. To keep clients satisfied in these situations, there is not much that can be done except the consideration of empowerment within the organization or finding a new client, which may not be a possibility. In this case, external empowerment is only successful if the project team agrees to value it and employs a teamwork approach that involves all stakeholders.

7.4 LEADERSHIP

Harrison Owen (1991) portrays doing business as "Riding the Tiger to Somewhere," where we know enough about tigers to be aware that riding them must be done very carefully. Jumping off is an invitation to lunch. Trying to stop the beast is not advisable because we all know what happens when you get a tiger by the tail. Riding the tiger is ludicrous when considering that somehow we might control our mount. The environmental industry fits well within Owen's analogy for environmental businesses and government agencies — like it or not, we are riding a tiger to somewhere. Despite all of the chaos during the ride, we attempt to predict and control our environment, but often to no avail because of technical uncertainties, poor communication skills and processes, inability to work as teams, and inability to collectively "ride the tiger" to the same location. Often, one entity heads north and another entity south, which certainly upsets the tiger and creates havoc in our jobs.

Leadership is critical in order for teams to work well together, and arrive at a predetermined location together within a reasonable time frame, i.e., ride the tiger gracefully. For the purposes of this section, there are two fundamental leadership models outlined. In the first leadership model, there is a single person who plays the role of the leader for the entire project. This is the most common leadership model that has been in place for many years. Football provides an excellent analogy for this leadership model. In football, the quarterback is the leader while the ball is in play, and is the key individual on the team that everyone looks to for leadership and direction. In addition, the quarterback needs everyone on the team in order to win the game; however, the quarterback determines the next play unless the coach steps in. The quarterback is the primary team player accountable for the team's success or failure. This leadership model is commonly used when an agency or a business attempts to complete a primary environmental investigation or remediation task, such as completing a remedial investigation/feasibility study (RI/FS). The second leadership model is less common, where the entire team shares in the leadership role at one time or another. Owen's (1990) leader model in his book *Leadership Is* uses the game of soccer as an analogy:

Whoever has the Ball is Leader
Ball Hogs Die
Never Oppose Force with Force
Play the Whole Field
Cooperate in Order to Compete
Honor the Opposition

This leadership model is effective if there are multiple entities working toward a primary goal, such as protecting human health and the environment, or streamlining the environmental process. In the soccer leadership model, it is the team that is accountable rather than the single individual in the football leadership model. This is not to say that some soccer players are not better than others. However, the less gifted soccer players do get to be the leader when the ball is in their portion of the field. In relationship to accelerating cleanup, different leadership models can be used; however, the skills, attitudes, and flexibility required for the two leadership models are different. The football leadership model can grow egos if we are not careful, and the soccer leadership model permits no egos, and encourages chaos if we do not know how to play the "game."

For complex field investigations, selecting the football leadership model may be desirable, depending on the ability and skills of the individuals on the field team (i.e., staff need specific direction in order to do their job efficiency and effectively). Alternately, the soccer leadership model may be desirable if the field team will be more efficient, effective, or because they function better in open and accountable environments. In either case, the field team leadership model is primarily an *internal* leadership concept originating from within one entity, such as a contractor.

External leadership is related to multiple entities working together, such as responsible parties and the EPA, where leaders from both entities are selected to guide their team to their destination. The football leadership model has been highly ineffective and inefficient in completing environmental efforts in reasonable amounts of time in these situations. However, using Owen's leadership model, external leadership can shift from two football teams battling each other, to a single soccer team, made up of the various entities, competing not against each other, but along side each other. In other words, using Owen's model, leaders share in the responsibility and accountability in order to overcome whatever competition there is to compete against, such as a technical uncertainty encountered during remedial actions. However, this type of leadership model is rare because of the way environmental regulations are designed to be enforced. An enforceable regulatory system is necessary in order to force polluters violating laws to act in cases where they are unwilling to do so. Nonetheless, in cases where there are no adversarial relationships between entities, a more open leadership model can be applied. This type of leadership model is most common in cases where a government agency, such as the military, works with another government agency, such as the EPA, to protect human health and the environment. In addition, under voluntary actions the soccer leadership model can be applied because it is not an enforceable action. However, all of the entities involved must follow the same type of leadership model. In cases where a soccer leadership model is applied, and there is one dissenting entity who wants to "play football," the soccer game will be most unenjoyable, and the entities will usually end up playing football in the end.

The type of leadership model employed, and its relative effectiveness, affects how quickly decisions are made and how streamlined the environmental process will be. Unfortunately, there are no referees to help control the environmental process. If leaders want to take low blows, they do, and they often receive one back in retaliation. Therefore, effective and nonconfrontational leadership is important if environmental professionals are to fast-track and streamline environmental processes. While we will not be literally eaten by a tiger due to inadequate leadership, we may end up having to go to court in some cases where leaders cannot get along, which in most cases wastes time and money and slows down protection of human health and the environment.

7.4.1 Leadership Skills

Effective leaders understand the "big picture" and are committed to the mission and visions of projects and/or programs. Not only do they see the technical aspects, they embrace the human side of decision making, understand regulatory compliance issues, apply a consistent approach when leading, accept diversity, foresee the end of projects, and get help when they need it in order to maintain performance. For external leadership involving multiple entities working together, excellent leadership skills are especially important. Leadership also involves considering what people need to keep them motivated and want to be at work. In *The One Minute Manager* (Blanchard and Spencer 1981), praise and reprimands are part of what effective office leaders use to increase performance. For example, leaders take on the responsibility of helping others learn from their mistakes, and offer praise when they do well. Effective leaders are also consistent. Consistency means that fundamental rules, principles, and values are applied across the board, and there is no room for favoritism. In turn, effective leadership will encourage loyalty, teamwork, and trust.

Dr. Stephen Covey's work goes beyond fundamental leadership skills and into "principle centered leadership," which is focused on maintaining the delicate balance between work and home, inspiring coworkers to free their creativity, breaking the endless cycle of "putting out fires" at work, and starting to work toward preventing crises before they happen. Covey also describes the "Seven Habits of Highly Effective People" which are the basic building blocks of human effectiveness, which relate directly to quality leadership. Covey is an excellent source for improving leadership skills.

Leadership and Management

Leadership is more than management. True leaders give meaning and purpose to jobs, whereas management controls and tracks people, tasks, and progress. Instead of managing people, leaders are caretakers of people, allowing them to grow and prosper from their successes and failures. An appropriate analogy is that of a gardener. A gardener manages weeds by pulling them out, but nurtures the flowers by watering and fertilizing them, protecting them, and encouraging them to be the best they can be. The gardener uses his/her technical knowledge, intuition, values, and principles to grow flowers, enjoy the garden, and take pride in its success. If a leader is similar to a gardener, then an appropriate analogy for a manager is that of an armed guard. Armed guards have explicit rules they are determined to follow; even though they may not agree with the rules, they live by them because that is their job. If anything out of the ordinary is observed, they are suspicious and deal with it, as necessary, to remain in control and follow the rules by the book. As environmental professionals, for example, we can be leaders of people and managers of money. For without effective leadership, teams fall apart, budgets cannot be easily managed, and goals and objectives are not met within a reasonable time frame. Leadership skills are also important not only for the benefit of others, but also to maintain self motivation. Lastly, leadership skills must be learned and practiced over time in order to be effective.

A critical component of effective leadership is the desire to be a leader, as well as growing the right skills. In hierarchies, people are often forced into becoming leaders of large environmental projects, even though all they really want to do is the technical work. In these situations, "people leadership" is often abandoned in favor of "people management," which can have the effect of impeding progress and result in following less proactive and more traditional pathways in order to complete projects. In other words, the armed guard is now in charge of the garden — and if the rule is to water the flowers every other day, then drought conditions may impact the garden because there are no exceptions to the rules. Most importantly, we all have leadership capabilities, but they go dormant unless we choose to embrace them. Therefore, it is important for people to look inside themselves and complete leadership training, and for others to offer honest advice and tips so they can grow their leadership capabilities or focus on more technical aspects of their job. Agencies, organizations, and businesses should ultimately have the people that display excellent leadership capabilities, and want to be leaders, in leadership positions and ensure that they receive training in order to maximize their leadership capabilities. The result of letting these people be leaders and training them is a greater potential for efficient and effective completion of environmental work, accelerated schedules, less conflict, and wiser use of resources.

Leadership is also about principles. Philip Howard (1994) uses this analogy: "Principles are like trees in open fields. We can know where we are and where we go. But the path we take is our own." So too in environmental leadership, the direction taken at toxic waste sites is based in part on the principles expounded by the project leaders. The influence of principles on projects can have a direct effect on how quickly sites are investigated, cleaned up, and closed. Traditional principles follow traditional pathways, which can be relatively slow and are relatively conservative compared to newer approaches, for example, real-time data collection vs. having all analysis sent to a laboratory. Accelerated cleanup as an environmental principle has the opposite effect and expedites projects.

The scope of this book includes descriptions of fundamental aspects of leadership and some of the primary skills required to be an effective leader in the environmental industry. There are many skills and models developed by the authors referenced in this chapter, such as Owen (1990; 1991), Block (1993), Blanchard and Spencer (1981), Stephen Covey, and Tom Peters, to name a few, that are excellent resources for learning more about leadership. Understanding the information in these resources is important in order to become an effective leader and maximize leadership skills, and to consider principles related to conducting environmental work.

The Power of Cynics

The Tom Peters Group (TPG) Leadership Series *The Leadership Challenge*™ *How to Get Extraordinary Things Done in Organizations* (Kouzes and Posner 1993) presents information from Kanter and Mirvis, *The Cynical Americans,* outlining three common personality types in the workplace. In their findings, cynics are said to be the most common personality type:

- 43% are cynics
- 41% are upbeat
- 16% are wary

The wary are the smallest population, but they can be swayed to either the cynical or upbeat point of view. Since there are more cynics than upbeat personalities, it is often an uphill battle for the upbeat personalities to persuade the wary to follow their line of reasoning. In relation to the topic of accelerated cleanup and decision making, upbeat personalities are more inclined to be proactive, and expedite projects to increase performance and save money. Upbeat personalities are more likely to accept higher levels of potential failure in order to quickly achieve project goals and fast-track projects. The cynics are less likely to agree with proposals that are out of the ordinary, and more likely to follow traditional pathways where there is less potential for making mistakes or failing. The wary are not sure which way to go, but listen to the others in order to make their own decisions. One verbal cynic in a room of 50 can set the tone and carry the day (Block 1993), and thus the power of cynics is realized.

When taking a proactive stance such as streamlining or accelerating the environmental process, leaders must guard their credibility when faced with the power of cynics, and carefully choose how they will address the concerns that are raised or limitations set on a proposed action. The leader must be able to lead teams following principles of *choice* (Block 1993). In group settings where the leader must address the concerns of a cynic, Block states that the leader should

- Acknowledge the other's opinion
- State the choice for faith and commitment in the face of the person's reservations
- Invite the same choice from the other person

A choice of faith and commitment over reservations does not promise or guarantee success; however, it is a credible approach for being proactive and allowing the team to weigh the alternatives and act as a group for accelerating and streamlining actions. However, Block also contends that there is at least partial truth in what cynics say, because they have history to recall upon. In addition, we all have reservations and we can often understand the cynics' point of view. The power of the cynic is therefore strong, especially if they are in a supervisory capacity, which can negate efforts to improve performance.

Job Satisfaction

People who feel good about themselves produce good results (Blanchard and Spencer 1981). People who feel good about themselves in the workplace enjoy their jobs. Leaders need to be concerned with job satisfaction and consider measuring the overall level of job satisfaction in order to improve performance and help individuals excel in their jobs and operate more efficiently and effectively. In the *Job Satisfaction Challenge* employee manual by Brander and PRC Environmental Management, Inc. (1995), the concepts of job satisfaction, the measurement of job satisfaction, and the reasons for caring about job satisfaction are described. The seven reasons for caring about job satisfaction include

1. A step toward an organizational partnership
2. A means of growing and honoring people
3. A method of linking happiness to the "bottom line"
4. A commitment to an open process
5. A way to create a safer work environment
6. An opportunity to cultivate spirit through storytelling
7. An opportunity to put [goodwill] into the organization.

Leaders that consider job satisfaction important and implement actions necessary to measure, maintain, and improve job satisfaction can more easily implement environmental programs and encourage the concepts of accelerated actions, streamlined decision making, and increased environmental performance. In addition, job satisfaction relates directly to worker safety in the field, and the number of days staff spend away from work if injured or ill. Insurance company studies show that a high level of job satisfaction is the single biggest reason for low worker injury and time spent away from work (Gice 1995). Maintaining project schedules and accelerating project schedules depends heavily on staff being on site and working safe. Job satisfaction is important for staff to be efficient and effective, to minimize replacement of staff, and to foster good working relationships with other people and entities — all of which are helpful in increasing performance and expediting the environmental process.

7.5 TRUST

The matrix that holds communication, teamwork, and leadership together is trust. Accelerated cleanup, innovative actions and ideas, and streamlined decision making are much more likely to be seriously considered in work environments that have high internal and external trust factors. The environmental industry as a whole has relatively low external trust factors between different government, private, nonprofit, and community entities. Internal trust factors vary depending on the individual organization or agency and how it is administered. External trust factors also vary. For example, community trust factors may vary based on different class, ethnic, and cultural backgrounds. Yet trust remains a critical element in the ability to fast-track and streamline environmental projects. If trust can be maintained at a high level where one entity is not attempting to take advantage of another entity, such as two responsible parties negotiating a settlement, the time and effort required to make decisions is substantially reduced. In addition, opportunities to streamline and accelerate environmental processes are accepted more readily in trusting relationships.

The difficult task facing environmental professionals is determining how to evaluate, build, and maintain trust on environmental projects. Trust is rather a fragile commodity; as soon as you see someone taking advantage of your good will, it vanishes (Chapple 1993). According to Chapple, building and maintaining trust is synonymous with having *faith* in someone, where the building blocks of trust are listening and honesty. This illustrates the intangible, yet powerful, characteristics of trust.

Trust in relation to listening involves hearing what the external and internal counterparts have to contribute to projects, their goals and objectives, their vision and mission, their expectations for a project or program, and any information they feel is important from an individual and organizational perspective. Honesty involves being up front with findings, providing the entire plan rather than just part of it, proposing realistic alternatives and ideas, comparing expectations, identifying the limitations of a project or program, and responding to questions in the same spirit they are asked. Together, listening and honesty can increase the level of trust between persons and entities if they truly desire to have a trusting relationship.

The experience of many professionals substantiates that honesty has been especially ignored in the environmental industry in order to gain something from somebody else. As a result, trust between agencies and responsible parties is often low, and the tenor of relations is sometimes even disrespectful. The impact of this dilemma has been to significantly slow protection of human health and the environment, substantially increase the amount of legal services required to complete projects, and increase the cost of environmental actions, which are directly and indirectly subsidized by the public. Increasing trust factors can have quite the opposite effect, providing benefit, safety, and savings to the public.

Trust also relates to accepting personal and professional diversity within environmental projects and organizations. Most would agree that administrative and technical understanding, interpretation, and regulatory approaches vary between individuals, organizations, and programs. Realizing and accepting the fact that there are other technical approaches, interpretations, and alternatives that need to be considered is an important step in limiting future conflict and maintaining trust factors when choosing one method or approach over another. Most importantly, the goal of accepting personal and professional diversity is not to drop your own way of thinking; rather, it is to understand other ways of thinking in order to improve communication, generate win–win solutions, and maintain trust factors. In the end, one approach generally is selected and an honest assessment of the choices should be completed in order to make the decision.

In cases where there are relatively low trust factors, improving trust can be extremely difficult, and project teams or individuals may be adversely impacted if their counterparts do not share the same level of trust. In these situations, there are three basic choices:

1. Focus on the trust building blocks to rekindle and increase trust factors. This, for example, could be accomplished using a formal partnership model to break down trust barriers.
2. Revamp the entire project team from all sides, starting with fresh individuals who are not dominated by the same trust issues, and increasing the overall level of trust through building the blocks of listening and honesty.
3. Continue the project following the current approach, which is likely to be slow and tedious, and recognize and work around low trust factors as well as possible. This is the most common approach used, which unfortunately provides the least improvement in performance.

7.6 PEOPLE TRAINING AND INTERVENTION

There is a virtual plethora of training and intervention services available to improve communication, teamwork, leadership, and increase trust factors. People training relates to preventative training sought in order to improve skills and knowledge necessary to avoid delays, miscommunication, and conflict, and improve success and performance. An example of this type of training would be the Massachusetts Institute of Technology Environmental Policy Group's *Negotiating Environmental*

Agreements. In this training, regulators, attorneys, consultants, and business executives, for example, learn more effective ways to negotiate environmental permits and agreements, which has a direct bearing on how we as environmental professionals communicate with our counterparts. Less environmentally specific examples of people training are Tom Peters Leadership Challenge courses where leadership skills are taught, from a generic perspective, in order to help improve leadership capabilities, bring out creativity, and help organizations be more successful by investing in their employees. These types of training are helpful because leadership skills are valuable for increasing performance no matter what industry they are applied in.

Intervention services relate to having to address issues after they have happened. Investing in intervention services is similar to a married couple seeking marriage counseling in order to help their marriage or simply improve their relationship. This is not to say that unless you get help, you will end up divorced! It is possible to remedy your own issues, and in fact many organizations and teams resolve their own communication, teamwork, leadership, and trust issues through their own intervention and determination. However, investing in professional help may expedite the learning process, resolve issues more quickly, and more quickly improve teamwork and levels of trust.

More importantly, it should be recognized that nontechnical training and intervention may be helpful to improve project performance and expedite reaching the project goals and objectives ahead of time. It is also important to realize that because there are so many training and intervention services available, the claims made may not all be true, and not all of them will be able to meet the mark or level of satisfaction you desire. Careful comparisons between services and training is recommended to help limit resource expenditures and time, and ensure that the desired results are realized.

7.7 REFERENCES

Alessandra, T. 1996. The Platinum Rule. Warner Books, New York.

Blanchard, K. and Spencer, J. 1981. The One Minute Manager. Berkley Books, New York.

Block, P. 1993. Stewardship — Choosing Service Over Self-Interest. Berrett-Koehler Publishers, San Francisco.

Brander, D.L. and PRC Environmental Management, Inc. 1995. Job Satisfaction Challenge. Helena, MT, p 48.

Chapple, T. 1993. The Soul of a Business — Managing for Profit and the Common Good. Bantam Books, New York.

Directed Business Concepts (DBC). 1997. *ah ha!* Business Improvement System. Calgary, Alberta, Canada.

Gice, J. 1995. The Relationship Between Job Satisfaction and Workers Compensation Claims. CPCU Journal. September, pp. 178-183.

Howard, P.K. 1994. Death of Common Sense. Random House, New York.

Kouzes, J.M. and Posner, B.Z. 1993. The Leadership Challenge Workshop. The Tom Peters Group Learning Systems, Palo Alto, CA.

Marston, W.M. 1928. The Emotions of Normal People. Harcourt, Brace, New York.

Owen, H. 1990. Leadership Is. Abbott Publishing, Potomac, MD.

Owen, H. 1991. Riding the Tiger — Doing Business in a Transforming World. Abbott Publishing, Potomac, MD.
Robinson, D.G. and Robinson, J.C. 1995. Performance Consulting. Berrett-Koehler Publishers, San Francisco.
Tannen, D. 1986. That's Not What I Meant! How Conversational Style Makes or Breaks Your Relations with Others. William Morrow, New York.
Tannen, D. 1990. You Just Don't Understand — Women and Men in Conversation. William Morrow, New York.

CHAPTER **8**

Strategic Planning and Lowering Environmental Costs

The purpose of this chapter is to outline and connect the concepts necessary for developing environmental strategies that accelerate cleanup and streamline decision making, as well as to summarize cost-saving measures, approaches, and initiatives that can be used to efficiently investigate, cleanup, and close sites. Strategy and cost-saving measures are often considered in the beginning of projects in order to define the pathway for eventual site closure. The concepts and elements of strategic planning and cost reduction are presented by linking the information outlined in the previous chapters of this book related to site characterization, cleanup, decision making, and site closure. In addition, the concept of revisiting strategy throughout a project is outlined where site conditions and actions are observed and measured in order to monitor performance and integrate new information into project decision making. As a result, environmental strategies can be adjusted for optimal performance, and cost-saving measures can be expanded or redesigned in order to more efficiently run environmental systems and projects.

The information in previous chapters represents the processes, theories, regulations, ideas, interactions, technologies, etc., that are fundamental for preparing and modifying strategic plans, not to mention the concepts of acceleration and streamlining. In addition, the information in previous chapters is important for developing and identifying cost-saving measures. Linking this information together will not only aid in the preparation of strategic plans and identification of cost-saving measures, but will also help ensure that regulatory requirements are met, and more importantly, that human health and the environment are protected. These aspects, and the advantages of developing a comprehensive site-specific strategy, cost-savings measures, and presentation of the Brownfields Initiative are divided into two main sections in this chapter.

8.1 STRATEGIC PLANNING

An environmental strategy is required for most projects in order to navigate through the technical and regulatory requirements, especially in cases where projects are relatively large and complex. In order to efficiently and effectively run environmental projects and work with external counterparts, it is essential to develop a strategy that links the various components of environmental projects together and provides a direction for the entire project team to follow in order to meet specific goals and objectives, including the final project *destination*. This type of strategy should not be confused with the technical strategies of projects, which are only one element of a comprehensive strategic plan. An example of an excellent information resource related to technical strategy development is the U.S. Environmental Protection Agency's (EPA) *General Methods for Remedial Operation Performance Evaluation* (1992). This resource outlines strategic considerations for groundwater pump-and-treat systems. This technical element of a strategy should be integrated into a comprehensive strategic plan which is described later in this section. Technical elements, for example, also include data quality, fluid behavior, monitoring strategies, pulse extraction pump-and-treat systems, criteria for measuring performance and efficiency, statistical applications, scale of measurements, etc. Projects that require a comprehensive project strategy are sites that are generally complex, and often includes evaluation and cleanup of mixed wastes and/or releases. Smaller projects may only require minimal strategic planning, which generally requires some level of technical strategy in order to implement cleanup actions.

Most importantly, a comprehensive strategic plan should be an *external* environmental strategy that (1) is designed to meet regulatory requirements, (2) protects human health and the environment, and (3) is developed by the entire project team. In addition, external strategies should be designed to follow the path of least resistance so that the direction taken will achieve the goals and objectives in a timely and cost-effective manner. The path of least resistance does not necessarily follow the path of least work; rather, it is a direction that minimizes conflict and the potential for litigation. The path of least resistance takes a proactive approach that integrates all project needs and concerns in order to fast-track, streamline, and reduce costs. In this context, an *effective* strategic plan moves beyond the confines of internal business or agency strategies, and integrates the needs and expectations of *all* stakeholders. In addition, effective strategic planning emphasizes win–win strategies that not only meet the goals and objectives of one entity, such as a responsible party or regulatory agency, but also meet the goals and objectives of the *entire project team*.

Internal organizational strategies do not need to be abandoned when developing an effective external strategic plan. Responsible parties must maintain their integrity, regulatory agencies must meet the spirit of the regulations they oversee, and environmental businesses must be profitable. All of these requirements must be integrated into a comprehensive strategic plan in order to efficiently and effectively conduct environmental work, especially if there is the desire to meet the needs and expectations of *society*. This section focuses on the development of external project strategies that integrate the goals and objectives of all parties involved. Certainly, this is not an easy task to accomplish. However, in terms of effort, external strategic plans can be developed and agreed upon with less overall effort than those following past strategic approaches that incorporate pursuing litigation with an "us and them" emphasis, as long as the project team has a relatively high level of trust for one another and there is effective leadership in place (see Chapter 7 for more information).

Strategic plans also have to be flexible so that they can be modified when new data are collected and changes become evident. Therefore, uncertainty must be addressed and built into strategic plans. Throughout the life of a project's life, the overall environmental strategy should be ultimately improved, focused, modified, or completely changed in order to efficiently and effectively meet the common goals and objectives agreed upon by the project team.

This chapter introduces the fundamental concepts of external strategic planning that is focused on accelerating cleanup, streamlining decision making, and reducing costs. There are many other resources on environmental strategies available through formal education, training, and publications. However, the majority of strategic development resources have historically presented strategic planning as an internal application. More recently, environmental strategies have become more focused on external strategic planning. For example, *Corporate Environmental Strategy: The Journal of Environmental Leadership* is a periodical focused on describing the connection between strategic environmental management and sound business strategies. Cases studies are used to illustrate basic and advanced concepts of sustainability and movement beyond regulatory and liability issues that motivated past environmental response actions. Another example is the University of California at Los Angeles (UCLA) Center for Clean Technology which integrates research and development of innovative technologies and risk management approaches with decision making, mutual trust, and cooperation in order to develop win–win environmental strategies.

Many other types of environmental resources help develop environmental strategies, i.e., they complement the "big picture strategy" rather than *being* the "big picture strategy." These types of resources are scientific, technical, and regulatory-related resources that advance the environmental industry as a whole. One example of a comprehensive resource useful for development of environmental strategies is the *Hazardous Waste Consultant*, which integrates information on technologies, remediation, investigation, regulatory developments, legal implications, and other subjects. Tracking these types of information resources is helpful in order to design site-specific strategic models that incorporate the most up-to-date environmental directions and technologies.

New environmental strategies are possible because of the current "relaxed regulatory setting" and because "corporations realize that proactive environmental initiatives [are] cheaper than reactive responses" (Williams 1996). For example, the September 15, 1994 final rule of the revised National Oil and Hazardous Substances Contingency Plan (NCP) clearly stresses that the goal of response actions is to achieve coordination and cooperation during oil-spill responses, rather than rely on a more rigid system of command and control (Hix-Mays 1995). Currently, there is an opportunity to follow more proactive environmental strategies that may efficiently and effectively operate, cleanup, and close sites. However, a comprehensive external strategic plan, developed using the most recent technology and regulatory information, may be needed in order to realize the full potential of the relaxed regulatory setting. A "new paradigm of cleanup" is upon the environmental industry. This paradigm has occurred because "our past efforts have been too costly and ineffective" as outlined in John Bredehoeft's 1996 editorial in *Ground Water*. Effective strategic planning is a fundamental component required to operate projects within this "new paradigm."

8.1.1 External Strategy Development and Implementation Process

External strategic planning and implementation can be achieved using several models and approaches. The purpose of this section is to focus on development and implementation of strategies that fast-track projects, streamline environmental processes, and reduce environmental costs. For this reason, strategic planning that involves the entire project team rather than a single entity is presented. The importance of the entire project team working together is outlined in Chapter 7. In this type of strategic plan, the need to conceal the desired goals and objectives is replaced by openly sharing project goals, objectives, needs, and expectations with all team members (i.e., the regulatory agency, other agencies, responsible party, etc.). In the past, strategic planning and implementation models have focused on the *internal* strategies of agencies, businesses, and organizations. These strategic plans generally emphasize outcompeting, enforcing, or controlling the external project counterparts in order to "win at all costs." In these models, the strategic plans are designed to be concealed rather than shared with the project team. Postulating what a counterpart's plans are has been used to develop countermeasures. An analogy would be that of two countries at war, where strategic plans are top secret because they detail how one entity plans to overcome another entity in order to win the war. In this model, only one side wins, and it is generally not the public in the case of environmental wars. Internal strategic plans of this type are not presented in this book because they are reflective of the environmental inaction experienced from the 1970s through the early 1990s. Furthermore, based on the environmental track record during this period, environmental professionals have already demonstrated that they thoroughly grasp how these types of strategies work, and are actually very *good* at them. In an attempt to move away from what is already known, this chapter describes a team strategic planning and implementation process that is in line with current team and participatory concepts that involve all stakeholders and help fast-track response actions. Depending on the project-specific needs and team preferences, the approaches presented in this chapter may be modified, expanded, or used "as is" in order to develop a site-specific external strategic plan that the project team can agree upon.

The external strategic planning and implementation process for environmental projects involves four principal phases. These phases are followed when developing and applying an external environmental strategy for a new project that employs the entire team. The fundamental four phases are illustrated in Figure 8-1 and include

STRATEGIC PLANNING AND LOWERING ENVIRONMENTAL COSTS

- **Revelation and Consensus** — The process where members of the project team share information on the project vision, mission, goals and objectives, and the regulatory process in order to gain consensus on project strategy and identify common ground.
- **Direction and Experience** — The process where common ground is identified and the group establishes a project *destination* (i.e., close site under an industrial cleanup scenario) that is based on experience, practicality, and case studies, all of which demonstrate that the destination is reasonable and achievable.
- **Development and Innovation** — The process where the elements of an environmental project are integrated into a comprehensive strategic plan, and innovative approaches/technologies are used to achieve the agreed-upon goals, objectives, needs, and expectations of the entire project team, including the public, when applicable.
- **Implementation and Determination** — The process where the strategic plan is implemented and progress is measured and observed to adjust or revamp the strategic plan in order to lead the project team to the predetermined destination (i.e., final site closure). Determination is often required in cases where the strategic pathway is arduous, complex, or initially unsuccessful. The predetermined *destination* should be flexible in part so that it may be adjusted due to unforeseen conditions.

Figure 8-1 shows the concepts, elements, and process used for developing and implementing external and team-based environmental strategies. As illustrated in the figure, development of a strategic plan requires following the above phases and also integrating the specific project elements that must be part of strategic plans. In order to quickly develop a strategic model, a checklist should be prepared that includes the primary and secondary project elements that support the overall project strategy. See Figure 8-1 for the basic, or "generic," project elements that are integrated into most strategic plans.

During the implementation phase of a strategic plan, observing and measuring progress is required in order to evaluate performance, consider new information, or implement a contingency plan that is needed to counter insufficient progress. It may be necessary to deviate from the original strategy in order to maximize progress toward the predetermined destination. For complex environmental settings and contaminant conditions, a reiterative process may have to be implemented in order to periodically adjust the strategy and account for new information, while still arriving at the predetermined destination. Clearly, predetermining "no action" as a destination when high levels of contamination are documented is an unlikely possibility in most cases. Instead, a reasonable destination must be agreed upon. A compilation of the site-specific elements that are integrated into strategic plans is described below from both large- and small-scale vantages.

8.1.2 Large-Scale Strategy Elements for Accelerating and Streamlining Environmental Projects

Large-scale strategic elements are the primary elements of a strategic plan that generally cannot be ignored or overlooked. They formulate the "big picture" for

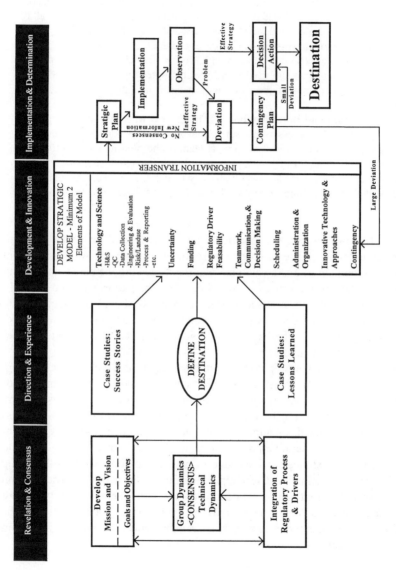

Figure 8-1 The four fundamental phases and external strategic planning process. The elements listed under the development of the strategic model are common elements. Site-specific requirements and conditions may include additional elements that can be addressed that are not on the list.

projects and the direction for reaching a final destination. In cases where they are overlooked or considered unimportant, they often become an issue and delay projects and increase costs. In some cases, additional elements may need to be included in order to address site-specific needs, such as community involvement, or site-specific elements that may be insignificant for other projects. For this reason, site-specific conditions and requirements should be considered when developing a checklist of primary elements included in strategic plans. The following are large-scale fundamental, or "generic," elements for developing and implementing an environmental strategic plan.

Regulatory Process and Flexibility

The regulatory process and guidelines that are followed or mandated for projects, such as the Comprehensive Environmental Response, Compensation and Liability Act (CERCLA), is the overall process that will be followed in order to resolve site contamination issues. When considering various project strategies, the regulatory process is important because it influences how decisions are made and what requirements must be met. Understanding the intent, enforcement procedures, and flexibility of the regulatory process is critical in order to develop the most advantageous pathway and define the sequence of events necessary to complete projects. For example, the project team may follow CERCLA as the governing regulatory process due to liability issues, and would need to identify the administrative needs, regulatory requirements, investigation reports, evaluation studies, and decision documents that are often completed under CERCLA. In turn, the flexibility of the regulatory process can be evaluated in order to efficiently navigate through the process. More importantly, it is possible to consider *eliminating* specific elements or documents in the process, in order to follow a more proactive approach for site restoration and closure, yet still meet the letter and spirit of the regulatory process. This could include, for example, using voluntary actions, undertaken by the responsible party, in order to fast-track cleanup and site closure. Likewise, the regulatory approach could be integrated with other regulatory programs, such as aspects of the Resource Conservation and Recovery Act (RCRA), in order to treat hazardous waste on-site by designating a Corrective Action Management Unit (CAMU), instead of having to address land disposal restrictions (LDR) and end up incinerating the hazardous waste off-site at a much higher cost.

Simply "plowing" through the regulatory process without planning how to address compliance issues is inefficient, because the team is then constantly backtracking in order to address less apparent compliance issues. The regulatory process and flexibility must be understood so that a plan can be developed for the overall project direction, and so that avenues for early response actions, site closure, and cost-saving initiatives can be taken advantage of. In most cases, the regulatory process and guidelines for environmental investigation and remediation are flexible, to the surprise of many, and do allow for creative problem solving as long as human health and the environment are adequately protected in a reasonable amount of time.

For example, when the Superfund Accelerated Cleanup Model (SACM) guidance was developed, changes in the NCP were not required. Prior to the development of SACM, environmental professionals could have followed the SACM approach for fast-tracking response actions all along, which some environmental professionals did in fact do. In this example, close evaluation of the *language* and *spirit* of the NCP, and other regulations and laws, reveals that the original goals of efficient and effective regulatory processes have not been followed, or have been ignored, in many cases during the development of project strategies. In most cases, this is due to liability concerns and/or the inability to work together as a team. In order to emphasize the flexibility of regulatory programs, EPA has more recently promoted models and guidance such as voluntary cleanup actions that emphasize acceleration, streamlining, and reasonable and practical approaches to environmental problems.

To illustrate this point further, an analogy may be useful. Tax laws are very complex; however, they allow for many deductions that can be used to reduce personal and business income tax liability. Minimizing tax liability by following these laws and the guidance of the federal government is certainly legal, and most would agree that paying only what you owe is fair. So too with environmental regulations, the regulatory process outlined in project strategies should focus on protecting human health and the environment and minimizing the cost and effort of projects by capitalizing on the flexibility of environmental regulations and guidance. Far too often environmental professionals, including regulatory personnel, go in the opposite direction, often resulting in overconservative project strategies. In these cases, it is sometimes reasoned or assumed that the environmental regulations and guidance prescribe a conservative rather than practical strategy when dealing with contaminated sites. However, most current, and even dated regulations and guidance mandate *reasonable* and *practical* solutions for environmental cleanup. Of course not all laws, regulations, and guidance fall into this category, yet a significant portion do. In the current regulatory climate, a practical project strategy can be emphasized in order to develop an effective external project strategy that fast-tracks and streamlines environmental response actions. The industry is "under the gun" to achieve more cleanup with fewer dollars. In order to accomplish this goal, a high level of regulatory evaluation is required and regulatory experts may need to be consulted in order to pinpoint regulatory flexibility related to site-specific issues. This is analogous to knowing tax laws sufficiently well so that what is paid is no more than what is owed.

Lastly, waivers for certain compliance issues may be applicable, depending on site-specific conditions, contaminants, location, and other factors, which can be used to negate specific "unreasonable" regulatory requirements. Waivers or exemptions should be integrated into strategic planning as options in order to efficiently and effectively protect human health and the environment. Following the same CERCLA example above, applicable or relevant and appropriate requirements (ARAR) must be evaluated in order to make use of other program and agency standards. In cases where there are prudent reasons that support noncompliance with some standards, EPA developed six ARAR waivers under CERCLA that can be applied under the following conditions (EPA 1989):

1. Interim measures waiver (i.e., used for early risk-reducing actions).
2. Equivalent standard of performance waiver (i.e., no significant benefit by complying with a standard).
3. Greater risk to human health and the environment waiver (i.e., more protective to waive the standard).
4. Technically impracticable waiver (i.e,. an unreasonable/unachievable technical standard).
5. Inconsistent application of state standard waiver (i.e., complying with the standard is inconsistent with other standards).
6. Fund-balancing waiver (i.e., it costs too much to comply with the standard).

In this example, it is clear that *practical* solutions can be applied if the solutions are protective, and properly proposed, planned, and evaluated. Similar circumstances can be applied to other regulatory processes at the federal, state, and local levels, and an option to consider waivers should be integrated into the regulatory elements of strategic plans. On a cautionary note, waivers are by no means easy to get agreement on. A clear and defensible argument for waivers is essential to reach consensus that a waiver should be approved.

Teamwork, Communication, and Decision Making

For external strategic plans, group dynamics are extremely important. Chapter 7 outlines the importance of teamwork, communication, leadership, and trust. These fundamental project components must be integrated into strategic plans in order for the overall project strategy to be effective. An approach such as partnerships may be proposed and agreed upon in order to work together cohesively throughout the duration of projects. By doing so, the need for internal project strategy development is minimized, and the opportunity to work together is maximized. As a result, common ground can be identified and focused on for completion of projects in a cost-effective and timely manner.

Most would agree that liability issues can cause havoc when attempting to institute or maintain a cohesive team effort or partnership — especially in cases when there are large cleanup costs at stake. In these cases, honest and fair consideration of *reasonable* alternatives is required, and those alternatives should be openly evaluated to ensure they are protective, cost-effective, and within the realm of regulatory flexibility, in order to avoid enforcement and legal actions that delay projects and ultimately slow protection of human health and the environment.

In addition, the decision-making process must be incorporated into strategic plans. Under most environmental regulations, the decision-making responsibility is relatively explicit, whereby the lead agency retains the majority of decision-making power, if not all of it. However, the project team can decide to work together and share the decision-making power, as long as reasonable and prudent decisions are made that reflect the needs and expectations of all stakeholders. The previous example of conducting a voluntary cleanup action is applicable, where decisions related to cleanup may be shared, negotiated, and agreed upon by the entire project team, including the public when appropriate.

In cases where there is a need for response actions to be enforced due to a lack of cooperation, shared decision making may not be an option. In order to avoid the additional cost and time of having to enforce actions, or in the case of the responsible party having to pay for potentially overconservative response actions, it is generally in the best interests of the entire project team to work together, find common ground, and share decision making, as long as high trust factors can be maintained. In cases where the public is involved, and there is an inherent lack of trust or significant issues emerge, the public should then be integrated into the environmental process in order to demonstrate that their concerns are being met. A high level of trust between all entities must be established and encouraged in order to foster agreement between all stakeholders and avoid conflict. Under circumstances where the project team *does* work well together, sharing the decision-making power can fast-track environmental actions and decision making. Therefore, the process used to develop the strategic plan should include the elements of teamwork, communication, leadership, and trust, along with who, how, and when decisions will be made for the project. For the sake of argument, the above "ideal" model may not always work. In these cases, trust and communication are likely issues. Refer to Chapter 7 for more information on trust and negotiating win–win solutions.

Future Land Use

Inclusion of land use, and consideration of potential future land use in project strategies is essential because land use can drive the overall level of cleanup that will be required at contaminated sites. Most often, risk-based industrial or recreational action levels are substantially higher than action levels designed for residential land use. Up-front planning and consideration of land use should be integrated into strategic plans in order to focus project investigation efforts and decision making, and streamline the overall environmental process by using less stringent nonresidential cleanup criteria, if protective. In addition, using nonresidential cleanup criteria will lower the overall cleanup costs and offer more technically practicable alternatives that can be considered for remediation because of technological limitations sometimes associated with lower residential cleanup criteria.

The EPA, Department of Defense (DoD), and Department of Energy (DOE) all have proactive viewpoints related to early determination of land use. In EPA's *Land Use in the CERCLA Remedy Selection Process* (1995), the importance of reasonable environmental decision making is linked together with land use in the following ways:

- Future land use assumptions allow the baseline risk assessment and feasibility study to focus on developing practicable and cost-effective remedial alternatives. These alternatives should lead to site activities that are consistent with the reasonably anticipated future land use. However, there may be reasons to analyze implications associated with additional land uses.
- Remedial action objectives developed during the remedial investigation/feasibility study (RI/FS) should reflect the reasonably anticipated future land use or uses.
- Land uses that will be available following completion of remedial actions are determined as part of the remedy selection process. During this process, the goal

of realizing reasonably anticipated future land uses is considered along with other factors. Any combination of unrestricted uses, restricted uses, or use for long-term waste management may result (EPA 1995).

The EPA directive provides additional language that strongly advocates and guides project teams to use realistic assumptions regarding land use and early consideration of land use. EPA recommends that project teams work with local land use planning authorities, appropriate officials, and the public, as necessary, early in the environmental process in order to consider reasonable future land uses. This may include extra efforts to reach out to and consult with segments of the community that are not ordinarily reached by conventional communication vehicles. Clearly, EPA is advocating that strategic planning fully integrate land use in the investigation and alternative analysis phase of CERCLA-regulated projects.

Under the Base Realignment and Closure (BRAC) program, DoD (1995) has developed a similar proactive approach regarding land use and streamlining the environmental process. For example,

- Risk assessment is a tool used to determine risk-based remediation goals and target areas for remediation, as well as areas requiring no further action. At BRAC installations, property reuse designations provide a unique opportunity for linking risk-based remediation goals to future land use. The outcome of the risk assessment will facilitate remedy selection and the identification of areas that meet the property transfer criteria of CERCLA §120(h).
- Prior to assessing risk, the risk assessors who will be involved in writing or evaluating risk assessments at closing installations should hold meetings to define assumptions they will use in their calculations. This should include the impact of planned reuse (DoD 1995).

Most importantly, under all environmental programs, an honest appraisal of future land use is important in order to determine what cleanup criteria will be used for remediating contaminated media. Assuming industrial cleanup levels in order to reduce cleanup costs does not meet the criteria of an honest appraisal and may not be protective of human health, especially if there is evidence to support that the site is or will be a residential setting. However, in some cases, it may be possible to control exposure through institutional controls, such as zoning or deed restrictions. In these cases, a strong argument would need to be presented that supports an alternative of this nature. A plan is also needed in order to ensure institutional controls are maintained in the future. Lastly, land use with respect to human health may not be the primary risk driver if there are major ecological concerns at contaminated sites, such as endangered species. At these sites the ecological concerns may drive cleanup criteria rather than human health concerns. Strategic plans must consider the impact of land use from a human health perspective, and often from an ecological land use perspective as well, in order to develop realistic assumptions and a reasonable future land use scenario. This element alone can have a significant effect on how quickly sites are remediated and closed, and how much it will cost to complete response actions.

Technology and Science Elements

On a larger scale, strategic plans should integrate assumptions and approaches related to the technical approaches and science that will be used to investigate and potentially remediate sites. By preplanning specific approaches, it is possible to fast-track projects by considering the methods or techniques ahead of time. For example, the project team could reserve the ability to identify operable units (OU) for segregating cleanup actions, identify investigation zones for focusing data collection efforts, or define critical release areas in order to focus funding on early response actions. The strategic plan should outline how and when these types of technical divisions may be applied at sites, and if they are necessary or prudent. Institutional controls are another example, where *consideration* of institutional controls must be integrated into the overall project strategy and, if applicable, included in the investigation and cleanup evaluation process in order to collect and evaluate data that may support their application.

More rarely, project teams may develop strategic plans that do not adequately emphasize the technical or scientific elements. Rather, they focus on other elements such as the regulatory process or land use. It is critical that the science and technology that supports environmental efforts be integrated into strategic plans. Disregarding the fundamental science and technical approaches can result in having to backtrack and redevelop a sound technical and scientific argument for project decisions. In other words, some level of scientific documentation is, for all practical purposes, always required in order to support environmental decisions. Likewise, the strategic plan should include how the project team will ensure that the proper technical and scientific methods will be properly applied, especially in cases were accelerated or streamlined approaches are proposed.

Funding Requirements and Allocation

Funding will often drive how quickly a site is cleaned up and closed. The strategic plan should integrate the funding needs and requirements in order to maintain the flow of resources needed to conduct proposed actions at sites. In cases where actions are accelerated, additional up-front funding is needed compared to traditional funding requirements in order to pay for early cleanup, focus investigation efforts, and support additional decision making, which is further outlined later in this chapter. From this perspective, the strategic plan should outline the funding source(s), mechanisms for receiving funding, and how the funds will be allocated. For government-funded environmental actions, the funding can be discussed relatively openly because tax dollars are used to pay for efforts, and generally there is a structured process for requesting and receiving funding. However, even under the best of conditions, federal, state, or local government funding may become unavailable due to shortfalls, unapproved budgets, or competition for funding. The strategic plan should outline the primary funding process, and then, if possible, outline contingency plans to replace lost funding. This, for example, could include lobbying for year-end federal dollars that need to be allocated or lost.

For privately funded projects, the potential funding requirements can also be discussed; however, the private party may have reservations for doing so. In cases where a million-dollar company has a $5 million potential liability and its insurance coverage is questionable, there is a strong likelihood that there will be insufficient funds to complete the work. For obvious reasons, this should be openly discussed. Larger private organizations may have substantial *total* resources, and may be able to pay for the response actions necessary to complete relatively large environmental projects. However, their *liquid* assets may be much less than their total assets, which can affect their ability to pay for environmental response actions. From this perspective, it is reasonable for project teams to consider cost-saving measures in order to minimize expenditures, maintain a team relationship with all stakeholders, and expedite protection of human health and the environment. Private organizations who feel that they are being taken advantage of have historically resisted voluntary response actions, which slows the environmental process and may increase costs. Regulatory enforcement is the primary option in order to ensure that unresponsive polluters pay for cleaning up their releases; however, the amount of time and resources needed to complete responses under these conditions is generally excessive. For this reason, frank discussions related to funding are essential in order to develop an effective strategic plan that will be paid for without extensive delays related to litigation. These discussions can include topics related to how much funding is available, what the anticipated project costs will be, and how funding will be allocated over the duration of the project. For those that have experience working with the private sector, open discussions related to funding may not come easily, especially if there are questions remaining regarding liability and trust. At a minimum, however, the anticipated project costs should be discussed in order for the liable party(s) to at least be able to plan for potential funding requirements, consider insurance liability issues, and communicate potential funding limitations that will impact the project. In turn, approaches that may help reduce costs and streamline the life of the project can be considered early in the project, rather than waiting until the bitter end. Initially, the anticipated project costs may be inaccurate because little is known about the site. Revisiting the anticipated project costs throughout the project can help improve cost estimates using the most current information and ideas, and this information can then be used to update the strategic plan.

In terms of considering cost-saving strategies, a comparison of voluntary cleanup actions that meets reasonable cleanup standards or criteria can be compared with the costs of enforced response actions. From this comparison, consensus on using less expensive alternatives may be achieved, and potential enforcement or litigation actions may be avoided. This is not an argument for limiting private liability costs associated with environmental cleanup for the sake of maintaining profits over protecting human health and the environment. Environmental laws explicitly identify when an entity is liable for environmental cleanup costs. Rather, this is an argument for private organizations to own up to their environmental liability, or portion(s) thereof, and take a *proactive* stance in order to *reduce* the overall cost of response actions through acceleration and streamlined approaches. Furthermore, this is not an argument for government agencies to consider big businesses as "deep pockets" to be taken advantage of in order to pay for overconservative response actions by holding the "enforcement hammer" over their heads. Rather, this is an argument for the government agency(s), responsible party(s), and the public to work together, while understanding funding *limitations*, in order to more efficiently complete projects. In turn, creative approaches and/or innovative technologies can be considered in order to help minimize the overall cost liability, whether government or private, and adequately protect human health and the environment. For these reasons, strategic plans should outline funding needs and allocation to the degree possible and as agreed upon by the project team.

Uncertainty Components

Uncertainty is an element of strategic plans that must be considered because of the inherent *lack* of certainty associated with hazardous waste sites. For this reason, throughout the life of a project it is necessary to measure and observe project progress, reevaluate strategic plans, and develop contingency plans or strategies that may be needed in order to address unforeseen conditions. It is important to identify what is known about projects, and also what is *not* known about sites when developing a strategic plan. Assumptions on what is not known can be used to reasonably anticipate strategic deviations, similar to engineering deviations integrated into the observational approach described in Chapter 6. Realizing that uncertainty is a guaranteed condition at hazardous waste sites no matter how much investigation is completed allows for the consideration of "possibilities," and the development of contingency strategies for potential unforeseen conditions. The project team should integrate measuring and observing performance and progress elements that can be used to trigger alternative project strategies. From this effort, a more effective strategy can be implemented to help accelerate and streamline projects.

Scheduling Considerations

Scheduling is an important element of strategic plans. The schedule for milestone events, pivotal decisions, and major deliverables should be agreed upon early in a project's life. This could also include contingency schedules for early site closure if limited or zero contamination is encountered at sites. In more complex cases, scheduling should include combining tasks that various project counterparts are responsible for in order to integrate efforts and more efficiently complete projects (see the Navy's integrated scheduling approach in Chapter 3 for more information). The actual dates selected for scheduled activities should drive performance for early completion of major tasks and subtasks, and identify creative ways to meet scheduled events and deliverables within the allocated budgets. An overconservative schedule has the tendency to encourage individuals to take the entire amount of time allocated rather than getting the job done sooner. However, proactive schedules should be realistic in that the resources and time available are *reasonable* in order to complete the effort on time. In turn, the need to continually revise schedules and reevaluate the resources needed to complete efforts is minimized, and project team members and support staff will not have to continually work overtime in order to complete tasks.

Innovative Technologies and Approaches

Innovative technologies and approaches are mechanisms that can be utilized in order to expedite actions, streamline decision making, and reduce costs. This book, for example, compiles a lengthy list of how to apply, consider, and implement innovative technologies and approaches, and Chapter 6 provides a clear distinction between innovative technologies and innovative approaches. Strategic plans should include consideration of innovative technologies and approaches so that the project team can evaluate these alternatives and compare them with conventional alternatives, in order to support the overall project strategy, or components thereof, such as following EPA's Superfund Accelerated Cleanup Model or a Brownfields approach. In addition, this element is helpful in cases where innovative technologies will be considered in the project investigation and cleanup phases. The strategic plan should outline up-front planning and the team's desire to use acceleration approaches and/or innovative technologies. As a result, up-front planning can be used to "break the ice" and streamline evaluation and selection of innovative alternatives. EPA's Brownfields Initiative is presented later in this chapter, because it is a potential strategy or innovative approach that can be used to expedite response actions at many contaminated sites across America. The Brownfields Initiative and the associated state voluntary cleanup programs are examples of "big picture" innovative approaches that be used to accelerate and streamline response actions.

8.1.3 Small-Scale Strategy for Accelerating and Streamlining Environmental Projects

Environmental strategies must also consider the small-scale elements of strategic plans that are geared toward the day-to-day operation and decision-making requirements of projects. From the project team perspective, some or most of these elements may be implicit components of strategic plans, rather then explicit components. However, under some circumstances, small-scale elements may need to be elevated at the project team level in order to make decisions, discuss information, or consider new project directions. Most importantly, the small-scale elements should be conducted in accord with the large-scale strategic elements and the overall focus of the strategic plan. The following are fundamental or "generic" small-scale elements of environmental strategic plans.

Technology and Science Elements

The validity of data relies on how technologies and science are used, and validity is needed in order to complete investigation and cleanup of contaminated sites. Not only must data validity be quantified through quality control procedures, but initial planning is also required in order to maximize data useability and respresentativeness. One example is the level of data quality, or analytical level, required to make project decisions. The EPA analytical level varies depending on the intended purpose. In order to identify potentially responsible parties (PRP) and complete risk assessments, EPA level IV or V data are usually required for National Priorities List (NPL) sites, which are expensive to analyze and validate and require additional effort to collect. EPA level I or II data, however, are less expensive to analyze and collect, and may include field testing analytical analyses. These data may be used for basic site characterization, but generally not for PRP or risk assessment efforts. In this example, the technology and science elements of data collection and testing should be considered ahead of time and *implicitly* integrated into the overall strategic plan in order to keep costs reasonable, streamline the environmental process, and ensure that data collected will be useable and representative. In some cases, only *limited* level IV/V data should be collected, and should be supplemented with lower level analytical data that are easier to collect and less expensive to analyze.

To illustrate this further, another example of why strategic planning should integrate small-scale strategic elements of technology and science is the argument for and against collecting dissolved and total metal samples from monitoring wells. Heated debates have developed over attempting to use dissolved or total metal analyses for multiple purposes. Developing a strategy that provides the *framework* and not necessarily the *details* of how small-scale technologies and scientific methods will be considered throughout the life of the project is helpful in order to efficiently and effectively meet project goals and objectives. In this example, typically risk assessments require filtered metal analyses, and drinking water analysis requires total metal analyses. In order to address project data needs of the future, these types of technology and science issues should be integrated into the strategic plan so that later on delays are avoided by not having to, for example, revisit the site and collect additional samples. Most importantly, the day-to-day technology and science issues for projects must be in accord with the overall "big picture" strategy so that data collected are useable and representative.

Administrative and Organizational Elements

Administrative and organizational elements are essential for smooth operations and efficient information transfer. Organizational talents and tools are the components of these small-scale strategic elements. From a talent perspective, those responsible for multiple tasks should possess, at a minimum, basic organizational skills. On larger and more complex projects, these types of skills grow in importance and can affect how well a project is accelerated and streamlined. Comprehensive and effective money management is often part of the organizational skills that are needed to run large and complex projects. Money management can be aided by using computer-based money management tools that combine project schedule information, task descriptions, and lead personnel for various project components in order to track progress and realize goals.

Tools that help organize and store information include databases and geographical information systems (GIS). For large projects, a GIS approach is almost mandatary based on the success of past GIS efforts for data storage, data relation, and data presentation. More recently, global positioning systems (GPS) have become the state-of-the-art approach for locating and logging site information. Together, GPS, GIS, and/or databases are tools that can streamline environmental projects. Smaller projects may not require an elaborate system in order to organize and compile findings. However, larger projects can benefit from these types of tools, and prior planning is helpful in order to maximize their use.

An example of a custom tool that can be used to help organize and administer projects is project story boards (Figure 8-2). Story boards are large-scale exhibits that illustrate the investigation, remediation, evaluation, and decision-making elements of projects. These tools act as a "visual action plan" and are beneficial for tracking large- and small-scale project tasks and completing projects within the scheduled time period and budget. Story boards can augment strategic plans. Theoretically, after competing each task identified on a story board, the project is completed and/or a final decision has been made. Due to uncertainty, however, story boards and strategic plans must be periodically reevaluated, or adjusted, in order to ensure they are reasonable and prudent.

From an administrative perspective, the need to store and report information should be integrated into strategic plans. If there is a strong likelihood that the project will end up in litigation, then the level of documentation must be meticulous and the language in documents clear. For the purposes of accelerating cleanup and streamlining decision making, litigation is not considered an optimal avenue or choice, and for that reason is not discussed in this section. In cases where the stakeholders can agree on project strategy and direction, the administrative elements considered may include developing *reasonably* sized reports for outlining project findings. There are numerous examples of past environmental efforts that have generated huge documents that are both bewildering to read and almost impossible to comprehend. An effective strategy can include elements that encourage reducing the size and complexity of deliverables, yet ensure they still contain sufficient information needed to support decision making.

Lastly, the details of projects must be addressed on a regular basis. Empowering individuals with the ability to oversee and track project details enables project teams to work more efficiently and effectively. Therefore, consideration of the administrative and organizational elements in the strategic plan is helpful for streamlining regulatory processes.

Teamwork, Communication, and Decision Making

In Chapter 7, the aspects of teamwork and communication were outlined because human interaction is essential in order to complete work on environmental projects. From a small-scale perspective, one-on-one communication and trust are essential. Recognizing the importance of maintaining positive and effective relationships are part of developing a comprehensive strategic plan. Poor interaction between project players can impact performance and impede progress. Fostering teamwork, effective communication, leadership, and high trust factors help ensure that professional relationships are maintained. These aspects of a project strategy cannot be ignored without the risk of having to resolve significant conflict on future project issues. More information on this subject is presented in Chapter 7.

FIGURE 8-2. PROJECT X STORY BOARD

Project Requirements and Data Gaps (examples)	Lead	Action & Reference	Field Tasks	Comments	Scope of Work	Uncertainty	Date Completed (Insert Columns)
<u>Air</u> • Sources • Sites/phase of effort <u>Soil</u> • Sources • Sites/phase of effort <u>Groundwater</u> • Sources • Sites/phase of effort <u>Local Conditions</u> • Existing data/documents • Land use • Environmental setting • Community Issues	Name the person, entity, or team that is responsible for each project requirement and data gap.	List the general action proposed that will address each data gap and project requirement. Each action should be referenced to a document or project team decision in order to link past decision making with future decision making requirements.	List the number of samples or wells, etc. that are anticipated, proposed, or installed, and other field related information to be conducted or used in the effort. Other examples of information include aquifer testing, geophysics, pilot testing, etc.	Clarify and link, if necessary, each data gap and project requirement with other project data gaps and requirements in order to correlate various field and office efforts. Case studies supporting the action/approach proposed may be referenced supporting actions and lessons learned.	Track in-scope and out-of-scope tasks, data gaps, and project needs in this column for budget negotiations. Indicate whether the out-of-scope tasks where approved, or if they had to be implemented/ acted upon because of a potential failure.	Briefly summarize the uncertainty related to primary actions, data gaps, and project requirements. If possible, list assumptions used in the site conceptual model and potential contingency plans considered or implemented.	Divide this column in to several rows in order to list the date when data gaps and project requirements are completed. A sequential format should be used to show progress during the effort. Example headers could include: • Field work • Data/site evaluation • Report/document • Final decision • Response action • O&M • Closure
<u>Work Plans, QC, H&S, CR, Databases, etc.</u> • Plans & documentation	This should include a contact in order for the project team to be able to follow up with questions.	(Consider reports, plans, and engineering evaluations as project requirements.)		List the acceleration and/or streamlining strategy, model, approach in this column. The anticipated savings or benefit may also be listed.			Track progress on the story board and the project schedule. Completion dates should correspond with the project schedule completion dates.
<u>Evaluation</u> • Characterization rpts. • Alternatives evals. • Risk assessments • ARARs studies • Regulatory process needs							
<u>Communication</u> • Teamwork & partnership • Trust & Leadership				The project schedule may need to be referenced in this column also.			When all columns are filled out with either a date or "not applicable" indicator, the project should be closed or long-term monitoring and/or O&M should be in place.
<u>Decision Making</u> • Permit modification • Decision documents							
<u>Cleanup & Response Actions</u> • IRAs, Removals, RD/RA • ICM & corrective actions • Site closure/O&M				Other information as necessary.			

Figure 8-2 Example contents of a project storyboard.

Scheduling Considerations

Scheduling project activities and tracking results is essential for keeping the project team abreast of the latest development and for identifying the project tasks completed. Schedules, similar to strategic plans and project story boards, should be updated periodically. The strategic plan can provide the first "roundtable" project schedule, and how and when the schedule will be updated. Periodic updating of the project schedule is needed in order to assess the need to get help or delegate effort, and to evenly share work loads. From a small-scale perspective, developing an approach to revise and review the project schedule is essential in order to keep projects running smoothly.

8.1.4 Strategic Plan

A strategic plan may consist of a single illustration that outlines the general strategic approach or model. This type of plan works well for smaller projects, comprised of small project teams, where much of the large- and small-scale strategic planning is implicit in running the project. In these cases the strategic plan is discussed in *concept,* or a flow chart is drafted to illustrate the main project efforts, and the project team works together to identify uncertainty and resolve issues. More complex projects, such as major DOE facilities that require restoration, generally require a comprehensive strategic implementation plan in order to outline the large- and small-scale elements of the strategic model, and the approach, timing, and implementation of both efforts. In these cases, there usually the large numbers of people that participate on the project teams, each person handling or overseeing various components of the project. Coordination of this many people and maintaining teamwork and communication necessary to efficiently run such projects dictates the need for a comprehensive strategic plan. Projects that fall in between these two sizes of projects require a plan, or model that information-wise falls somewhere in between. Most importantly, the strategic plan is a vital document that the project leaders periodically review and use to help redirect future efforts.

In summary, a strategic plan should be comprehensive; however, the size of the document must be reasonable, and generally should allow it to be easily read in a short period of time. The information should be clear and concise, and illustrations and tables should be used to convey the information presented. An environmental strategic plan that is developed from an external perspective should include the following components, not necessarily in this order:

- Mission and vision statements and identification of a final project "destination"
- Revision dates and information supporting past revisions
- Goals and objectives
- A project team consensus statement for progress and teamwork/partnership
- Description of the overall strategy and regulatory program approach, such as Brownfields, CERCLA, RCRA, voluntary actions, etc.
- A list of large- and small-scale project elements needed to support the strategic plan
- A project story board listing sites/areas of concern, etc., and a project schedule

- A description of uncertainty and assumptions considered that support the strategic plan, and a listing of potential contingency approaches or directions
- Graphical displays showing integration of regulatory programs, technical issues, project elements, and team player and/or contractors interaction
- A description of the agreed-upon decision-making processes that will be used by the project team
- Site-specific and program information that supports innovative technologies and approaches, and acceleration, streamlining, and cost-reduction initiatives
- Other information needed to support the strategic plan.

8.1.5 Brownfields and Environmental Strategic Planning

An example of a large-scale strategy that can be used to encourage a team approach or partnership with stakeholders is EPA's Brownfields Initiative, and the associated state voluntary cleanup programs. The following is an introduction to the Brownfields Initiative, pilot projects, success stories, and how the Brownfields strategy can be integrated into private sector efforts. The information presented on the EPA's and state Brownfields programs was compiled from the EPA Headquarters, Region I, and Region V home pages on the World Wide Web. More detailed information is available from the EPA, including the EPA home page (http://www.epa.gov; see also Appendix A).

Background

Many sites across the country once used for industrial/commercial purposes have been abandoned or are under-used, some are contaminated, and some are merely perceived to be contaminated. A report from the General Accounting Office (GAO: Community Development, Reuse of Urban Industrial Sites, June 1995, GAO/RCED-95-172) finds that "as states and localities attempt to redevelop their abandoned industrial sites, they have faced several obstacles, including the possibility of contamination and the associated liability for cleanup . . . This situation is caused largely by Federal and state environmental laws and court decisions that impose or imply potentially far-reaching liability. The uncertain liability has encouraged businesses to build in previously undeveloped nonurban areas called 'greenfields' where they feel more confident that no previous industrial use has occurred." The National Environmental Justice Advisory Council (NEJAC) has likewise "determined that there exists a compelling need to address issues of economic development and revitalization of America's urban [and rural] communities." The NEJAC has requested that EPA "provide leadership in stimulating a new and vigorous national public discourse over the compelling need to develop strategies for ensuring healthy and sustainable communities in America's urban [and rural] centers and their importance to the nation's environmental and economic future."

Increasingly, prospective buyers of property, real estate developers, and lenders hesitate to acquire urban land that may possess some level of contamination. The reason is that discovery of contamination could subject the owner to liability under federal and state environmental laws, especially CERCLA. Instead of pursuing these urban properties, or "brownfields," buyers often choose to develop "greenfield" sites in suburban and rural areas. This trend not only encourages the use of undeveloped land, often leading to suburban sprawl, increased traffic congestion, and the destruction of habitat, but also restricts the redevelopment of urban areas, thus limiting economic growth. Since many brownfield sites are located in poor and minority communities, their existence raises environmental and economic justice concerns as well.

One of the most important reasons why developers shy away from purchasing brownfields is that environmental laws assign liability to a broad range of parties — including the present owner. Buyers are nervous that they might get stuck paying for the cleanup of a previous owner's pollution. Current owners also fear potential future liabilities related to unknown site contamination. While EPA does not intend for innocent parties to pay, the agency nonetheless maintains that a comprehensive liability scheme is critical to ensure that polluting parties assume responsibility for their pollution. At this point, the challenge facing government agencies is how to encourage redevelopment/economic growth in these urban areas without compromising public health and environmental protection. The Brownfields Initiative is a program aimed at understanding and addressing this challenge.

There are many factors that influence a decision to purchase property and redevelop it, but concern about environmental liability is often serious enough to preclude a real estate transaction from moving forward. Some of these other factors include location, financing constraints, suitability of property to purchaser's purpose, local infrastructure supports, availability of a trained labor pool, local land use policy, and taxation considerations. Fear of liability from environmental contamination is only one factor among many that affects such development decisions.

Brownfields Applies to Many Sites

It is nearly impossible to determine the exact number of brownfields throughout the U.S. EPA estimates that there are probably 500,000 sites or more with a high potential for some contamination, based on previous industrial or commercial use. Many urban areas have thousands of abandoned and potentially contaminated properties. The majority of these sites are located in urban areas where heavy manufacturing and other industrial activities have occurred. A brownfield might be found at an abandoned industrial yard or at a facility currently in operation. But brownfields can also can be small residential or commercial lots where some form of contamination is suspected to exist even though no such information may be currently available. To address these issues and the vast number of sites, EPA's efforts under the Brownfields Initiative can be grouped into four broad and overlapping categories

1. Providing grants for brownfields pilot projects
2. Clarifying liability and cleanup issues
3. Building partnerships and outreach among federal agencies, states, municipalities, and communities
4. Fostering local job development and training initiatives

Uncertainty and Brownfields Redevelopment

The predominant characteristic associated with brownfields redevelopment is uncertainty. Uncertainty permeates virtually every aspect of the process, making potential buyers, developers, and lenders hesitant to get involved. Consider the following:

- There is usually uncertainty about whether contamination exists on the property, and if so, to what extent developers and lenders might be held liable for cleanup costs.
- Standards for environmental cleanup vary from state to state, and these in turn may differ from federal standards.
- The question of "how clean is clean?" remains hotly debated.
- It is sometimes difficult to gauge the costs associated with cleaning up and redeveloping a brownfield site, but there can be significant transaction costs (legal fees) and potentially substantial cleanup costs.
- Assessing what kind of contamination exists on the site can also be costly.
- As the purchaser of a brownfield site, it may be extremely difficult to secure financing. Lending institutions are wary of becoming involved with potentially contaminated property without some degree of government sign-off on liability.
- Communities that contain brownfields are often not integrated early into the brownfields planning process.

Brownfields Initiative Helps Integrate Environmental Programs

Environmental justice and pollution prevention (P2) are two of the many issues that are linked to brownfields, in addition to CERCLA. In urban areas, many lower income or minority communities are concerned about multiple pollution sources that may expose residents to unacceptable amounts of environmental contamination. These communities want plans that promote both economic redevelopment and environmental justice. Brownfield redevelopment must proceed in a way that does not compromise the environmental health and well-being of local residents. Pollution prevention (P2) promotes the development of more efficient manufacturing practices that generate less pollution and ultimately lower pollution-control costs. Consideration of P2 is an important part of a brownfield redevelopment project. P2 not only results in lower manufacturing costs, but it also promotes cleaner and more efficient redevelopment of a site. P2 practices encourage sustainable land use redevelopment and reduce the likelihood of a redeveloped site becoming another brownfield.

EPA's Brownfields Assessment Demonstration Pilots

As part of EPA's Brownfields Initiative, the Brownfields Assessment Demonstration Pilots are designed to empower states, communities, and other stakeholders to work together in a timely manner to prevent, assess, safely cleanup, and reuse brownfields. EPA has awarded cooperative agreements to states, cities, towns, counties, and tribes for demonstration pilots that test brownfields assessment models, direct special efforts toward removing regulatory barriers without sacrificing protectiveness, and facilitate coordinated public and private efforts at the federal, state, and local levels. As of 1996, EPA funded 76 Brownfields Assessment Pilots. Of those pilots, 39 are national pilots selected under criteria developed by EPA Headquarters and 37 are regional pilots selected under EPA Regional criteria.

The brownfields assessment pilots test cleanup and redevelopment planning models, direct special efforts toward removing regulatory barriers without sacrificing protectiveness, and facilitate coordinated environmental cleanup and redevelopment efforts at the federal, state, and local levels. The funds are to be used to bring together community groups, investors, lenders, developers, and other affected parties to address the issue of cleaning up sites contaminated with hazardous substances and preparing them for appropriate, productive use. The pilots serve as vehicles to explore a series of models for states and localities struggling with such efforts. In addition, the pilots focus on EPA's primary mission of protecting human health and the environment. Likewise, it is an essential piece of the nation's overall community revitalization efforts. EPA works closely with other federal agencies through the Interagency Working Group on Brownfields and builds relationships with other stakeholders on the national and local levels in order to develop coordinated approaches for community revitalization.

An example pilot project is EPA Region I's State of Rhode Island Brownfields Pilot. The state is focusing its project on the communities in the watersheds of the Woonasquatucket and Blackstone Rivers. Decades of industrial use have left many properties contaminated, which has led to under-use, vacancy, and decay at these sites. There are 53 known brownfields in the watersheds, but inventories of contaminated sites are incomplete and not cross-referenced with development potential. Sites not on the state's inventory may pose threats to the health of the surrounding communities. The stigma of possible contamination has impacted local economies and has impeded the implementation of a major project to create a greenway along the Woonasquatucket River. The state is proposing development of a model ecosystem-based program to bring the vacant or under-used contaminated properties in the communities back to beneficial use. The key to the success of the state's ecosystem-based approach will be the coordination of revitalization efforts being conducted by multiple state and local agencies with jurisdiction in the watersheds. The state's goal is to develop a statewide brownfields marketing strategy based on the results of the EPA demonstration pilot. Two critical pieces of information needed are (1) a degree of certainty in the level of contamination that will allow accurate estimates of cleanup costs; and (2) the financial incentives necessary to support marketing of the Brownfields. Once the environmental concerns are known, the State's Economic Development Corporation can market the sites. Activities planned as part of this pilot include

- Conducting a regional survey of both ecosystems to determine likely candidates sites for further assessment, including investigation of regional demographics, aerial photographs, and interviews of local community groups to determine areas of concern.
- Facilitating roundtable meetings of all stakeholders to prioritize identified sites based on community concerns, potential beneficial reuse, and environmental threat.

- Assigning specific contact persons to reach out to affected communities.
- Assessing four priority sites already identified and others that may be identified by municipal and community groups.
- Developing a baseline property survey checklist.
- Developing community relations planning guidance.

Linking CERCLA, Brownfields, and State Environmental Programs

The law at the heart of the brownfields is CERCLA, which governs not only cleanup but the liability associated with inactive contaminated sites. CERCLA was enacted to address the most serious hazardous waste sites, such as the well-known Love Canal site in New York. The law adopted a broad liability scheme to ensure that parties responsible for polluting would be held accountable for their actions. While Superfund has had success in terms of addressing the "hotspots," by identifying PRPs, facilitating over 280 NPL cleanups, and coordinating 2000 emergency removals, the law also has been indirectly responsible for generating much of the fear associated with small-scale industrial property transfers. Indeed, it has served to discourage exploration and redevelopment of sites possessing only minimal levels of contamination or perhaps no contamination at all. Although most low-level contamination sites probably will never be subject to a Superfund investigation and cleanup by EPA, the specter of liability nonetheless serves as a major disincentive for potential buyers and lenders to get involved in such properties.

EPA and state governments recognize that successful brownfield redevelopment can best occur when state and local governments, community groups, and the private sector work together to solve problems that combine environmental cleanup, urban planning, and economic redevelopment. A major reason that redevelopment projects are often halted is due to fear of liability in clean-up costs. To address this, EPA has formed partnerships with states to more effectively communicate liability issues, increase understanding of environmental risks, and educate local governments, communities, and the private sector about brownfields redevelopment. In recent years, states have developed voluntary cleanup programs that are designed to provide liability protection to private parties that clean up brownfield sites. Even though the federal government may have little or no legal interest in most brownfield sites, EPA is taking steps to diminish the fear of liability under federal laws by modifying existing agreements with states that have voluntary cleanup programs. Through these modified agreements, EPA fully supports state cleanup programs and essentially pledges that the successful cleanup of a site under the state's program will also satisfy EPA.

State Voluntary Cleanup Programs

In response to the enforcement-intensive federal Superfund law, states have begun to implement their own voluntary cleanup programs, which are often more responsive than the federal program. This relationship between state and federal legal authority is often confusing, however. While a party might be in compliance with a state voluntary program, there is still a risk or fear of potential legal action taken by EPA or a third party suit under CERCLA. In an effort to reduce some of the uncertainty, EPA Region V, for example, has entered into numerous Superfund Memoranda of Agreements (SMOA) with states to assure that agencies will not take action at a facility in compliance with a state voluntary cleanup program. The SMOA formalize EPA's approach to the state's voluntary cleanup program and essentially pledge that successful participation in the state program will also satisfy EPA. Illinois, Indiana, Wisconsin, and Minnesota have signed such an agreement with EPA Region V. Similar changes to the agreements with Michigan and Ohio are anticipated in the near future. Most states are taking the lead in terms of quantifying the liability issues associated with Brownfield redevelopment. This is why it is critical to contact your state if you have specific questions about the state program. Many states have written their own laws specifying the liability scheme, enacted voluntary cleanup programs, outlined cleanup standards, and provided financing opportunities. Voluntary cleanup programs established within EPA Region V are summarized as examples of how some state voluntary cleanup programs operate. Opportunities to use these types of state programs should be thoroughly considered when developing strategic plans.

Illinois Voluntary Cleanup Program — The Illinois Environmental Protection Agency (IEPA) created its voluntary cleanup program, titled the Pre-Notice Program, in 1989. The program was developed under the authority of the state's Environmental Protection Act, since Illinois has no Superfund program per se. The program offers technical assistance for cleanup of contaminated sites and lessens the threat of future enforcement action. An initial prepayment of $5,000 or half of anticipated costs is required, and a review and evaluation of services agreement has to be signed, that stipulates that all work done on site must be carried out in a manner approved by IEPA. Currently, there is no financial assistance available through the program. By September of 1995, 480 sites had entered the program, with 125 of those being released with a "clean letter" from the state, indicating no further corrective or remedial action is necessary.

Through EPA's site assessment dollars, IEPA is targeting publicly held sites and tax-reverted properties for the program. To date, IEPA has performed four assessments in Chicago and three in East St. Louis, and will soon begin field work on three additional sites in East St. Louis. IEPA estimated they will spend $300,000 in 1996 on Brownfield assessments. The assessments IEPA will perform are not typical Phase I/Phase II audits. Instead they will perform a hybrid assessment based upon the Superfund Site Investigation and the Phase II guidances in order to leave the NPL-scoring option open. "As clean" levels are set by IEPA's new state-wide Tiered Approach to Corrective Action process. The state encourages cities to propose sites for the Pre-Notice Program after demolition of structures on the property.

Indiana Voluntary Cleanup Program — The Indiana Department of Environmental Management (IDEM) started its Voluntary Remediation Program (VRP) in July 1993. Since that time, 81 sites have entered into the program with 5 sites receiving "Certificates of Completion." Participants must submit a $1,000 check and an application describing the history of the site, including the results of a Phase I site assessment. A Voluntary Remediation Agreement must be signed that includes provisions for cost recovery of IDEM time and effect on the site. IDEM will issue a Certificate of Completion and the Governor's office will issue a covenant not-to-sue upon successful completion of a remediation. There is currently no financial assistance available.

As a recipient of a Brownfields pilot grant through its Site Assessment program, IDEM has performed three Brownfield assessments, one in each of the northwest Indiana cities of Gary, Hammond, and East Chicago. Two additional sites have been sampled in smaller communities; both sites have potential purchasers and redevelopment plans. IDEM performs a hybrid assessment that is neither a traditional Superfund assessment nor a Phase II. Their "as clean" levels are set at the same levels as the VRP industrial reuse levels. IDEM is conducting community outreach and education meetings or workshops throughout the state to market and explain its brownfields program and services. Through its Site Assessment program it is developing a brownfield application package that will not only allow IDEM to select sites upon which it will conduct the assessments, but will also help the communities determine which sites in their communities would be more suitable for brownfields redevelopment efforts. Finally, IDEM will utilize state funds to target and assess sites for immediate removals under state authority, concentrating on the smaller communities where it believes it can achieve higher results with the state's limited resources.

Michigan Voluntary Cleanup Program — The Michigan Environmental Response Act (MERA) of 1982 established the means for the state to fund contaminated site cleanups and recover costs from responsible parties. In 1990, the Polluters Pay amendments to this act established strict joint and several liability for potentially responsible parties and provided for administrative orders and covenants not-to-sue for use at brownfields. In 1995, the state cleanup law was again amended and is now known as Part 201 of the Natural Resources and Environmental Protection Act. The amendment was specifically designed to encourage the reuse of brownfields by changing the liability standard to causation, allowing parties to purchase contaminated property without liability after completing a baseline environmental assessment and providing for due care in the reuse of the property, and providing for use-based cleanups as well as a lower cleanup standard.

The state cleanup authority was supported with a $425 million bond issue, the Environmental Protection Bond, established in 1989. A total of $45 million of the bond was specifically designated for brownfield redevelopment through the Site Reclamation Program. The Site Reclamation Program, which started operation in 1992, provides a total of $35 million for grants to local units of government to investigate and clean up sites of contamination where a developer has been identified. The remaining $10 million is used to provide grants, also to local units of government, for Brownfield site assessment. Site assessments can include Phase I and phase II site assessments and limited remedial investigations. Grants cover 100% of eligible costs. Fifty-six grants for $22.6 million have been issued. Nine grants for a total of $6 million have been completed and have generated approximately $50 million in capital investment and 400 jobs. EPA brownfield site assessment funds are a valuable supplement to this program and have allowed assessments to take place at sites and in communities that otherwise would not have received assistance. In 1995 the Department of Environmental Quality (DEQ) performed 10 EPA site assessments in Detroit and plans another 6 throughout the state. The DEQ estimates that at least 45% of its workload is brownfields-related.

Minnesota Voluntary Cleanup Program — The purpose of Minnesota's Voluntary Investigation and Cleanup (VIC) program is to investigate and remediate contaminated land and bring it back into productive use by providing technical assistance and liability assurances. The VIC program provides five broad categories of written liability assurances which include Technical Assistance Approval Letters; No Action Letters or No Action Agreements (No Action Agreements include a covenant not-to-sue); Off-Site Source Determination Letters or Agreements; No Association Determination Letters; and Certificates of Completion (partial and full cleanups).

Since 1988, 670 sites have entered the VIC Program and 393 have been cleaned up, found acceptable for purchase, development, or refinancing, or transferred to other regulatory programs. There are no application fees; however, oversight costs are recovered by billing on a quarterly basis at a rate of $75 to $85 per hour. Under VIC program oversight, more than 2500 acres of industrial and commercial property have been returned to productive use, providing environmental and economic benefits to communities without cost to Minnesota's taxpayers. The VIC program is an integral part of two funding programs created to address investigation and cleanup of contaminated sites: the Contamination Cleanup Grant Program, administered by the Minnesota Department of Trade and Economic Development, and the Tax Base Revitalization Account, administered by the Metropolitan Council.

The Contamination Cleanup Grant Program provides funds to investigate and clean up contaminated sites, thus offering a greater opportunity to convert contaminated property into a marketable asset. The Tax Base Revitalization Account provides grants for properties in the Twin Cities metropolitan area for polluted land cleanup. To be eligible for either program, a site must have a Response Action Plan (cleanup plan) approved by the State, typically through the VIC Program. Under the oversight of the VIC Program, an 18-month pilot was implemented in conjunction with EPA to demonstrate the potential effectiveness of using state-based voluntary cleanup programs to resolve the status of Comprehensive Environmental Response, Compensation and Liability Information System (CERCLIS) sites that have not yet undergone sufficient characterization to prioritize them using the Hazard Ranking System (HRS) scoring process. Thirty sites were involved in the pilot program. This program proved to be substantially more efficient and cost-effective than the traditional assessment process of Superfund. It was the recommendation of Minnesota Pollution Control Agency (MPCA) staff that EPA consider duplicating this program in other states based on the success of this pilot program.

The MPCA Preremedial Superfund Program interacts with the VIC program to aid in the redevelopment of orphaned contaminated sites. This program is funded by EPA and has just over $100,000 to conduct Brownfields investigations on properties with high redevelopment potential, but where real or perceived levels of contamination limit redevelopment activity. These properties generally have no viable responsible party or voluntary redevelopment party, and may be tax delinquent or abandoned. The MPCA works with local units of government to identify appropriate sites. MPCA staff conducts Phase I and limited Phase II investigations with the expectation that, by removing some of the uncertainty around contamination issues, redevelopment activity will be stimulated. Interested developers will sign up with the VIC program to complete the site investigation and the redevelopment process. These voluntary parties may be asked to reimburse the MPCA for some of the preremedial Phase I and Phase II activity expenses.

Ohio Voluntary Cleanup Program — Ohio's Voluntary Action Program was established in June of 1994. A person undertaking a voluntary action to clean up his or her property may contract with consultants and/or contractors to perform investigations and cleanup activities. When the property has been cleaned up and meets the standards for its specified reuse, the owner must contract with a certified professional and certified laboratory to prepare a No Further Action letter and supporting documents to send to Ohio EPA. A certified professional also must prepare a request for a variance from particular standards, if needed. Ohio EPA then can issue a covenant not-to-sue based upon this No Further Action letter. Participants may be charged for technical assistance provided by Ohio EPA on a fee-for-service basis, as well as required to pay for insurance of covenants not-to-sue. Financial assistance for sites is available in the form of low-interest loans administered by the Chamber of Commerce. The Property Revitalization Board serves as a "clearinghouse" for this and other available financial incentives. Sites with groundwater contamination are precluded from entering the program. Numerous sites have entered the program, with three sites receiving covenants not-to-sue.

Wisconsin Voluntary Cleanup Program — Wisconsin's voluntary cleanup program was established in 1978 under the Hazardous Substance Spill Law. The Spill Law established notification requirements, responsibility for environmental investigation and cleanups, and a hazardous substance spill fund. Since 1978, more than 90% of all sites cleaned up in Wisconsin have been cleaned up through this voluntary process. To enhance the Spill Law so that it encourages environmental cleanups by parties not responsible for contamination, the Wisconsin legislature developed the Land Recycling Law, which became effective in May 1994. This law was designed to provide liability exemptions to encourage environmental cleanups, revitalize rural and urban areas, and return property to the tax base. The Land Recycling Law exempts municipalities from cleaning up contaminated property acquired through tax delinquency or bankruptcy court order if the original discharge was not caused by the municipality; exempts lenders from responsibility to clean up contaminated property they acquire through foreclosure, if they meet certain conditions; exempts purchasers from future liability when contaminated property is cleaned up by the purchaser, when they meet certain conditions; delegates authority to political subdivisions to negotiate and recover costs for cleaning up property they own, if it was contaminated by past owners, operators, or transporters (including landfills); and provides the Department of Natural Resources (DNR) the authority to file a superior lien for state-incurred cleanup action costs, except on residential properties.

Under the Land Recycling Program, purchasers (i.e., current innocent landowners or those purchasing property) of contaminated property will be granted limited liability under the Spill Law for past releases on their property if they investigate and clean up the entire property with DNR oversight. There are currently 38 properties in the Land Recycling Program. After successfully cleaning up historically contaminated industrial sites in New Berlin, Cellular One received the first "Certificate of Completion" under Wisconsin's new Land Recycling Program in September 1995. (Cellular One entered the Land Recycling Program in May 1995, and the cleanup was complete less than 4 months later.) The DNR Land Recycling Program also developed a Brownfields Environmental Assessment Pilot and sought participation by municipalities statewide that have potentially contaminated properties. Under this federal- and state-funded pilot, DNR staff will conduct preliminary assessments at abandoned, tax delinquent, or bankrupt properties with development potential to determine if contaminants are present. Eleven communities were chosen to participate in this innovative pilot. These communities, with DNR's help, can then market the properties for cleanup and development to get them back on the tax rolls with the possibility of returning more than 210 acres to productive community use.

U.S. EPA Region V and State Environmental Program Success Stories

The following are EPA Region V, and state voluntary cleanup program Brownfields success stories.

Illinois Success Story — The John Deere Plow Works site located in Moline was used as a manufacturing plant for farm machinery until 1986. The site is located in the Mississippi flood plain, where the groundwater is between 8–15 ft deep. The plant owned and operated several underground storage tanks (UST) in at least 15 buildings. Environmental audits indicated widespread site contamination from the USTs. John Deere Plow Works donated the property to the City of Moline in 1988 to redevelop the property into a new civic center. After joining the Pre-Notice Program, approximately 35,000 gal of contaminated groundwater were delivered to the City of Moline Water Pollution Control Plant. The site sits on top of 15–25 ft of demolition fill that is a source of low-level polynuclear aromatic hydrocarbons (PNA) and volatile organic compounds (VOC). In 1991 under the Pre-Notice Program, IEPA agreed with the plans to redevelop the site, but required monitoring of the groundwater. IEPA's oversight costs totaled $26,877, of which $4,341 paid for personnel, $10,304 for an IEPA contractor, and $12,079 paid for analytical costs. The City of Moline transferred ownership of the cleaned up property to the Quad City Civic Center Authority. This $40 million public works project contributes several million dollars annually to the Quad City areas each year, and the new civic center also provides some 1200 parking spaces to the riverfront district.

Indiana Success Story — C&M Plating was a metal plating facility in Roanoke that was closed in the late 1980s. EPA removed the plating chemicals and other hazardous substances that remained at the site. The buildings, which included the main plating building and office, a wastewater treatment building, and an abandoned gas station that was used for storage, remained with the site along with much of the plating equipment. The structures were deteriorating, the walls and floors appeared to be contaminated, and indications of unauthorized entry by children were evident. The local community wanted the potentially hazardous site removed from their neighborhood and local business people were interested in developing this downtown property. A cooperative effort to remediate the site as a brownfield project was agreed to by EPA, the state, and the local governments of Huntington County and City of Roanoke. A four-phase approach was adopted. EPA decontaminated the structures on the site. The local community using volunteer labor and contractors demolished and disposed of the buildings. IDEM removed the underground storage tanks, and EPA removed the concrete slabs beneath the buildings and excavated contaminated soil.

Minnesota Success Story — The Minneapolis Community Development Agency (MCDA) received the state's first Certificate of Completion under the Minnesota Land Recycling Act on May 18, 1994, for the cleanup of the Sawmill Run Property. This downtown Minneapolis property was formerly the site of a coal gasification plant and other industrial operations. The MCDA entered the VIC program in September 1991. Response actions were completed in 1 year, at which time more than 23,000 tons of contaminated soil, and 40 barrels of paint and solvents were removed, treated, and/or disposed of. The cleanup action also included the recovery of petroleum floating on the watertable and groundwater monitoring to assure that contamination levels decrease over time and will not affect the Mississippi River. The $1.3 million cleanup paves the way for a $12.1 million upscale redevelopment project that includes riverfront townhomes.

Ohio Success Story — Kessler Products manufactured extruded PVC polymer products from 1964 until 1994 on 15 acres in Boardman, Ohio. Kessler Products had previously tried to sell the facility as an operating concern; however, because of soil contamination at the site, no one would buy it. The manufacturing equipment was sold and moved to another factory. The facility remained empty. The company tried to sell the property, but still no one would buy it because of the potential liability associated with the soil contamination. As part of EPA's Brownfields Initiative to return urban land to productive use, EPA and the State of Ohio worked to evaluate the level of contamination. The state issued a covenant not-to-sue under the Voluntary Action Program to Kessler Products. With these liability uncertainties resolved and the Kessler Product's commitment to bring the site into compliance with the clean up goals of the Interim Voluntary Action, Kessler Products was able to negotiate a sale of the property to Modern Builders. Modern Builders Supply, a wholesale distributor of building supplies since 1944, already manufactures vinyl doors and windows at a plant in Boardman near the Kessler site. They will employ

47 new workers at their new site. Ronald N. Kessler, President of Kessler Products, said, "the program was a good solution to a problem we had been facing for a while. We had been looking for a speedy, cost effective way in which to clean up the property, but none of the existing programs provided us with one. The Voluntary Action Program allowed us to establish definite clean-up levels and gave us a process to follow. Once we knew what we had to do, the voluntary action took less than two months. Now that this clean-up is complete, I am going to be turning my attention to developing the adjacent property."

Wisconsin Success Story — The former Balko Plant and Jump River Sites are located on 8.8 acres along Highway 27 at the entrance of the City of Ladysmith. Balko, Inc. manufactured boat and utility trailers since the 1950s. K. D. Balko (a different company) bought the buildings in 1992, manufacturing the same products. When K. D. Balko declared bankruptcy in 1994, 18 employees lost their jobs and the building sat vacant and tax delinquent. In mid-1995, the Scott Equipment Company, a farm equipment manufacturer, leased the 18,000 ft^2 plant with an option to purchase and hired 14 new employees. Scott Equipment is still in the building. The affiliated Jump River property is vacant land with no prior development. In 1992, the Wisconsin Department of Natural Resources found that Balko, Inc. had been dumping paint solvents and other chemicals onto the ground near a loading dock. DNR began and continues to take enforcement actions against Balko, Inc. It is unknown how much contamination remains on the site. Financing for the cleanup will be provided by the Rusk County Economic Development Revolving Loan Fund. If additional funds are needed, the City of Ladysmith also can provide a revolving loan fund. When the site is cleaned up, Scott Equipment plans to add 7500 ft^2 and then 20,000 ft^2 to the plant, and hire up to 20 additional employees. The investment in these two sites is expected to add $1.3 million in county revenues. The extension of city water and sewer will create development potential along the entire northwestern edge of the City of Ladysmith. Nearly $1 million will be invested in Scott's new plant and infrastructure, which will be used for future marketing efforts undertaken by the Ladysmith area. Getting these properties back to productive use is a pivotal part of a long-range, comprehensive development effort.

Michigan Success Story — The City of Detroit received an $850,000 site remediation grant to conduct a remedial investigation and feasibility study at the former Detroit Rockwell property located on Fort and Clark Streets. The investigations indicated subsurface soil contamination by PNAs, benzene, toluene, ethlybenzene, and xylene (BTEX), and other organic contaminants; possible off-site migration of contaminants; and the presence of five underground storage tanks, all with confirmed releases, which were removed. The investigation has been completed, with a final remedial action plan under development by the City's environmental consultant. Additionally, sampling will be conducted when excavation for the development's truck bays begin. Motor City Intermodal Distribution, the proposed developer, plans to build a 250,000 ft^2 warehouse that will include 10,000 ft^2 of office space. The new center will combine marine, truck, and rail modes of transportation with warehousing facilities at one central location near Detroit's riverfront. The operation has been approved as a foreign trade zone. The development will expand the developer's current business. The proposed expansion will also result in the direct addition of 108 new jobs, including managers, professionals, operatives, skilled craftspeople, clerical staff, and laborers. The new developer has received protection from liability under an Administrative Order by Consent negotiated between the state and the City of Detroit.

Brownfields and Strategic Planning

The EPA Brownfields Initiative is currently in the development and testing phase. Pilot projects are being used to demonstrate the Brownfields Initiative and redevelopment possibilities, and demonstrate the success of the program in communities. Brownfield success stories are becoming more common, such as those provided in this section, which suggests that a Brownfields strategy can help facilitate an accelerated response action and stimulate economic growth. The EPA Brownfields program is geared toward demonstration and offers a limited number of grants for pilot projects, which may not integrate well into most environmental strategic plans easily, or not at all, if not pilot grants are available. However, in light of the federal government's Brownfields program, and the opportunity to clean up idle or abandoned properties, private sector businesses can integrate the concepts of Brownfields into environmental strategic plans if there is an *economic* driver for revitalizing or selling real estate.

One example is Brownfields Remediation, LLC of Cary, NC. In addition to helping property owners reduce their liability through site restoration under, for example, voluntary cleanup actions, Brownfields Remediation, LLC also developed a Brownfields decision-making process that emulates the spirit of Brownfields by purchasing contaminated properties and cleaning them up for resale (Figure 8-3). Their approach uses a relatively simple model of purchasing properties where they are both the environmental contractor and the property owner. Using a Brownfields redevelopment strategy, it is important that they own the property in order to limit their overall liability. Their strategy focuses on the transition between cleaning up properties, if necessary, and selling the property for redevelopment purposes. After the original owner conducts an initial investigation and results support a redevelopment alternative, they can limit their liability following a Brownfield site. The final owner, after the site is cleared or cleaned up, has a useable property for development purposes. The interim property owner limits its liability through agreements, SMOA, etc., with the state or EPA in order to pay for or finance the cleanup. Most importantly, the strategy focuses on identification of an economic driver that creates an incentive for remediating contaminated sites. Cost-wise, this can be illustrated as

[Estimated Final Property Value] – [Cleanup Costs] – [Debt] = A Reasonable Profit

Cost-saving measures, such as property tax abatement based on environmental problems, can further enhance the applicability of this type of strategy in order to create an incentive through potential increased profitability or financing. Idle or

STRATEGIC PLANNING AND LOWERING ENVIRONMENTAL COSTS 315

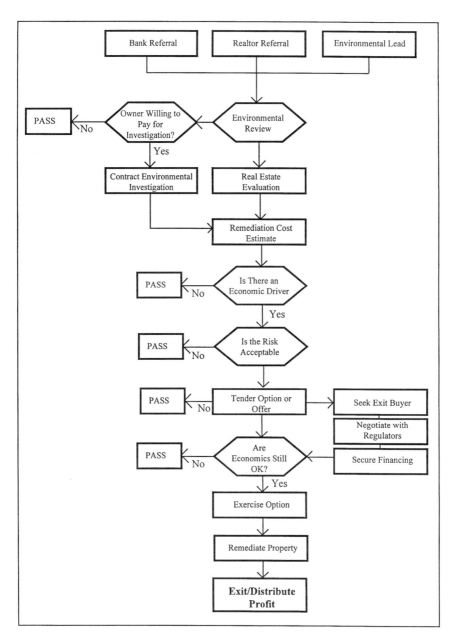

Figure 8-3 A Brownfields strategy developed specifically for private sector organizations to clean up idle property and redevelop it for economic growth. (Courtesy of Brownfields Remediation, LLC, Cary, NC.)

abandoned sites that follow a Brownfields model of this type are generally not highly contaminated, or the types of contaminants encountered can be dealt with relatively easily, such as petroleum. The majority of idle or abandoned sites, as referenced earlier in this section, fall into this category. An example would be issues associated with petroleum contamination at gas stations. This model can succeed in cases where significant cleanup costs are not necessary, or in cases where significant cleanup costs do not supercede the final "clean" property value. In some situations, the Brownfields strategy may be the best avenue because of funding limitations. In turn, businesses and communities can realize direct economic gains by redeveloping Brownfields properties, risk to human health and the environment can be lowered, and through limited liability, opportunities become available for property owners to finance, cleanup, and sell properties.

8.2 LOWERING ENVIRONMENTAL COSTS

"Cost is a central factor in all Superfund remedy selection decisions" (EPA 1996). In addition, cost is a central factor for all environmental programs — and keeping costs reasonable, while still being protective, is important. Guidance, such as *The Role of Cost in the Superfund Remedy Selection Process* (EPA 1996) is an example document that describes the role of cost in screening remedial alternatives, performing detailed analyses and remedy selections, and waiving statutory requirements because of cost. Understanding the role of cost in environmental decision making is important in order to reduce environmental costs and propose/select protective alternatives that meet regulatory requirements as practicable. In this chapter, approaches for lowering costs are described from a qualitative perspective. A detailed or quantitative financial analysis of costs, approaches, and alternatives is possible, but is beyond the scope of this book. Site-specific conditions, coupled with the project team's ability to work together and implement cost-reducing measures, affect the overall costs of environmental projects. For this reason, a qualitative approach is presented, which in turn can be used to evaluate site-specific conditions and needs, and project teams can decide whether it is necessary to quantitatively compare costs for decision making.

In general, there are no short cuts, simplistic comparisons, or magical formulas that forgo uncertainty in comparing costs, methods, and approaches. As with engineering cost opinions, a substantial effort may be needed in order to determine the most cost-effective alternative or approach. Furthermore, when integrating the cost implications of a partnership, for example, or conducting a cost–benefit analysis of "high and low trust factors" within project teams, the task becomes much more arduous than comparing remediation alternatives. In this example, the only way to quantitatively evaluate the actual cost savings may be to compare two very similar projects *after they have been completed,* assuming that two very different approaches were employed regarding teamwork and trust. Following this approach, it may be possible to estimate a savings for future projects. Even so, other uncertainty can cloud cost comparisons. Therefore, each project, and each cost-saving approach, must be considered individually in order to minimize the overall cost of projects. The following are generic examples, methods, approaches, and techniques that can be used to lower costs of environmental projects.

Technology-Based Approaches

Technology-based cost-reduction approaches are the most common mechanisms applied to reduce costs. Examples include investigation tools, such as field testing kits, on-site analytical laboratories, real-time data collection techniques/tools, geophysics, GPS, and many other investigation related innovative technologies (see Appendix B for more examples). In general, innovative investigation tools can be used to more quickly investigate sites, reducing the overall project cost by limiting the use of traditional investigation methods and number of hours needed to complete efforts. In addition, investigation tools can cost less to operate and provide similar data needed to characterize sites and make decisions.

Interpretive tools that reduce the cost of projects include computer applications and modeling techniques, databases, GIS, etc. These types of applications can streamline data storage and interpretation of complex settings, aid in the evaluation and presentation of remedial alternatives, or reduce the amount of time necessary to organize and interpolate site data. Remedial processes and technologies are commonly used to reduce costs by reducing the amount of time required to clean up sites or by lowering operation costs. Examples of existing and innovative remedial technologies are listed in Appendix C.

There are many technological approaches available that reduce costs and a number of different approaches that can be used together that have a significant impact on lowering costs. The most difficult task concerning technology-based approaches is wading through the various vendors and applications that are available in order to make the best decision for who and what to use. Keeping abreast of the latest technological advances is therefore important in order to maximize cost savings.

Acceleration and Streamlining Approaches

Quite often, when accelerating and streamlining projects, the up-front costs of investigation and response actions are greater than those costs associated with following conventional approaches (Figure 8-4). However, the overall cost of projects can be significantly lowered using an accelerated or streamlined approach. For example, in Chapter 4, several examples of DOE success stories are presented that include the actual cost savings resulting from accelerating and streamlining projects. Examples of acceleration and streamlining approaches include early actions under EPA's SACM approach, Brownfields Initiative, presumptive remedies, integrated assessments, or the RCRA site stabilization initiative (see others listed in Chapter 2). Other federal and state agencies have also developed acceleration and streamlining approaches, such as DoD's BRAC program and DOE's Streamlined Approach for Environmental Restoration (SAFER) program (see Chapters 3, 4, and 5 for more examples). In addition, private sector entities have developed approaches, such as the observational approach, sequential risk mitigation, and the Risk-Based Cleanup Approach (RBCA), for example (see Chapter 6 for more examples). Most importantly, there are many approaches that can be considered or implemented in order to accelerate and streamline actions. In turn, the overall cost of projects can be

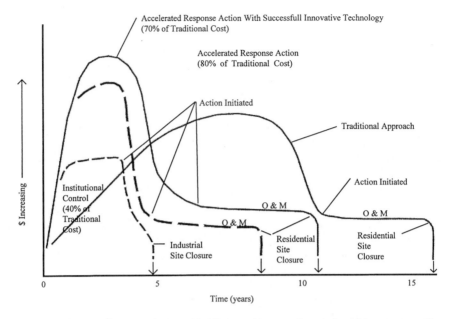

Figure 8-4 A qualitative comparison of funding requirements for accelerated response actions and traditional approaches. Accelerated response actions generally require more up-front funding with less total funding needed than traditional approaches, which generally have a higher total cost. Innovative technologies, approaches, and institutional controls can further reduce the overall cost of accelerated response actions and fast-track site closure.

reduced as long as the project team is able to implement the environmental program efficiently and effectively, and implement the approach using sound technical and practical judgment, which some environmental professionals have coined as "environmental common sense."

Contracting Strategies

Contracting strategies should be used to ensure that the technologies, acceleration, and streamlining approaches are efficiently conducted. As outlined in Chapter 3, the activities contracted should be solicited by using the most cost-effective contracting vehicle. For example, in cases where there is substantial uncertainty or an observational method is being applied, a flexible contracting strategy is beneficial for minimizing costs. A fixed-price contract, in this situation, would likely make it difficult for the contractor to estimate cost, which usually increases the overall cost of projects due to uncertainty which they include in the cost estimate. In addition, it would be burdensome and costly for the contract administrator to continually address change orders in this situation. In situations where the level of certainty is relatively high, however, a fixed-price contract will likely lead to the most cost-effective action. Cost reduction should therefore integrate consideration of the type of contracting mechanism or vehicle. By doing so, the most beneficial relationship between the client and contractor can be established in order to realize the most reasonable cost of projects.

Cost Practicality

Cost practicality is an essential element of decision making and implementation of response actions at environmental sites. The practicality of alternatives should be evaluated in order to segregate the reasonable alternatives from the unreasonable alternatives in relationship to primary projects elements, such as the investigation and remediation efforts. The extremely high costs of restoring sites to pristine conditions must be weighed against alternatives that will provide the same or perhaps even a lesser level of protection using significantly less capital. Groundwater contaminated with dense free phase solvents is perhaps the most common situation where costs can be extremely high in order to remove contamination. In this case, the technical impracticability would also likely apply when evaluating cleanup alternatives along with cost practicality (see Chapter 2).

Under residential land use scenarios, some or even all of the cleanup alternatives may be extremely cost-prohibitive, depending on the site conditions and the type and level of contamination encountered. In addition, the potential exposure routes for groundwater or soil contamination must be evaluated in order for cost-reduction opportunities to emerge, such as institutional controls. Rather than treating soil or groundwater down to "technically expensive cleanup levels," it may be feasible to use institutional controls in order to ensure pathways and exposures to contaminants are blocked, so that human health and the environment are protected. For example, city water service could be provided in order to eliminate drinking water hazards that stem from citizens using domestic water supply wells. This approach may save substantial amounts of money, while still providing the same level of protection as long as ecological concerns are addressed and long-term plume migration is not an issue. In turn, a more cost-effective approach can be implemented in order to address the groundwater contamination, such as intrinsic remediation or plume stabilization, etc., which may not result in restoring the site to pristine conditions. Thus, in some cases, reducing costs may include evaluation of alternatives that *do not* provide the same level of cleanup, but *do* provide an adequate level of protection.

8.2.1 Reducing Remediation Costs

Cutting Remediation Costs But Not Corners is a paper written by Dennis Trip of Applied Science & Technology, Inc. of Ann Arbor, MI. His paper was published in the *Michigan Lawyers Weekly* (Trip 1995) and is presented in its entirety with permission from the publisher. Trip's paper outlines how costs can be lowered for projects requiring cleanup.

> Remediation procedures and techniques are becoming increasingly more specialized and more costly for business and property owners. However, many of today's remediation costs can be minimized, without jeopardizing effectiveness, by gaining a better understanding of remediation procedures and the various options available at the different stages in the process. The following guidelines provide a concise yet comprehensive overview of the leading factors that can impact the cost of remediation for soil or groundwater contamination at each phase of a "typical" program. More

importantly, the treatment of each cost factor also includes a review of frequently available alternatives that can dramatically reduce the ultimate cost of a remediation program without detracting from its effectiveness.

Identifying the Problem

The first step in any remediation process is the proper analysis and identification of the contamination problem, both its exact nature and extent. The more thorough the initial analysis the less likely that costly surprises will surface at a later stage. The techniques employed to make these determinations will depend on the magnitude of the problem. For example, on large sites requiring extensive sampling it is often more cost effective to conduct an on-site analysis using a mobile laboratory, especially if plant operations must be shut down during this sampling process. Although bringing a mobile lab on-site is not inexpensive, in certain situations (such as those requiring shutting down operations) it can be very cost effective by reducing the amount of down time. The alternative is to ship samples to an off-site lab and wait for results to come back, which can add days or weeks to the process depending on the size of the site and the number of samples required.

A more focused initial site investigation by the environmental consulting firm will also help to keep costs to a minimum. If the consultant concentrates the site investigation on only the issue or issues that compelled the property owner to seek assistance in the first place, then there is no need to undertake broad screening and very costly investigative procedures, such as a priority pollutant analysis. If an analysis is capable of identifying traces of more than 100 different compounds, it is quite likely that some other contaminants will be found. While this generates additional work for the consulting firm, it only creates unanticipated and often unnecessary added costs for the owner.

Determining the Right Remediation Method

Once the magnitude of the problem has been identified, the next step is the development of a remediation plan which includes a feasibility analysis. This analysis is intended to determine the most technically feasible and cost effective method for remediating a particular site and should be carefully reviewed by the property owner since it often presents several available remediation options. A well prepared feasibility analysis should include a discussion of the proposed remediation's impact on the business beyond just the capital costs. It should weigh such factors as the trade-off between time and cost and an analysis of the sensitivity of cost estimates. For example, should a business delay or draw out the remediation process and costs over several years [or accelerating the process], if possible? [Or,] can the process be financed from corporate capital or will it require bank financing?

A well designed feasibility analysis will typically take into consideration such factors as:

Type of soils. Certain soils respond better to particular remediation techniques. For example, if the soils on a contaminated site are primarily sandy, then certain remediation methods such as vacuum extraction or bioremediation might be appropriate depending on the type of contaminant. On the other hand, if the soil has a high clay content, then excavation and disposal, thermal desorption or incineration may be better suited to the task.

Contaminant location. The location of the contamination relative to surface features, such as buildings, utilities, etc., may limit the available remedial options. For example, soil contamination located under a building may make an *in situ* method much more cost effective than excavation and *ex-situ* treatment. If the contamination is located at or below the water table, a combination of remediation techniques (vacuum extraction, air sparging and pump-and-treat) may be required to accomplish the remediation. Or, on the other hand, the nature of the contaminant (e.g., PCBs) may be such that de-watering and excavation and disposal is the only feasible option.

Nature of the contamination. Some contaminants such as hydrocarbons, including gasoline and diesel oil, generally can be dealt with easily and inexpensively. If necessary, they can even be dug up and moved to a Type 2 landfill. Other contaminants, such as chlorinated hydrocarbons, however, cannot be hauled to a landfill without costly pre-treatment. Similarly, some remediation methods are better suited to certain types of contaminants. For example, bioremediation has proven to be very effective with petroleum gasoline hydrocarbons such as diesel fuel and heating oil products, but less effective with chlorinated solvents and PCBs.

Amount of contaminated material. Some remediation techniques are more cost effective for larger volumes of contaminants, such as bioremediation or low temperature thermal desorption. With bioremediation, contaminants are consumed or detoxified by micro-organisms, which are either naturally occurring or genetically engineered. Since this method typically is employed on the site, it eliminates the cost and liability of transferring contaminants from one location to another. Low temperature thermal desorption is another on-site remediation technique, but is best suited for petroleum hydrocarbons. Utilizing a unit resembling a modified asphalt plant, the contaminated soil is heated and volatile compounds are driven off and captured in the vapor stream where they are destroyed. Since the decontaminated soil is returned to its original site, this process also eliminates the liability issue and the need to bring in clean fill dirt. Other methods, such as excavation and disposal are generally more cost effective for smaller scale projects than for sites with large volumes of contaminated soils.

Time required to remediate the site. If the contaminated site is also the location of an operating industrial plant or other business which may be disrupted by the clean-up process, then the time required to accomplish remediation becomes a critical element in the decision as to which technique to adopt, assuming that several alternatives are appropriate. Often the more expensive remediation option can be the most cost effective, if it requires the least down-time. Conversely, the least costly remediation method could prove to be the most expensive technique, if it necessitates shutting down or disrupting operations for extended periods. Although the actual time required for various remediation techniques is highly site and contaminant dependent, some of the more commonly used procedures can be ranked according to the following typical time requirements (going from least time to most time): excavation and disposal — days to several weeks; vacuum extraction — months to years; low temperature thermal desorption — days to weeks; and bioremediation, which may require years at a minimum. In a situation where only one remediation technique is feasible, business owners can sometimes still reduce costs if they can take steps to modify their operations to facilitate the remediation, which in turn could minimize the time and expense of a clean-up procedure.

Time of year. In Michigan, for example, the time of year can also impact the cost of remediation because of the state's Frost Laws. These regulations place weight restrictions on the size of a load that a truck can haul during the Spring thaw. During a six to eight week period usually in March and April, trucks are limited to hauling only partial loads which can add to the time and expense of any remediation procedure requiring the removal and transfer of site materials. Savings achieved by delaying the transfer process a few weeks until the Frost Laws have expired can often compensate for the added inconvenience.

Use of remediation specialists. For more specialized remediation techniques such as bioremediation, vacuum extraction or soil vitrification it can be more cost effective to retain a firm that specializes in a particular procedure. By bringing in a specialized remediation provider with extensive knowledge and experience rather than taking time to become familiar with new equipment and procedures, the lead consultant on the project can save the client time, money and grief. After all, it is always less expensive to remediate a site correctly the first time than to go back and correct oversights.

Conclusion

A good environmental consulting firm will include the foregoing factors as well as others in their feasibility analysis and provide the property owner with various remediation options, if applicable, and recommend the approach that is most cost effective. But the final decision ultimately must be made by the owner. If he or she is aware of the remediation options available and how they may affect day-to-day operations, the less costly that decision will be.

8.2.2 Remedial Performance Evaluations

Remedial performance evaluations are necessary in order to monitor cleanup progress, and if necessary, adjust the remedial system in order to maintain or improve performance. In addition, performance evaluations can also be employed early in projects in order to track investigation and decision-making progress (see Chapter 3 for examples). This section, however, focuses on the remedial operation performance evaluations that can be used to

- Evaluate whether the remediation will meet its proposed timeliness
- Evaluate whether the remediation will stay within budget (EPA 1992).

In terms of cost reducing approaches, maintaining a relatively high level of performance during remediation helps maintain and reduce cleanup costs. Tracking and evaluating performance is therefore important. In EPA's *General Methods for Remedial Operation Performance Evaluation* (1992), generic principles for formulating site-specific performance evaluations and strategies related to groundwater contamination remediation are described. This EPA document is a valuable resource that can be used to help develop groundwater remediation performance strategies through the application of rigorous methods that are outlined by EPA. By completing performance evaluations, it is possible to reconsider original assumptions used to develop the remediation system, and if necessary, adjust the remedial system in order to increase performance, meet remediation goals, and ensure project budget limitations are met. EPA's recommendations focus on groundwater remediation, and this section provides a brief overview of their work. The fundamental concepts presented for groundwater remediation may be adapted for other remedial actions, such as soil remediation. However, the specific *methods* and *protocols* presented below may *not* apply to other media beyond groundwater. EPA's methods include computational, statistical, graphical, and theoretical methods. These are described below and are further described in EPA's document (1992).

- **Computation methods**. Computational indicators of performance range from simple arithmetic to highly sophisticated transport models incorporating chemical, physicochemical, and biological reaction rates. The use of a particular method should be justified by the conditions to which it is being applied. Such justification should be supported by the appropriate quantity and quality of requisite data. Computer solute fate and transport modeling is one example (see also Chapter 6 on computer modeling).
- **Statistical methods**. Most site documents lack statistical evaluations, and many site reports present inappropriate simplifications of data sets, such as grouping or averaging broad categories of data without regard to the statistical validity of these simplifications. There are important uses of statistics in the evaluation of remedial performance and these include analysis of variance (ANOVA) techniques. ANOVA techniques may be used to segregate errors due to chemical analyses, sampling procedures, and the intrinsic variability of the contaminant concentrations at each sampling point. These include *correlation coefficients*, which may be used to provide justification for lumping various chemicals together (such as total VOCs), or for using a single chemical as a class representative, or by linking sources of similar chemical behavior. *Regression equations*, which may be used to predict contaminant loads based on the historical records and supplemental data, and may be used to test cause-and-effect hypotheses about sources and contaminant release rates. *Surface trend analysis techniques*, which may be used to identify recurring and intermittent (i.e., seasonal) trends in contour maps, or groundwater elevations and contaminant distributions, which may be extrapolated to source locations or future plume trajectories.
- **Graphical methods**. Graphical methods of data presentation and analysis have been used heavily in groundwater flow and geochemical evaluations. Plots and contour maps offer much information related to site conditions, such as velocity computations, distance relationships, and concentration data that are needed to estimate the overall remedial performance.
- **Theoretical**. General indicators of physicochemical reaction potential include the cation-exchange-capacity (CEC), alkalinity, and acidity of subsurface samples. Theoretical relationships link CEC, alkalinity, and acidity to the reserve of reactants, and consequently to the potential products. Theoretical equations have also been developed and tested for various wastewater treatment unit designs. In addition, batch sorption isotherms and static microcosms can be used to generate

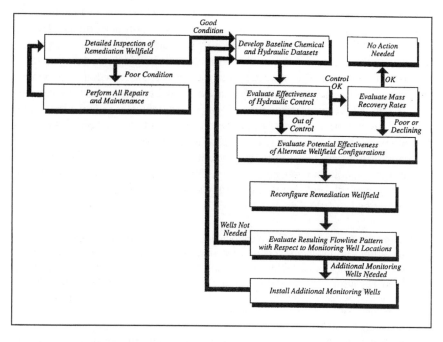

Figure 8-5 Outline of key efforts needed to select and use monitoring and reporting requirements for evaluating remedial performance and optimizing groundwater extraction systems. (Source: EPA. 1992. General Methods for Remedial Operation Performance Evaluation. EPA/600/R-92/002.)

estimates of limiting reaction rates of specific contaminates. It is most common to evaluate data from such experiments by comparison to an equation that has been derived from theory, as opposed to the use of empirical relationships.

General protocols for performance evaluations of groundwater remediation summarized by EPA (1992) include (1) monitoring and reporting criteria, (2) violation reporting and response criteria, and (3) design and operation modification criteria. For the purposes of this section, Figures 8-5, 8-6, and 8-7 illustrate the "skeletons" for which site-specific charts can be developed in order to identify key activities for comprehensive remedial performance evaluations. Overall, the methods and protocols developed by the EPA are useful for monitoring and evaluating remediation performance, specifically for groundwater contaminant remediation. In turn, cost-reducing strategies and methods can be developed and employed at sites undergoing restoration using the evaluation data. In other words, performance data is an important method needed to access where cost-reducing measures can be implemented. More detailed information is presented in the EPA guidance.

8.2.3 Property Tax Reduction

Perhaps one of the most applicable, yet unused, avenues for offsetting environmental investigation and cleanup costs is property tax abatement based on environmental

STRATEGIC PLANNING AND LOWERING ENVIRONMENTAL COSTS 325

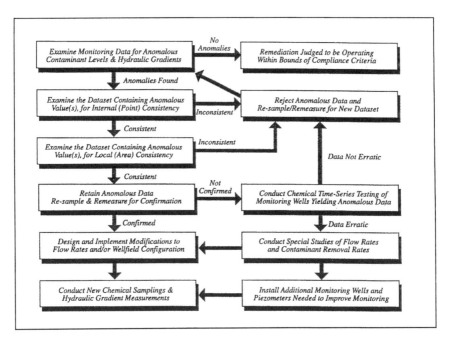

Figure 8-6 Key efforts needed to identify, verify, and correct inadequate performance of the remediation. (Source: EPA. 1992. General Methods for Remedial Operation Performance Evaluation. EPA/600/R-92/002.)

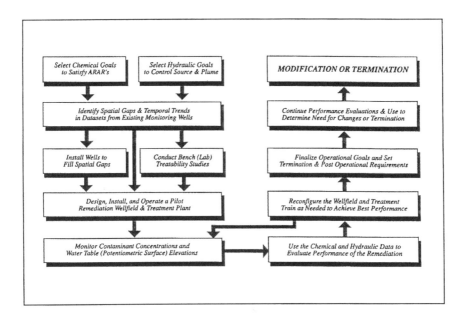

Figure 8-7 Key efforts needed to prepare for, implement, and verify modifications of the remediation. (Source: EPA. 1992. General Methods for Remedial Operation Performance Evaluation. EPA/600/R-92/002.)

factors. Fair market value is generally used to determine the property tax assigned to commercial properties. Fair market value should include assessment of environmental problems associated with industry-wide issues, such as asbestos or toxic waste, because these types of problems can substantially reduce property values. In general, the presence of hazardous substances and the associated environmental *stigma* impacts the salability and cash flow of real estate. As outlined in Randy Airst's 1994 paper, literally millions of properties may be subject to internal or external environmental problems that should be considered when assessing the fair market value of properties. In theory, the tax reduction realized because of environmental problems can be reinvested into the property site for restoration activities, or raising profitability needed to help pay for restoration costs. This approach can help subsidize environmental work needed to be completed at sites that otherwise would not be conducted because of limited cash flow and/or inadequate insurance coverage. In addition, property tax abatement enables property owners to address contamination issues that may hinder property sale, transfer, or financing.

Internal factors that affect property values include (1) presence of hazardous contaminants/USTs on site, (2) pending regulatory actions against the property, (3) tort law suits based on toxic environmental factors, and (4) violation of an environmental statute(s) that cancels a permit/license vital to the operation and economic viability of the business (Airst 1994). Depending on real estate comparables, income adjustments, and/or replacement values, property tax appeals can be argued and litigated successfully, especially in cases where there is a bona fide comparable that can be used to demonstrate a reduced fair market value. Most tax assessors do not have the information or the technical background required in order to fairly assess property with environmental problems. Furthermore, most county and local tax assessing agencies do not have the resources needed to confront, and technically defend, their evaluation of environmental problems assigned to properties, if any.

Property tax appeals generally are possible through informal, formal, and litigation procedures. While the informal process may not always grant tax appeals, at the litigation level, courts have repeatedly supported arguments for substantially reducing property tax based on the presence of environmental problems. Examples of litigation are outlined in Airst's paper. The avenue best suited for appealing property tax depends upon the county and/or local tax assessing system and the appeals process. A successful appeal will depend greatly on the ability of the protestor to provide a complete and comprehensive assessment of comparable properties, income adjustments/cash flow analysis, and/or replacement values. Moving through the informal, formal, and litigation phases of the appeal process, and offering a sound argument related to property tax reduction because of an environmental problem, a substantial tax savings for large industrial and commercial sites is possible. Most importantly, tax abatement is not based on the cost of investigation or cleanup. Rather, tax abatement is based on the environmental stigma attached to a property which has resulted from an external or internal environmental problem.

8.2.4 Increasing Efficiency by Understanding What Motivates People

Individual and team efficiency plays an important role in how much time it takes to complete tasks and projects. In most cases the more time it takes, the more it costs to complete the effort, which in turn can have the effect of increasing the overall cost of projects. Those staff and professionals that are highly efficient can usually expedite projects and more easily meet deadlines. Unfortunately, for most large project teams, it is seldom common that all team members are high achievers. Generally, only a smaller percentage of team players are gifted with the innate ability to produce quality results and deliverables in relatively short periods of time. These people are analogous to the people who run 4-minute miles. There are simply not that many of them.

This is not to suggest that there are few hard workers in the environmental industry! While there may be a few individuals who cannot pull their fair share of the work load, most staff and professionals give their best when it comes to hard work. The question becomes, why are some people more gifted, or able to produce results, compared to others? Some highly efficient people are extraordinary people; however, there are many people who efficiently produce results because they enjoy what they do. In other words, they are motivated to complete work efficiently because they *enjoy* their job, they are good at it, and they are motivated to do more of the same work.

For those that work for a large agency or company, all that is generally necessary to conclude that most people are not motivated is to ask a few people how their day is going and if they enjoy their job. Generally, there will be some staff who complain about the work they do. The majority of staff will not be overly unhappy with what they do, yet they are not overly excited to get back to their "post." The staff who truly enjoy their jobs do not have the time to tell you about it, because they are too busy helping the office or agency keep up with schedules! Overall, there are a lot of people who want to enjoy their work more. People are motivated by challenge, and specifically, challenges that allow them to do the work they enjoy. Empowering and encouraging people so they can address their need to be challenged at work fosters a higher level of efficiency. In some cases training may be needed to help expedite the learning process if a new area of expertise must be learned by staff. However, if the individuals enjoy what they are doing, they will likely complete the task(s) more efficiently than if they are doing something they do not enjoy doing. In some cases, no matter what challenge is offered, some individuals may not excel. Perhaps they are in the wrong profession, or should consider other areas of focus. In addition, not all work can be challenging or enjoyable. Striking a balance between these types of tasks is therefore important in order to maintain efficiency. Those that find enjoyment in their work will focus on getting tasks completed more quickly, doing a quality job because they care, and will offer their services in hope of doing more work in the future. Under these types of conditions, high efficiency can reduce costs. Therefore, developing and nurturing a project team who, in general, enjoy their work can help projects stay on schedule and within budget, and potentially accelerate response actions.

8.3 REFERENCES

Airst, R.L. 1994. External Environmental Factors May Justify a Reduction in Real Estate Taxes. Journal of Multistate Taxation, Vol. 4, No. 1, April, pp. 1-4.

Bredehoeft, J. 1996. A New Paradigm for Cleanup? Ground Water, July-August, Vol. 34, No. 4, p. 577.

Hix-Mays, R. 1995. Revised NCP Stress Planned, Coordinated Spill Response Strategy. Environmental Solutions, Vol. 8, No. 3, March, pp. 21-24.

Trip, D.E. 1995. Cutting Remediation Costs But Not Corners. Michigan Lawyers Weekly, No. 20, March 27.

U.S. Department of Defense (DoD). 1995. BRAC Cleanup Plan Guidebook. Revised edition, Fall.

U.S. Environmental Protection Agency (EPA). 1989. Overview of ARARs Focus on ARAR Waivers. OSWER Pub. 9234.2-03/FS.

EPA. 1992. General Methods for Remedial Operation Performance Evaluation. EPA/600/R-92/002.

EPA. 1995. Land Use in the CERCLA Remedy Selection Process. OSWER Dir. 9355.7-04. Memorandum from Elliott P. Laws OSWER to selected EPA personnel dated May 25.

EPA. 1996. The Role of Cost in the Superfund Remedy Selection Process. OSWER Pub. 9200.3-23FS, EPA/540/F-96/018.

Williams, R.S. 1996. Creative Remediation Consulting in Today's More Relaxed Regulatory Environment. Remediation Management, Vol. 2, No. 5, pp. 32-37.

APPENDIX **A**

Environmental Resource Guide

Appendix A contains a list of selected environmental resources available to environmental professionals. Some resources have limited access and a few charge for their use. Most of the resources originate from the U.S. government and include a variety of topics, including general environmental information, access to environmental databases, viewing or ordering environmental documents, information on innovative technologies, current environmental regulations and policy, aerial photography/map resources, climate data, environmental hotlines, networks for information transfer, and links to literally hundreds of other information environmental resources, databases, and services. For example, *Access EPA* can be downloaded or viewed online, and allows the user to access EPA information compiled on numerous environmental resources and databases, many of which are not listed in this appendix and otherwise would be difficult to find.

With the expansion of the Internet, the number of environmental resources grows daily. International environmental resources also are available on the Internet by searching, or by linking with them through U.S. Internet addresses. The organizations and agencies listed in Appendix A were contacted prior to the publication of this book. The author makes no guarantee they will be accessible in the future, or that their addresses or telephone numbers are current. Inclusion of an environmental resource does not in any way represent an endorsement of the resource, organization, or agency. The environmental resources include the divisions shown below with subjects listed in alphabetical order. The majority of the resources included are from the federal government. Only limited state and private sector resources are listed.

A-1 U.S. Government — Environmental Protection Agency
A-2 U.S. Government — Military
A-3 U.S. Government — Environmental Protection Agency Bulletin Boards/Databases
A-4 U.S. Government — Military Bulletin Boards/Databases
A-5 U.S. Government — Department of Energy
A-6 U.S. Government — Department of Energy Bulletin Boards/Databases
A-7 U.S. Government — Other Sources/Bulletin Boards/Databases
A-8 Private and State — Sources/Bulletin Boards/Databases/Internet Addresses

A-1 U.S. GOVERNMENT — Environmental Protection Agency

Source	Number or Address	Comments
Access EPA (comprehensive link to additional environmental resources)	Download: http:\\earth1.epa.gov: 80\access\ View: http:\\epawww.ciesin.org. org\national\epaorg\epaorg4.html	Access EPA is EPA's compilation of environmental resources. It includes links to public information tools, major EPA dockets and duplication fees, clearinghouses and hotlines, national and agency records management programs/policies, major EPA environmental databases (including some none-EPA databases) and how to access, future database development, National Library and information services, directory of state environmental libraries and services, state environmental contacts, EPA scientific models, and information on the National Environmental Supercomputing Center. A portion of this information is provided in this appendix. Access EPA is major source for linking up with many environmental resources.
Air Docket	202-382-7548	Provides public access to regulatory information that supports the Clean Air Act.
Air Pollution Emissions Control Technology Center Hotline	919-541-0800	Component of EPA's Air Toxics Strategy. The hotline provides information to state and local pollution control agencies on sources of emissions of air toxics.
Air RISC Hotline	919-541-0888	Information on health effects, exposure, and risk assessment of toxic air pollutants.
Air Quality Standards Library/Office of Air Quality Planning and Standards	919-541-5618	This office develops national standards for air quality, emission standards for new stationary and mobile sources, and emission standards for hazardous pollutants. Information on specific standards available. Library focuses on air pollution and control technology, including costs, chemical technology, minerals, and statistics.
Asbestos Ombudsman Clearinghouse	703-305-5938 800-368-5888	Asbestos abatement.
Best Available Control Technology and Lowest Achievable Emission Rate Clearinghouse	919-541-0800	Air pollution control technology related to new source review permitting requirements.
CERCLIS Helpline	800-775-5037	Provides information on CERCLA sites.
Clean Air Hotline	800-296-1996	Provides information on air quality questions.
Clean Lakes Clearinghouse	202-382-7111	Clean Lakes Clearinghouse collects, organizes, and distributes information on lake restoration, protection, and management for researchers, EPA personnel, lake managers, and state and local governments. The clearinghouse has a bibliographic database that can search specific topic areas.
Drinking Water Docket	202-475-9598	The docket contains comprehensive information on regulatory phases and will be expanded as new MCGs and MCLs are proposed. Information included Federal Register notices, letters, public hearing transcripts, National Drinking Water Advisory Council materials, public comments, technical documents, etc.

ENVIRONMENTAL RESOURCE GUIDE

Resource	Contact	Description
Emission Factor Clearinghouse	919-541-5285	Air pollution emission factors for criteria and toxic pollutant from stationary sources, area sources, and mobile sources.
Environmental Justice Hotline	202-260-6357 800-962-6215	Environmental protection of people of color and low income communities.
Environmental Research Laboratory Narragansett (ERLN)	401-782-3025	EPA's center for marine, coastal, and estuarine water quality research. General information cover aquatic toxicology, biological oceanography, biomedical science, coastal research, fisheries biology, marine biology, marine ecology, marine organisms. Also, water use designation and quality criteria for estuarine and marine water and sediment, and environmental assessment of ocean discharge.
EPA Acid Rain Hotline	202-233-9620	Air emission monitoring, permitting, allowance tracking, and conservation information.
EPA/Ada Laboratory	405-436-8651	Publications from the EPA/Ada Laboratory.
EPA Headquarters Library	202-260-5921	Provides comprehensive information services including environmental policy hazardous waste, air, water, environmental law, solid waste, toxic substances, and test methods.
EPA Home Page on the World Wide Web	http://www.epa.gov	EPA's home page is a vast resource of environmental information periodically updated with new material. Search capabilities and EPA office links available. Also has the Government Information Locator Service (GILS).
EPA Listserver	listserver@unixmail.rtpnc.epa.gov	EPA provides a mailing list that distributes information to you by electronic mail, such as the daily table of contents of the Federal Register. To join, send an email message to "SUBSCRIBE TO LISTSERVERS" at their email address (on left). Leave the *subject* field blank and enter this *message*: "Subscribe EPA-Waste [your 1st name] [your 2nd name]". After you subscribe, you can request a list of all EPA listservers by sending an email to "LISTS" with "LISTS" in the body of the message.
EPA Public Information Center	202-260-7751	Links the public to EPA resources.
EPA Technology Innovation Office (EPA-TIO)	703-308-8827 703-308-8800	Comprehensive program disseminates information on innovative technologies. Helps identify incentives to use of innovative technologies and move new technologies from laboratories and vendors into waste cleanup programs. Innovative Technology for Site Characterization and Remediation Recent Releases is a publication sent out periodically to update environmental professionals on new information.

A-1 (continued) U.S. GOVERNMENT — Environmental Protection Agency

Source	Number or Address	Comments
Fax-On-Demand Service	OSW documents: 202-651-2060 CEEPO documents: 202-651-2061 OERR documents: 202-651-2062	Use handset from fax machine and dial the fax number (on left) for the type of document you want. Follow the voice prompt for ordering options. You may order up to three documents per call. After you are finished, confirm your selection by pressing "1" or "2" to re-enter the document number(s). Press "#" to end document entry and press "start" on your fax machine to begin transmission of documents to your fax machine. Return handset to its cradle. The requested documents should be automatically downloaded to your fax machine.
Federal Facilities Docket Hotline	800-548-1016	Provides the name, address, NPL status, agency, and region for Federal Facilities on the docket. Facilities include those reported as RCRA TSDF or those with CERCLA release or potential.
Federal Remediation Technologies Roundtable	703-308-8827 703-308-8800 (See NTIS for publications)	Provides federal information exchange and a forum for joint activity for development and demonstration of innovative technologies for hazardous waste remediation. Cost and performance of innovative technologies information is available. Representatives from all major federal agencies involved.
Ground-Water and Drinking Water Resource Center	202-260-7786	Distributes groundwater and drinking water publications and maintains a bibliographic database on Office of Groundwater and Drinking Water documents.
Ground-Water Fate and Transport Technology Support Center	405-436-8603	Provides technical support on groundwater remediation technologies and maintains a catalogue of various groundwater remediation technologies.
Hazardous Waste Ombudsman Program	202-260-9361 800-262-7937	RCRA emphasis hazardous waste.
INHALE	216-575-6040	National Healthy Air License Exchange Air Quality and Pollution Credit information.
National Air Toxics Information Clearinghouse	919-541-0850	Air toxics and the development of air toxics control programs.
National Center for Environmental Publications and Information (NCEPI)	513-891-6561	Stores and distributes a limited supply of most EPA publications, videos, posters, and multi-media materials.
National Pesticide Telecommunication Network	800-858-7377	General pesticide information.
National Radon Hotline	800-767-7236	Information on radon health effects and testing.
Non-Point Source Information Exchange	202-260-3665	Non-point source water pollution and other issues related to water.

ENVIRONMENTAL RESOURCE GUIDE 333

Name	Contact	Description
OSWER Directives (Superfund Document Center)	202-260-9760	EPA document center for OSWER/Superfund directives.
OSWER Pipeline Integration	703-603-8802	Information on EPA's regional Accelerated Response Centers and communication links to developing fast-tracked cleanup programs.
OUST Docket	202-260-9720	Provides documents and regulatory information on RCRA Subtitle 1 UST program.
Pollution Prevention Information Clearinghouse	202-260-1023	General information on pollution prevention.
RCRA Docket and Information Center	202-260-9327	Indexes and provides public access to all regulatory materials supporting actions under RCRA and Office of Solid Waste publications
RCRA, Superfund, and OUST Hotline	703-412-9810 800-424-9346 http://www.EPA.gov/epaoswer/hotline	Provides regulatory assistance with RCRA, CERCLA, UST, pollution prevention programs. Hotline training modules available for RCRA and EPCRA (text files available on the World Wide Web).
Records of Decision Hotline	703-603-8881 703-271-5400	Information on final decision documents developed under CERCLA.
Risk Communication Hotline	202-260-5606	EPA service for technical assistance for communicating with the public about environmental risks.
Risk Reduction Engineering Laboratory (RREL) and Site Superfund Videotape Library	201-535-2219	Provide videotapes on a number of EPA produced documentaries and on specific SITE program demonstrations.
Robert S. Kerr Environmental Research Laboratory Library (RSKERL)	405-332-8800 On-line information: http://www.epa.gov/ada/Kerrlab.html	RSKERL is EPA's center of expertise for investigation of subsurface environments. Example areas of research include chemical contamination of groundwater and the mathematical and computer modeling of the movement of groundwater and the influence of various contaminants in environments. Comprehensive databases maintained such as Ground Water On-line and computer modeling information and software.
Safe Drinking Water Hotline	800-426-4791	Information on the Safe Drinking Water Act and amendments.
Solid Waste Information Clearinghouse and Hotline (SWICH)	800-67-SWICH on-line service: 301-585-1404	Comprehensive information on solid waste management.
Stormwater Hotline	703-821-4823	Information on stormwater runoff.
Superfund Docket	202-260-3046	Provides access to Superfund regulatory documents, Federal Register notices, and RODs.
Superfund Health Risk Technical Support Center	513-569-7300	Provides EPA risk assessors, state agencies, and EPA contractors technical assistance with chemical-specific questions, support, and review.

A-1 (continued) U.S. GOVERNMENT — Environmental Protection Agency

Source	Number or Address	Comments
Superfund Innovative Technology Evaluation Program	Demonstrations 513-569-7696 Emerging Tech: 513-569-7665 MMTP 702-798-2373 (see ATTIC and VISITT)	A comprehensive source of new characterization and cleanup technologies. Information on SITE demonstrations, emerging technologies, and the SITE Monitoring and Measurement Technologies Program (MMTP) available.
SW-846 Information Service	703-821-4789	Provides information on SW-846.
Technical Information Exchange (TIX)	908-321-6860	Provides technical environmental-related videocassettes.
Title 40 of the Code of Federal Regulations	http://www.epa.gov/epacfr40/	Title 40 available on-line for free. Not searchable, but can be downloaded by section in portable document format for display and printing found in the CFR.
Toxic Substances Control Act Information Service	202-554-1401 202-554-1404	General information on Toxic Substances Control Act and PCBs.
Volatile Organic Compounds/Reasonable Achievable Control Technology (VOC) Clearinghouse	919-541-5625	This clearinghouse is intended to control VOC emissions by facilitating the exchange of technical data and experience to control emissions.
Water Resource Center	202-260-7787	Groundwater and drinking water documents and videotapes.
Watershed Information Resource System	202-833-8317 800-726-5253	Lake restoration, management, and protection.
Wetlands Protection	703-525-0985 800-832-7828	Value and function of wetlands and options for protection.

ENVIRONMENTAL RESOURCE GUIDE

A-2 U.S. GOVERNMENT — Military

Source	Number or Address	Comments
Air Force Center for Environmental Excellence (AFCEE)	210-536-4038 210-536-1110 210-536-3066	Provides environmental restoration, compliance, panning and pollution prevention, technology transfer, military construction management, and facility design. Also develops innovative environmental technologies and provides Air Force documents not available from NTIS.
Army Environmental Hotline	410-671-1699	Information on environmental policy, references, points of contact, stewardship, laws and regulations for Army personnel.
Defense Technical Information Center (DTIC)	703-274-3848 Internet: http://www.dtic.dla.mil/	Source of technical DoD documents and information.
DoD/National Environmental Technology Demonstration Program (DNETDP) and National Test Site (NTS) Program	Navy: 805-982-1299 Army: 410-671-1560 Air Force: 904-283-6291 EPA: 202-260-2583	DNETDP provides locations for innovative technology demonstrations and comparative evaluations, and transfers innovative remediation technologies from research through fullscale operations. NTS transfers application of innovative water, sediment, and soil cleanup technologies and information related to the fate and transport processes of DoD contaminants and how they relate to cleanup technologies.
Environmental Technology Transfer Committee (ETTC)	410-671-1575 619-553-5475 703-325-0314	Committee provides an information exchange of technology related to cleanup, compliance, conservation, pollution prevention, implementation of technologies, coordination of demonstrations. Multi-agency organization.
Environmental Update	See address	Newsletter containing information on Army environmental programs. To receive the newsletter write: Army Environmental Center SFIM-AEC-PA Bldg. E-4461T Att. Mike Cast Aberden Proving Grounds Maryland, 21010-5401
Explosive Ordnance Disposal Technology Center	301-743-6817	Information and technology on UXO.
Legacy Program	703-325-8525	Legacy is program that funds projects for DoD. Under some conditions, it is possible to obtain funding for demonstration of an innovative cleanup technology, etc., if the project meets certain criteria.

Name	Phone	Description
National Defense Center for Environmental Excellence (NDCEE)	814-269-6820 814-269-6491	Provides information on environmentally acceptable solutions for government and industry manufacturing. Center identifies, evaluates, demonstrates, and transfers manufacturing processes.
Navy Environmental Leadership Program (NELP)	619-545-1125 803-743-6057 904-270-6730	NELP demonstrates and implements new and innovative methods/management/technologies for Navy environmental compliance, restoration, pollution prevention, and natural resource protection. Information is available on NELP initiatives that expedite actions and reduce costs.
Naval Facilities Engineering Service Center	805-982-5751 805-982-2640	Provides technical and program information related to all ongoing Naval environmental projects worldwide. Conducts research, consulting, and field engineering services. Multi-agency interaction.
PRO-ACT	800-233-4356	Air Force's environmental clearinghouse and research service.
Remediation Technologies Screening Matrix/Comprehensive Guide to IR Site Remediation Technologies	Navy: 805-982-3020 Air Force: 904-283-6244 Army: 410-671-1575 EPA-TIO: 703-308-8749	Information source for innovative and conventional technologies used for site remediation and environmental cleanup. Information available to assist decision makers responsible for screening technologies at sites. Large tabular format available for reviewed technologies. An ETTC initiative.
Small Business Innovation Research (SBIR)	800-225-3842 703-697-1481 See also the Defense Technical Information Center	Funds small businesses to do research and development of innovative concepts to solve specific defense-related scientific or engineering problems. Helps commercialize innovative technologies.
Tri-Service Project Reliance	601-634-3723 805-982-1294 904-283-6244	Integrates the science and technology projects of the Research and Development Laboratories of the Army (6.1), Air Force (6.2), and Navy (6.3A) into a single program. Develops new technologies and transfers technology development information to help eliminate duplication and reduce environmental cleanup and technology development costs.
U.S. Army Environmental Center (USACE)	410-671-3348 410-671-4811	Provides a centralized management, oversight, coordination, and execution of Army-wide environmental programs. Also provides technology transfer and information on environmental technologies and demonstrations not available through NTIS or DTIC with CETHA or AMXTH numbers.
U.S. ARMY Engineer Waterways Experiment Station (WES) and Army Corps of Engineers Environmental Engineering Division	601-634-2504 601-634-3723 601-634-3703 WES documents: 601-643-2856	Provides less expensive, more rapid and effective technologies for site characterization and remediation of contaminated media. In addition, documents with WES numbers not available from NTIS are available from the Environmental Engineering Division.

ENVIRONMENTAL RESOURCE GUIDE 337

A-3 U.S. GOVERNMENT — Environmental Protection Agency Bulletin Board/Databases

Source	Access or Operator	Comments
Alternative Treatment Technology Information Center (ATTIC) and Office of Research and Development (ORD)	Access: 513-569-7610 or 513-569-7700 Operator: 513-569-7272 (see also Fedworld)	Comprehensive alternative hazardous waste technologies information.
Bioremediation in the Field Search System (BFSS)	See CLU-IN below	PC based database of information on the use of bioremediation at waste sites across the country. Over 450 cleanups/studies outlined for location, media, contaminants, cost, and performance.
Case Study Data System	Operator: 703-308-8646	The data system for PC stores and retrieves over 200 studies for case-specific information to support rule and guidance development activities affecting facility siting, corrective actions, and closure. Topics include floodplains, disposal technology, treatment, and environmental affects.
Clean-up Information Bulletin Board System (CLU-IN)	Access: 301-589-8366 Internet: http://clu-in.epa.gov Operator: 301-589-8368	A comprehensive Superfund response activities, hazardous waste correction action information source for environmental professionals needing technological and regulatory information, access to databases, announcements on the Commerce Business Daily related to remediation.
Cost of Remedial Action Model (CORA)	Operator: 703-478-3566	PC-based and designed for the environmental professional, it is an "expert" model to help recommend remedial actions at Superfund sites and also estimate the cost of cleanup actions.
DIALOG Database	Operator: 800-3-DIALOG	Contains files relevant to hazardous waste such as Enviroline, Pollution Abstracts, Energy Science and Technology, National Technical Information Service (NTIS), and others.
Emission Factor Bulletin Board	Access: 919-541-5742	Air emission information.
Environmental Security Technology Certification Program	703-697-9107	Program demonstrates and validates innovative environmental technologies for DoD's most urgent environmental needs. The technologies must be projected to pay the investment in 5 years through cost savings and improved efficiency.
Envirosense	Access: 703-908-2092 Operator: 703-908-2007 Internet: http://es.inel.gov	EPA's Office of Research and Development and Enforcement Compliance and Outreach.
EPA Envirofacts	Access: http://www.epa.gov/enviro/html/ef_home.html	A comprehensive relational database combines data from RCRIS, CERCLIS, TRIS, PCS, and FINDS.

A-3 (continued) U.S. GOVERNMENT — Environmental Protection Agency Bulletin Board/Databases

Source	Access or Operator	Comments
Federal Facilities Environmental Leadership Exchange (FFLEX)	Access: 202-401-5730 Operator: 202-260-4640	Full spectrum environmental bulletin board system (such as compliance, technical assistance, and voluntary cleanup).
Ground Water Remediation Technologies Analysis Center (GWRTAC)	Access: http://www.gwrtac.org Operator: 800-373-1973	GWRTAC compiles, analyzes, and disseminates information on innovative groundwater remediation technologies for government, private sector, and public parties. Interactive case study and vendor information databases available on the web site.
Hazardous Waste Superfund Collection Database	See CLU-IN or EPA Headquarters Library Operator: 202-260-3021	On-line database contains a special collection of hazardous waste documents located throughout the EPA library network. References, abstracts, OSWER policy and guidance directives, legislation, regulations, and nongovernment books.
Innovative Treatment Technologies: Annual Status Report Database (ITT Database)	See NCEPI and CLU-IN	Contain site specific information on approximately 400 sites documented in the 7th Annual Status Report on Innovative Treatment Technologies EPA 542-R-95-008. Database is searchable and can generate reports.
International Cleaner Production Information Network	Access: 703-506-1025 Operator: 202-260-3161	Pollution prevention, source reduction, recycling.
National EnviroText Retrieval System	Operator: 703-603-8877	Access for retrieving environmental information.
Office of Air Quality Planning and Standards Technology Transfer Network (TTN)	Access: 919-541-5742 Internet: telnet://ttnbbs.rtpnc.epa.gov Operator: 919-541-5384	Comprehensive air quality planning and standards information transfer.
Office of Research and Development (ORD)	Access: 513-569-7610 Operator: 513-569-7272	General information and technology transfer. SITE Program administration is through the ORD.

ENVIRONMENTAL RESOURCE GUIDE 339

Online Library System (OLS)	Access: 919-549-0720 Operator: 513-569-7183 Telnet: telnet:\\epaib.rtpnc.epa.gov	Online access to many publications, papers, documents, etc., via EPA's main computer system.
Risk Reduction Engineering Laboratory and Treatability Database	Fax request: 513-891-6685	Request database on diskette that contains information on removal and destruction of over 1,200 chemicals in aqueous and solid media.
Record of Decisions System (RODS)	Operator: 703-603-8889	Provides a comprehensive on-line information database containing information on Superfund RODs for hazardous waste cleanup sites nationwide.
Risk Reduction Engineering Laboratory (RREL) Treatability Database	Operator: 513-569-7408 (see also CLU-IN)	Database provides treatability data for the removal or destruction of organic and inorganic chemicals in aqueous and solid media. 1,207 compounds and 13,500 data sets. PC-based.
Robert S. Kerr Laboratory Databases	See EPA information sources	See EPA information sources.
Soil Transport and Fate Database and Model Management System	Operator: 405-332-8800	The database includes information on approximately 400 chemicals as well as models for predicting the fate and transport of hazardous constituents in the vadose zone. Chemical properties, toxicity, transformation, bioaccumulation available. PC-based system.
Technology Transfer Network	Access: 919-541-5742 Operator: 919-541-5384 Internet: http://ttnwww.rtpnc.epa.gov	EPA's Office of Air Quality, Planning and Standards.
Vendor Facts	Operator: 202-512-2250 (see also CLU-IN)	Comprehensive database of field analytical and characterization technologies. Available on diskette for PC.
Vendor Information System for Innovative Technologies (VISITT)	Operator: 800-245-4505 (see CLU-IN and ATTIC)	Comprehensive database of innovative remedial technologies. Available on diskette for PC.
Waste Water Information Exchange	Access: 800-544-1936	Information on waste water programs.

A-4 U.S. GOVERNMENT — Military Bulletin Boards/Databases

Source	Access or Operator	Comments
Defense Environmental Electronic Bulletin Board System (DEEBS) (see also DENIX)	Access: 703-695-7820 (see also DENIX)	System serves as a centralized communication platform for disseminating DERP information pertaining to DoD's meetings, training, cleanup, and technologies. Access to 800 number available and access to other data networks.
Defense Environmental Network and Information Exchange (DENIX)	Access: 800-637-0958 Operator: 217-373-4517 Internet: http://www.dtic.mil/envirodod/envdocs.html	Comprehensive electronic bulletin board for environmental exchange among DoD environmental professionals, contractors, and DoD-sponsored federal and state employees. An excellent resource for DoD environmental information.
Defense RDT and E On-line System (DROLS) Defense Technical Information Center	Operator: 703-274-6871	Bibliographic database provides information on DoD's research and technology development in separate areas including: Research Work, Unit Information System, Technical Report Database, and Independent Research and Development Database.
DLA Hazardous Material Information System and Hazardous Technical Information Services	Operator: 804-279-5168 800-848-4847	Electronic database of Material Safety Data Sheets (MSDS), transportation, label, and disposal information for DoD personnel.
Environmental Technical Information System (ETIS) Army Corps of Engineers	Operator: 800-USA-CERL, ext. 652 Support: 217-333-1369	System provides help for conducting analyses to document environmental consequences of DoD activities. ETIS subsystems include information exchange on chemicals, regulations, hazardous materials, and hazardous wastes.
Installation Restoration Data Management Information System (IRDMIS)	Operator: 410-671-1655	Database supports technical and managerial requirements of the Army's Installation Restoration Program and other environmental efforts. Maintained by the Army Environmental Center. Requires special software.

ENVIRONMENTAL RESOURCE GUIDE 341

A-5 U.S. GOVERNMENT — Department of Energy

Source	Number or Address	Comments
Center for Environmental Management Information	800-736-3282 202-863-5084 301-903-7924 workshops: 800-831-8333 TIE Quarterly: 703-231-3572	Comprehensive service develops and transfer new technologies that are safer, faster, more effective, and less expensive than current methods. Also provides information on DOE environmental management.
DOE Technology Information Network	http://www.dtin.doe.gov/	Network source for DOE environmental technologies.
DOE Office of Environmental Management (EM)	http://www.em.doe.gov/	Comprehensive source for DOE environmental information and programs, and links to DOE sites.
DOE's Office of Environmental Policy and Assistance	http://www.eh.doe.gov/oepa	Internet information source for DOE environmental policy.
Environmental Technologies from DOE	http://iridium.nttc.edu/env/env_doe.html	Source for DOE environmental technologies. See also the DOE Technology Information Network page.
OSTI Department of Energy Documents	See address at right.	Order DOE documents with OSTI numbers from: OSTI U.S. Dept. of Energy Oak Ridge, TN 37801
Pacific Northwest National Laboratory (PNNL) Environmental Technology Division (in association with Battelle)	http:\\etd.pnl.gov.2080.etdtext.html	Information on development, demonstration, implementation of innovative solutions on various DOE technical subjects specific mostly to the Pacific Northwest.
Remedial Action Program Information Center (RAPIC)	423-576-6500 Fax: 423-576-6547	Comprehensive service for searching and ordering DOE and other environmental documents for DOE personnel and their contractors. RAPIC maintains a bibliographic database of technical literature. Prepares an annual directory of worldwide environmental restoration contacts.
Strategic Environmental Research and Development Program (SERDP)	703-696-2121	A joint, multi-agency effort that funds environmental research, development, demonstration of innovative technologies and programs. SERDP provides information and help to government and private organizations in developing technologies.

A-6 U.S. GOVERNMENT — Department of Energy Bulletin Boards/Databases

Source	Number or Address	Comments
Energy Science and Technology Database	Operators: ITIS: 615-576-1222 DIALOG: 800-334-2564	A multidisciplinary database of worldwide references from basic to applied sciences and technical research literature. Available through the Integrated Technical Information System (ITIS) or DIALOG Information Services (see EPA databases).
Environmental Technologies Remedial Actions Data Exchange (EnviroTRADE)	Operator: 301-903-7930 Internet: http://ramah.geoid.sandia.gov/et.html	A system designed to help facilitate the exchange of environmental restoration and waste management technologies. International information on waste management technologies, organizations, sites, activities, funding, and contracts.
Environmental Technology Information System (TIS)	Operator: 208-526-0614 800-845-2096	System provides information on waste cleanup technologies. On-line advice for screening technologies based on site-specific input information. Special software required.
New Technology from DOE (NTD)	Operator: 615-576-1222 (see ITIS)	Maintained as part of the ITIS, the system is designed to disseminate information about DOE research results that have potential for commercialization. Included are technology descriptions, patent status, secondary applications, literature citations, and DOE information. VT-100 emulation required.
ProTech and the Technology Catalogue	Protech: 301-903-7961 Technology Catalogue: 301-903-7449 Internet: http://texas.pnl.gov:2080/webtech/menu.html	Macintosh platform provides descriptions of technologies supported under the Integrated Demonstrations (IDs) Program. Provides detailed technical cost performance data on deployable technologies advanced by DOE.

ENVIRONMENTAL RESOURCE GUIDE

ReOpt: Electronic Encyclopedia of Remedial Action Options (formerly Remedial Action Assessment System (RASS) Technology Information System)	Operator: 509-375-3765	The system provides information collected from EPA, DOE, and other sources on remedial action, technologies. Data include diagrams, descriptions, engineering and design parameters, technology and regulatory information, and other information. PC and Macintosh based. New releases/upgrades offered.
Research in Progress (RIP) Database	Operator: 615-576-9374 (see also DIALOG)	The database contains administrative and technical information on all unclassified current and recently completed research projects funded by DOE. VT-100 emulation required.
Technology Connection Program	Operator: 970-248-6722 Internet: http://www.doegjpo.com	A directory of vendors for identifying commercial technologies. Includes both foreign and domestic cleanup businesses that are organized by problem type (soil, groundwater, etc.) within categories of characterization, treatment, and extraction or delivery.
Technology Integration System Support (TISS)	Operator: 615-435-3173 (see also TIS)	Provides help with development of new environmental technologies. Database includes environmental technologies, points of contact, DOE documents, vendor information, procurement, and other information.
Waste Management Information System (WMIS)	Operator: 615-435-3281	Provides a comprehensive resource for the explanation and selection of appropriate technologies for handling hazardous, mixed, radioactive, or remedial action wastes. Information on TSD capabilities, waste profiles, and other information.

A-7 U.S. GOVERNMENT — Other Sources/Bulletin Boards/Databases

Source	Number or Address	Comments
Air Resources Laboratory NOAA Environmental Research Laboratory	301-427-7684	The laboratory performs weather research to understand and predict human influences on the environment, especially those involving atmospheric transport and dispersion of pollutants.
Agency for Toxic Substances and Disease Registry (ASTDR)	http://astdr1.atsdr.cdcgov:8080/	Hazardous Substance Release/Health Effects database containing information on the release of hazardous substances from emergency events, Superfund sites, and health effects.
Air Pollutants, Asbestos, Safety/Consumer Product Safety Commission Hotline	800-638-2772	Provides information on consumer safety and guidelines on exposure to formaldehyde, asbestos, and air pollutants. They offer copies of studies and other related documents.
Code of Federal Regulations	http://www.pls.com:8001/his/cfr.html	Internet address of a complete, searchable CFR database back to 1992.
EnviroNetworks	gopher://envirolink.org:70/11/.environetworks	Access to other environmental computer networks.
FEDWORLD	Access: 703-321-3339 Operator: 703-487-4608 Internet: http://www.fedworld.gov	Allows access to more than 100 federally operated on-line computer systems, 8 of which are environmentally related under a single source. Sponsored by NTIS.
Government Printing Office Federal Register On-line	http://www.access.gpo.gov/su_docs/aces/aaces002.html	Internet address of a Federal Register database, updated daily, dated from January 1994.
Internal Revenue Service:	Internet: http://www.irs.ustreas.gov	Information on USTs related to IRS tax information.
Marine Environmental Response Division	202-267-0518 202-267-2611	Data is available on laws relating to the protection of the marine environment, incidents involving releases of oil or other hazardous substances, and federally funded spill response operations.
Office of Marine Safety, Security, and Environmental Protection U.S. Coast Guard		
Material Safety Data Sheets (MSDS)	http://hazard.com	Conduct searches for MSDS sheets by chemical or manufacturer.
National Technical Information Service (NTIS)	Ordering: 703-487-4650 Information: 703-487-4600 (see also DIALOG)	A comprehensive source of government technical documents and publications available to the public. NTIS is available on-line through DIALOG (see EPA databases).
National Oceanic and Atmospheric Administration (NOAA)	http://www.esdim.noaa.gov/	NOAA environmental information services.
NOAA Ocean Pollution Data and Information Network (OPDIN)/CCRO National Oceanographic Data Center	Operator: 202-673-5539	Access to ocean pollution data and information generated by 11 participating federal departments and agencies. OPDIN provides a wide range of products and services to researchers, managers, and others who need data and information about ocean pollution.

ENVIRONMENTAL RESOURCE GUIDE 345

Organization	Contact	Description
Occupation Safety and Health Administration	Internet: http://www.osha.gov/ http://www.osha-slc.gov/	Provides regulatory and technical guidance on the Hazard Communication standard, HAZWOPER standard, asbestos, compliance, and indoor air quality. All of Title 29 of the CFR is available on line. Searching capabilities available.
RTK-NET	Access: 202-234-8570 Operator: 202-234-8494 Internet: http://rtk.net	Unison Institute and OMB Watch nonprofit organizations Right to Know Computer Network.
Tank Waste Information Network System (TWINS)	Internet: http://twins.pnl.gov:8001/refmain.htm	Provides tank waste information.
Thomas	Internet: http://thomas.loc.gov	A site created by the 104th Congress to provide free federal legislative information
U.S. Department of Interior Research Center	801-524-6112	Provides U.S. Department of Interior documents available from the Library of Salt Lake City Research Center.
USGS Distribution Branch	800-USA-MAPS	Land use maps and land cover maps are available for most of the United States. Land use maps refer to human uses of the land (housing and industry) and land cover maps describe the vegetation, water, natural surface, and construction on the land surface. The scale used ranges from 1:100,000 for a few maps in the western states to 1:250,000 for most other maps.
USGS Environmental Research	http://info.er.usgs.gov/research/environment/index.html	Provides an index of environmental research work conducted by the USGS.
USGS National Water Data Exchange (NAWDEX)	703-648-5677	NAWDEX is a confederation of federal and non-federal water-oriented organizations working together to improve access to water data. Information on sites for which water data is available, the types of data available, and the organizations that store the data is available from NAWDEX.
USGS Water Resource Information Center (WRSIC) Research Abstracts and Water Resources Division	703-648-6817 http:www.uwin.sin.edu/databases/wrsic/index.html	More than 265,000 international abstracts and citations searchable by key word. Be specific with key words. Also, National Water Conditions Monthly Update available for a summary of hydrologic conditions in the United States and southern Canada (free upon application).
USGS Office of Water Quality	703-648-6884	The National Water-Quality Networks Program describes and appraises the nation's water resources. The largest of these networks is the National Stream Quality Accounting Network (NASQAN), consisting of more than 400 sampling sites used to measure physical and chemical characteristics on a quarterly or bimonthly schedule.
Water Pollution Resources Superintendent of Documents Government Printing Office	202-783-3238	Water conservation and management books. The National Water Summary books on hydrologic events and the subscription service, Soil and Water Conservation News, are among the free selections.

A-8 PRIVATE AND STATE — Sources/Bulletin Boards/Databases/Internet Addresses

Source	Number or Address	Comments
Advanced Applied Technology Demonstrations Facility (AATDF) Rice University Consortium	713-527-4086 713-527-4700, ext. 3338 Internet: gopher://riceinfo.rice.edu:70/11/subject/	Provides help with selection, development, demonstration, and commercialization of advanced applied cleanup technologies at DoD sites. Strong industry connections and participation.
Air Pollution Control Association (APCA)	412-621-1090	The APCA collects and distributes information about air pollution and its control. It publishes the APCA Journal.
American Academy of Environmental Engineers (AAEE) Innovative Site Remediation Engineering Technology Monographs	410-266-3390	AAEE through a cooperative agreement with EPA's TIO and funding from DOE and DoD developed series of eight peer-reviewed state-of-practice engineering publications titled "Innovative Site Remediation Engineering Technology Monographs" that provide definitive engineering information on bioremediation, chemical treatment, soil washing/flushing, solidification/stabilization, solvent/chemical extraction, thermal desorption, thermal destruction, and vacuum extraction.
American Institute of Hydrology (AIH)	612-379-0901	AIH registers and certifies hydrologists and hydrogeologists.
American National Standards Institute (ANSI)	Internet: http://www.ansi.org	Source for ANSI information.
American Petroleum Institute (API)	Internet: http://www.api.org	Source for API information.
The American Society of Petroleum Operations Engineers (ASPOE)	http://www.bizinfonet.com/assn/aspoe.htm	A site providing information on the group, its members, its publications and special events.
American Society of Testing and Materials (ASTM)	http://www.astm.org	Source for ASTM protocol/directives and information.
Andrew W. Briedenbach Environmental Research Center Library	513-569-7703	Subjects in this library's collection are bacteriology, biology, biotechnology, chemistry, engineering, hazardous wastes, hydrobiology, microbiology, solid waste management, toxicology, water pollution, and water quality. Databases maintained here include BRS, CAS On-line, CIS, DIALOG, Dun & Bradstreet, Hazardous Waste Database, LEXIS/NEXIS, NLM, Toxline, and Toxnet. General collections include bacteriology, biology, biotechnology, microbiology, physics, solid waste management. This library's special collections cover the environment, Canada, legal issues, hazardous waste, and solid waste.

ENVIRONMENTAL RESOURCE GUIDE 347

Organization	Contact	Description
Association of State and Interstate Water Pollution Control Administrators (ASIWPCA)	202-624-7782	This association is comprised of chief water pollution control administrators from the states, District of Columbia, U.S. possessions and interstate agencies. It establishes objectives, policies and standards for state water pollution control.
Athens Environmental Research Laboratory (ERL) Library	404-546-3302	The Athens ERL Library provides information services covering a wide range of environmental and management subjects including aquatic toxicology, microbiology, biology, pesticides, chemistry, water pollution, engineering, and water quality. Databases maintained include CIS, DIALOG, and Ground Water On-line.
Cahner Manufacturing Marketplace Web Site	617-558-4373 Internet: http://www.manufacturing.net	Comprehensive web site with environmental products. 30,000 product categories, 50,000 manufactures and distributors, database of related associations, new product listings, trade show listings, classified ads, etc.
California Base Closure Environmental Committee (CBCEC)	Cal-EPA: 916-322-3294 Cal Toxic Control Board: 916-255-2012	The committee helps identify methods and technologies that promote accelerated restoration and expedited transfer of BRAC properties in California. Not officially a technical transfer program. However, CBCEC has produced valuable information related to investigation and cleanup available in publications and documents.
Clean Sites Inc.	703-739-1217 703-739-1240 703-739-1299	Company bridges the gap between private companies, government, environmental organizations, and community groups. Facilitates cooperative agreements to move projects forward with stakeholders and PRPs. Promotes technically and fiscally sound approaches to cleanup.
Counterpoint Publishing	800-998-4515 Internet: http://www.counterpoint.com	Source for environmental regulations on the internet and CD ROM. Federal Register with archives, CFR, state environmental and safety regulations, specialty titles, etc.
Environmental Engineering ENVENG-L listserver	listserv@vm.temple.edu	Subscribe at the email address to receive information on education, research and professional practice of environmental engineering. Environmental engineering topics will encompass the entire field, including water and wastewater treatment, air pollution control, solid waste management, and radioactive waste treatment.
Environmental Network System	Internet: http://ceres.ca.gov.env.html	Linking service to environmental resources.
Environmental Organizations on EnviroLink	Internet: http://envirolink.org/orgs/index.html	Comprehensive listing/links for environmental organizations and information from A to Z. Information covers over 130 countries.
Environmental Performance Cooperative, Inc.	215-666-8786	Helps facilitates rapid delivery of best available environmental performance and solutions at DoD facilities. Emphasis on UXO and explosive contamination.

A-8 (continued) PRIVATE AND STATE — Sources/Bulletin Boards/Databases/Internet Addresses

Source	Number or Address	Comments
GovCon	http://www.govcon.com/	Provides "yellowpages" of CBD and FAR via the internet. Must register to obtain information.
Halogenated Solvent Industry Alliance (HSIA)	202-223-5890	HSIA consists of producers, users, distributors, and equipment manufacturers in the halogenated solvent industry. It was established to develop constructive programs on legislative and regulatory problems involving halogenated solvents.
National Ground Water Association (NGWA)	800-551-7379 Fax: 614-898-7786 Internet: http://www.h2o-ngwa.org	NGWA provides comprehensive environmental education and conferences services. Maintains research databases and publishes Ground Water, Ground Water Monitoring Review, and other publications.
Petroleum Marketers Association of America (PMAA)	Internet: http://www.pmaa.org	A site that contains information about PMAA, and its activities, and links to other petroleum-related web sites.
Petroleum USENET	Newgroup: news:sci.geo.petroleum	A newgroup of particular interest for those working in the petroleum industry.
Pollution Liability Insurance Association (PLIA)	312-969-5300	PLIA is a reciprocal pool reinsuring pollution liability policies written by member insurance companies.
Public/Private Partnership	See EPA-TIO and Clean Sites Inc.	Organization seeks to pair government cleanup sites with private companies for in the field demonstrations of innovative technologies.
Recommended Internet Locations on the World Wide Web	http://www.einet.net/galaxy/community/the-environment.html	The Environment Home Page.
Recommended Internet Locations on the World Wide Web	http://www.eng.wayne.ed/~hwrm/winter.html (student Tom Crosby)	An exceptional student web site created by Tom Crosby in 1996. It contains links and information on USTs in U.S. An excellent starting point to find out about state UST programs (updated periodically).
Recommended Internet Locations on the World Wide Web	http://www.clay.net	Environmental professionals home page.
Recommended Internet Locations on the World Wide Web	http://www.imt.net/~dcouncil/env.html	Resource links for environmental professionals.
Recommended Internet Locations on the World Wide Web	http://www.iso14000.com	Information center for ISO 14000.
Recommended Internet Locations on the World Wide Web	http://www.webcom.con/~staber/welcome.ht	Environmental law information resource. Updated periodically.
Recommended Internet Locations on the World Wide Web	http://www.envirosearch.com	Searches, regulations, studies, and general environmental information.

Recommended Internet Locations on the World Wide Web	http://www.wef.org	The Air and Waste Management Association and the Water Environment Federation page.
Recommended Internet Locations on the World Wide Web	http://unisci.com	Compiles the latest research from American universities.
Society of Petroleum Engineers (SPE)	http://www.spe.org	A site that provides information on SPE, its members, its publications, and special events.
Urban Land Institute	202-289-8500	The Urban Land Institute consists of developers, architects, and public officials organized to provide accurate, unbiased information useful to land planners and developers. The organization publishes Land Use Digest, Project Reference File, Urban Land.
Western Governors Association (WGA)	303-623-9378 Fax: 303-534-7309	Nonpartisan organization of governors that develops and advocates environmental professional policy that reflects regional interests of 18 states, 2 pacific territories, and 1 commonwealth. Also helps commercialization and transfer of cost-effective waste management and cleanup technologies.

APPENDIX **B**

Completed U.S. EPA Site Monitoring and Measurement Technologies Evaluation Projects as of 1994

Developer	Technology	Technology Contact	EPA Project Manager	Waste Media	Applicable Waste Inorganic	Applicable Waste Organic
Analytical and Remedial Technology, Inc., Menlo Park, CA	Automated Volatile Organic Analytical System	D. MacKay 415-324-2259	Stephen Billets 702-798-2272	Water, Air Streams	Not Applicable	VOCs
Asoma Instruments, Moorpark, CA	Model 200 XRF Analyzer	Bob Friedl 805-529-7123	Harold Vincent 702-798-2129	Solids, Liquids, Slurries, Powders, Pastes, Films	Nonspecific Inorganics	Not Applicable
Bruker Instruments, Billerica, MA	Bruker Mobile Environmental Monitor	John Wronka 508-667-9580	Stephen Billets 702-798-2272	Air Streams, Water, Soil, Sludge, Sediment	Not Applicable	VOCs, SVOCs, PCBs, and PAHs
Dexsil Corporation, Hamden, CT (2 Demonstrations)	Environmental Test Kits	Steve Finch 203-288-3509	J. Lary Jack 702-798-2373	Soil, Sediment, Transformer Oils	Not Applicable	PCBs
EnSys, Inc., Research Triangle Park, NC (2 Demonstrations)	Penta RISc Test System	Aisling Scallan 919-941-5509	J. Lary Jack 702-798-2373	Groundwater, Soil	Not Applicable	PCP
Geoprobe Systems, Salina, KS	Geoprobe Conductivity Sensor	Collin Christy or Tom Christy 913-825-1842	J. Lary Jack 702-798-2373	Soil, Rock, Hydrogeologic Fluids	Nonspecific Inorganics	Nonspecific Organics
Graseby Ionics, Ltd., and PCP, Inc., Watford, Hertfordshire, England/West Palm Beach, FL (2 Demonstrations)	Ion Mobility Spectrometry	John Brokenshire 011-44-923-816166 Martin Cohen 407-683-0507	Eric Koglin 702-798-2432	Air Streams, Vapor, Soil, Water	Not Applicable	VOCs
HNU Systems, Inc., Newton, MA	HNU-Hanby PCP Test Kit	Jack Driscoll 617-964-6690	J. Lary Jack 702-798-2373	Soil	Not Applicable	PCPs
HNU Systems, Inc., Newton, MA	HNU Source Excited Fluorescence Analyzer-Portable (SEFA-P) XRF Analyzer	Jack Driscoll 617-964-6690	Harold Vincent 702-798-2272	Solids, Liquids, Slurries, Powders	Nonspecific Inorganics	Not Applicable

COMPLETED U.S. EPA SITE MONITORING AND EVLUATION PROJECTS

Company	Technology	Contact	EPA Contact	Media	Inorganics	Organics
HNU Systems, Inc., Newton, MA	Portable Gas Chromatograph	Ed Lazaruck 617-964-6690	Richard Berkley 919-541-2439	Air Streams	Not Applicable	VOCs, Aromatic Compounds, Halocarbons
Idetek, Inc. (formerly Binax Corporation, Antox Division), Sunnyvale, CA	Equate® Immunoassay	Richard Lankow 408-752-1353	Jeanette Van Emon 702-798-2154	Water	Not Applicable	Aromatic Hydrocarbons
MDA Scientific, Inc. Norcross, GA	Fourier Transform Infrared Spectrometer	Orman Simpson 404-242-0977	William McClenny 919-541-3158	Air Streams	Nonspecific Inorganics	Nonspecific Organics
Microsensor Systems, Incorporated, Bowling Green, KY	Portable Gas Chromatograph	N. L. Jarvis 410-939-1089	Richard Berkley 919-541-2439	Air Streams	Not Applicable	VOCs
Millipore Corporation, Bedford, MA	EnviroGard™ PCB Immunoassay Test Kit	Alan Weiss 617-275-9200	J. Lary Jack 702-798-2373	Soil, Water	Not Applicable	PCBs
Millipore Corporation, Bedford, MA	EnviroGard™ PCP Immunoassay Test Kit	Alan Weiss 617-275-9200	J. Lary Jack 702-798-2373	Soil, Water	Not Applicable	PCPs
MTI Analytical Instruments (formerly Microsensor Technology, Incorporated), Fremont, CA	Portable Gas Chromatograph	Mark Brunf 510-490-0900	Richard Berkley 919-541-2439	Air Streams	Nonspecific Inorganics	Nonspecific Organics
Ohmicron Corporation, Newtown, PA	Pentachlorophenol RaPID Assay	Mary Hayes 215-860-5115	J. Lary Jack 702-798-2373	Soil, Water	Not Applicable	PCPs
Outokumpu Electronics, Inc., Langhorne, PA	Metorex X-MET 920P XRF Analyzer	James Pasmore 800-229-9209	Harold Vincent 702-798-2129	Solids, Liquids, Slurries, Powders, Films	Nonspecific Inorganics	Not Applicable
Photovac International, Inc., Deer Park, NY	Photovac 10S PLUS	Mark Collins 516-254-4199	Richard Berkley 919-541-2439	Air Streams	Not Applicable	VOCs
SCITEC Corporation, Kennewick, WA	Metal Analysis Probe (MAP®) Portable Assayer	Mike Mullin 800-466-5323 509-783-9850	Harold Vincent 702-798-2129	Soil, Sediment, Filter and Wipe Samples	Nonspecific Inorganics, Lead	Not Applicable
Sentex Sensing Technology, Inc., Ridgefield, NJ	Scentograph Portable Gas Chromatograph	Amos Linenberg 201-945-3694	Richard Berkley 919-541-2439	Air Streams	Not Applicable	VOCs
SRI Instruments, Torrance, CA	Gas Chromatograph	Dave Quinn 310-214-5092	Richard Berkley 919-541-2439	Air Streams	Not Applicable	VOCs

Developer	Technology	Technology Contact	EPA Project Manager	Waste Media	Applicable Waste	
					Inorganic	Organic
TN Technologies, Inc., Round Rock, TX	Spectrace 9000 X-Ray Fluorescence Analyzer	Margo Meyers 512-388-9100	Harold Vincent 702-798-2129	Soil, Sediment, Filter and Wipe Samples	Nonspecific Inorganics, Lead	Not Applicable
Tri-Services, Aberdeen Proving Ground, MD	Site Characterization Analysis Penetrometer System (SCAPS)	George Robitialle 410-671-1576 John Ballard 601-634-2446	J. Lary Jack 702-798-2373	Soil	Not Applicable	Petroleum, PAHs, VOCs
Unisys Corporation, Eagon, MN	Rapid Optical Screen Tool	David Bohne 612-456-2339 Garry Hubbard 612-456-3721	J. Lary Jack 702-798-2373	Soil	Not Applicable	Petroleum, PAHs, VOCs
United States Environmental Protection Agency, Las Vegas, NV	Field Analytical Screening Program PCB Method	Howard Fribush 703-603-8831	J. Lary Jack 702-798-2373	Soil, Water	Not Applicable	PCBs
XonTech Incorporated, Van Nuys, CA	XonTech Sector Sampler	Matt Young 818-787-7380	Joachim Pleil 919-541-4680	Air Streams	Not Applicable	VOCs

Source: U.S. EPA, 1994, Superfund Innovative Technology Evaluation Program Technology Profiles Seventh Edition, EPA/540/R-94/526.

APPENDIX C

Remediation Technology Matrix

Inclusion of an environmental resource/vendor does not in any way represent an endorsement of the resource, vendor, agency, product, or services. In addition, the author makes no guarantee that the addresses or telephone numbers are current; however, the majority of vendors were contacted prior to the publication of this book.

STRATEGIES FOR ACCLERATING CLEANUP AT TOXIC WASTE SITES

Technology Type	Description	Vendor	Contact	Phone	Soil, Sed, Sldg	GW	Air	In Situ	Ex Situ	VOC	Chl. VOC	BTEX	Petr, Oil, Lubs	Met	SVOC	PAH	Pest	PCB	Asbst	UXO	Radio-nucleides	Ref. Source
Acid Extraction	A reagent-based process for treating radioactively contaminated fluoride bearing wastes. The process extracts radioactive elements for recycle or disposal as a volume reduced material.	ADVANCED RECOVERY SYSTEMS, INC. 1219 Banner Hill Road Erwin, TN 37650 USA	Steve Schutt	615-743-6186	Yes	No	No	No	No	No	No	No	No	No	No	No	No	No	No	No	Yes	VISITT
Acid Extraction	The acid extraction treatment system removes heavy metal contamination from soil. The AETS include a regeneration system to removal metals from the spent extracting and reclaim the acid for continuous reuse.	CENTER FOR HAZARDOUS MATERIALS RESEARCH 320 William Pitt Way Pittsburgh, PA 15238 USA	Stephen W. Paff	412-826-5321	Yes	No	No	No	Yes	No	No	No	No	Yes	No	No	No	No	No	No	No	VISITT
Acid Extraction	An extractive process that leaches heavy metals from contaminated soil, dust, sludge, or sediment with a proprietary aqueous leaching solution.	COGNIS, INC. 2331 Circadian Way Santa Rosa, CA 95407 USA	Eric Klein	707-576-6239	Yes	No	No	No	Yes	No	No	No	No	Yes	No	No	No	No	No	No	No	VISITT
Acid Extraction	A system to treat combined heavy metal and organic contamination via a two-stage process where the metals are extracted from the contaminated soil followed by bioremediation of the organic in a slurry bioreactor.	COGNIS, INC. 2331 Circadian Way Santa Rosa, CA 95407 USA	Eric Klein	707-576-6239	Yes	No	No	No	Yes	No	No	No	No	Yes	No	No	No	No	No	No	No	VISITT
Acid Extraction	This technology selectively removes regulated metal contaminants from soil by using a multistage extraction process which utilizes proprietary additives in an acid solvent to preferentially remove metal contaminants.	EARTH TREATMENT TECHNOLOGIES, INC. Dutton Mill Industrial Park 396 Turner Way Aston, PA 19014 USA	Troy Duguay	610-497-6729	Yes	No	No	No	Yes	No	No	No	No	Yes	No	No	No	No	No	No	No	VISITT
Acid Extraction	For treatment of metals, the soil is washed with hydrochloric acid. The solids are separated from final wash solution by gravimetric sedimentation. The acid extract is distilled for the recovery of excess metal.	IT CORPORATION 312 Directors Drive Knoxville, TN 37923 USA	Edward Alperin/ Stuart Shealy	615-690-3211	Yes	No	No	No	Yes	No	No	No	No	Yes	No	No	No	No	No	No	No	VISITT

REMEDIATION TECHNOLOGY MATRIX

Technology	Description	Company	Contact	Phone													Source
Acid Extraction	The technology involves the solubilization of the constituents of concern followed by the isolation of the various soluble elements into appropriate forms. The goal is to minimize the volume of hazardous/radioactive constituents for ultimate disposal.	LOCKHEED CORPORATION 980 Kelly Johnson Drive Las Vegas, NV 89119 USA	Ron May	702-897-3313	Yes	No	No	No	No	No	No	Yes	No	No	No	Yes	VISITT
Adsorption/absorption — In Situ	FORAGER(TM) Sponge is a porous open-celled sponge containing a specialized polymer having a selective affinity for certain species of metals and inorganics dissolved in aqueous solutions.	DYNAPHORE, INC. 2709 Willard Road Richmond, VA 23294 USA	Dr. Norman B. Rainer	804-672-3464	Yes	Yes	No	No	No	No	No	Yes	No	No	No	Yes	VISITT
Adsorption/absorption — In Situ	RECLAIM passive remediation is accomplished by exposing a porous polymer (proprietary "RECLAIM" material) to solution-phase and/or vapor-phase volatile organics. It adsorbs and concentrates VOCs based on partial pressures available.	ENVIRONMENTAL FUEL SYSTEMS, INC. P.O. Box 1899 709 Main Street Bandera, TX 78003 USA	Cal Chapman/ Jeff Keese	210-796-7767	No	Yes	No	Yes	No	Yes	Yes	No	Yes	No	No	No	VISITT
Air Sparging	The system operates by a combination of air injection below the water table, vapor withdrawal above the water table, and stimulation of the microbial community to increase bioremediation of less volatile compounds.	BILLINGS & ASSOCIATES, INC. 3816 Academy Parkway N-N.E. Albuquerque, NM 87109 USA	Dr. Gale K. Billings	505-345-1116	Yes	Yes	No	Yes	No	Yes	Yes	No	Yes	No	No	No	VISITT
Air Sparging	In situ air sparging involves the injection of air into the saturated zone within the areas of contamination. The VOCs dissolved in water are subsequently transported into air phase, within the radius of influence of an soil vapor extraction system.	ENVIROGEN, INC. 480 Neponset Street Canton, MA 02021 USA	Scott Drew	617-821-5560	No	Yes	No	Yes	No	Yes	Yes	No	Yes	No	No	No	VISITT
Air Sparging	The BioSparge System is utilized a designed system of gas injection wells combined with surrounding vapor extraction wells and a mobile surface treatment system to provide injection, capture and cleaning without gas venting and emissions.	HAYWARD BAKER ENVIRONMENTAL INC. 1130 Annapolis Road Odenton, MD 21113 USA	Derek Rhodes	410-551-1995	No	Yes	No	Yes	No	Yes	Yes	No	Yes	No	No	No	VISITT

Technology Type	Description	Vendor	Contact	Phone	Soil, Sed, Sldg	GW	Air	In Situ	Ex Situ	VOC	Chl. VOC	BTEX	Petr, Oil, Lubs	Met	SVOC	PAH	Pest	PCB	Asbst	UXO	Radio-nucleides	Ref. Source
Air Sparging	This technology involves using the trenched horizontal well installation process that was developed by HTI. Using trenched air sparging system to volatilize applicable compounds provides a much greater effective area of treatment than vertical wells.	HORIZONTAL TECHNOLOGIES, INC. 2309 Hancock Bridge Parkway (33990) P.O. Box 150820 Cape Coral, FL 33915-0820 USA	Donald R. Justice	813-995-8777	No	No	No	No	No	No	No	No	No	No	No	No	No	No	No	No	No	VISITT
Air Sparging	The groundwater well has two separate screens, the water rises inside the well due to negative pressure, fresh air is drawn through the water under negative pressure. Groundwater is drawn through the well shaft, is stripped of volatile contaminants.	IEG TECHNOLOGIES CORP. 1833 D Crossbeam Drive Charlotte, NC 28217 USA	Dr. Eric Klingel	704-357-6090	No	Yes	No	Yes	No	Yes	Yes	Yes	Yes	No	Yes	No	Yes	No	No	No	No	VISITT
Air Sparging	Air sparging system will use a network of air sparging well points, air blown into the well in the upper part of the well screen and then moves outward and upward through the groundwater. Air is usually injected at over the hydrostatic pressure.	IT CORPORATION 2925 Briar Park Houston, TX 77042 USA	John Mastroianni	713-784-2800	No	Yes	No	Yes	No	Yes	No	Yes	Yes	No	Yes	No	No	No	No	No	No	VISITT
Air Sparging	Quaternary Investigations, Inc. (QI) has developed a groundwater air sparging system that provides in situ enhancement of air (oxygen) levels in groundwater and volatilization of VOCs.	QUATERNARY INVESTIGATIONS, INC. 300 W. Olive Street Suite A Colton, CA 92324 USA	Tony Morgan	800-423-0740	No	Yes	No	Yes	No	Yes	No	Yes	Yes	No	No	No	No	No	No	No	No	VISITT
Air Sparging	Sparge Vac (TM) is an effective method for remediation of vadose-zone soils and groundwater containing volatile and semivolatile compounds. This technology is particularly effective when large quantities of groundwater must be treated.	TERRA VAC, INC. 1555 Williams Drive Suite 102 Marietta, GA 30066-6282 USA	Charles Pineo		Yes	Yes	No	Yes	No	Yes	Yes	Yes	Yes	No	Yes	Yes	No	No	No	No	No	VISITT

REMEDIATION TECHNOLOGY MATRIX

Technology	Description	Company	Contact	Phone														U.S. Department of Energy
Bioremediation	In situ bioremediation of groundwater and sediment contaminated with chlorinated hydrocarbons. The process simulates indigenous methanotrophic bacteria by injecting methane through horizontal wells.	SAVANA RIVER TECHNOLOGIES Westinghouse Savannah River Company 227 Gateway Drive, Aiken, SC 29803 USA	Christine Spanard		Yes	Yes	No	Yes	No	Yes	Yes	No	No	No	No	No	No	Serdp
Bioremediation	Bioremediation process that uses rhizosphere effects as a co-substrate to enhance contaminant degradation.	U.S. Army, CRREL 72 Lyme Road Hanover, NH 03755-1290 USA	Dr. C. M. Reynolds		Yes	No	No	Yes	No	Yes	Yes	No	No	No	No	Yes	No	
Bioremediation — In Situ Groundwater	Contaminated water and soil below the water table are treated by stimulating naturally occurring bacteria to biodegrade the contaminants in place. Biological activity is enhanced by the delivery of ABB-ES specially formulated nutrients.	ABB ENVIRONMENTAL SERVICES, INC. Corporate Place 128 107 Audubon Road Wakefield, MA 01880 USA	Jaret Johnson, P.E.	617-245-6606	No	Yes	No	Yes	No	Yes	Yes	No	Yes	No	No	No	No	VISITT
Bioremediation — In Situ Groundwater	Bio-Genesis uses naturally occurring, non-pathogenic and endemic to targeted organic compound. Bio-GT blend several different types of microbes that have each been selectively adapted and enhanced to degrade the hazardous contaminants.	BIO-GENESIS TECHNOLOGIES 7343 E. Camelback Road #D Scottsdale, AZ 85251 USA	Victor Coukoulis	602-990-0709	No	Yes	No	Yes	No	No	Yes	No	Yes	No	No	No	No	VISITT
Bioremediation — In Situ Groundwater	FyreZyme is a multifactoral liquid (aqueous) agent, combining a rich source of bacterial growth-enhancing agents, extracellular enzymes, and bioemulsifiers.	ECOLOGY TECHNOLOGIES INTERNATIONAL, INC. 1225 South 48th Street, Suite 2 Tempe, AZ 85281 USA	Pete Condy	602-985-5524	No	Yes	No	Yes	No	Yes	Yes	No	Yes	Yes	No	No	No	VISITT
Bioremediation — In Situ Groundwater	BIOPIM (Biological Process Integrated Method) is an in situ groundwater treatment system. Groundwater is pumped up with vertical extraction wells. Contaminants are biologically removed in a biological sand filter (BIOPIM).	ECOTECHNIEK B.V. Het Kwadrant 1 P.O. Box 1330 Maarssen, 3606AZ The Netherlands	J. Bouman	346-557-700	No	Yes	No	Yes	No	Yes	Yes	No	Yes	No	No	No	No	VISITT

Technology Type	Description	Vendor	Contact	Phone	Soil, Sed, Sldg	GW	Air	In Situ	Ex Situ	VOC	Chl. VOC	BTEX	Petr, Oil, Lubs	Met	SVOC	PAH	Pest	PCB	Asbst	UXO	Radio-nucleides	Ref. Source
Bioremediation — In Situ Groundwater	The electrokinetic bioremediation (or bioelectrokinetic remediation) technology for continuous treatment of groundwater or soil in situ, utilizes either electroosmosis or electrochemical migration (ion migration) to initiate or enhance insitu bioremediation.	ELECTROKINETICS, INC. LA Business and Technology Center Suite 155, South Stadium Drive Baton Rouge, LA 70803-6100 USA	Robert Marks/ Yalcin Acar/ Robert Gale	504-388-3992	No	No	No	No	No	No	No	Yes	Yes	No	No	Yes	No	No	No	No	No	VISITT
Bioremediation — In Situ Groundwater	The remediation approach for this technology combines the use of steam injection for the removal of dense nonaqueous phase liquids (DNAPL), coupled with the application of nutrients to groundwater to promote biotransformation of chlorinated VOCs.	ENSR CONSULTING AND ENGINEERING 1220 Avenida Acaso Camarillo, CA 93012 USA	Greg Smith	805-388-3775	No	Yes	No	Yes	No	Yes	Yes	No	No	No	No	No	No	No	No	No	No	VISITT
Bioremediation — In Situ Groundwater	Soil flushing and bioremediation in situ of groundwater involves the elutriation of organic and/or inorganic contaminants from soil for removal and treatment.	EODT SERVICES, INC. 10511 Hardin Valley Road, Building C Knoxville, TN 37832 USA	Paul Greene, Monirul Haque	615-690-6061	No	No	No	No	No	Yes	Yes	No	No	No	Yes	No	No	No	No	Yes	No	VISITT
Bioremediation — In Situ Groundwater	The technology is based on recovering contaminated groundwater; treating the groundwater in an aboveground fixed-film bioreactor with a combination of microbes, nutrients, and oxygen; and then reintroducing the groundwater to the subsurface.	ESE ENVIRONMENTAL, INC. 3208 Spring Forest Road Raleigh, NC 27604 USA	M. Tony Lieberman	919-872-9686	No	Yes	No	Yes	No	Yes	Yes	Yes	Yes	No	Yes	Yes	No	No	No	No	No	VISITT

REMEDIATION TECHNOLOGY MATRIX

Technology	Description	Company	Contact	Phone															
Bioremediation — In Situ Groundwater	GAIA-NET is designed to remediate remote, inaccessible subterranean groundwater in aquifers and porous media. GAIA-NET can be used for the remediation of military sites contaminated with hazardous, toxic, and radioactive materials.	GAIA RESOURCE, INC. P.O. Box 314 Chicago, IL 60690 USA	T.J. Lawrence	312-329-0368	No	No	No	No	No	No	No	No	No	No	No	No	No	No	No	VISITT
Bioremediation — In Situ Groundwater	An in situ biotreatment process has been developed which uses microorganisms to remove the H2S.	GEO-MICROBIAL TECHNOLOGIES, INC. East Main Street P.O. Box 132 Ochelata, OK 74051 USA	Daniel Hitzman	918-535-2281	No	No	No	No	No	No	No	No	No	No	No	No	No	No	No	VISITT
Bioremediation — In Situ Groundwater	In situ bioremediation is a method for treating groundwater and subsurface soils without excavating the impacted soils. The techniques utilized nutrients and a suitable electron acceptor.	GROUNDWATER TECHNOLOGY, INC. 100 River Ridge Drive Norwood, MA 02062 USA	Peggy Bliss/ Dick Brown	800-635-0053	No	Yes	No	Yes	No	Yes	Yes	No	No	Yes	No	No	No	No	No	VISITT
Bioremediation — In Situ Groundwater	In situ bioremediation of groundwater and saturated soil is achieved by developing a system of extraction and injection wells that deliver nutrients and oxygen and contain the plume of contaminants.	IT CORPORATION 312 Directors Drive Knoxville, TN 37923 USA	Maureen Leavitt	615-690-3211	No	Yes	No	Yes	No	Yes	Yes	No	No	Yes	No	No	No	No	No	VISITT
Bioremediation — In Situ Groundwater	A system of infiltration and recovery galleries is installed and connected to a bioreactor and an aeration system. The system is a stack of vertical trenches to recirculate groundwater in a closed-loop system.	KEMRON ENVIRONMENTAL SERVICES, INC. 2987 Clairmont Road, Suite 150 Atlanta, GA 30329 USA	James A. Novitsky	404-636-0928	No	No	No	No	No	No	No	No	No	Yes	No	No	No	No	No	VISITT
Bioremediation — In Situ Groundwater	In situ bioremediation of groundwater is accomplished through batch or continuous feed treatments using M-1000 microbial consortiums. The principle is based on the isolation of natural bacteria that have degradation capabilities to accelerate degradation.	MICRO-BAC INTERNATIONAL, INC. 3200 N. IH 35 Round Rock, TX 78681-2410 USA	Andrew Timmis	512-310-9000	Yes	Yes	No	Yes	No	Yes	Yes	No	No	Yes	No	No	No	No	No	VISITT

Technology Type	Description	Vendor	Contact	Phone	Soil, Sed, Sldg	GW	Air	In Situ	Ex Situ	VOC	Chl. VOC	BTEX	Petr, Oil, Lubs	Met	SVOC	PAH	Pest	PCB	Asbst	UXO	Radio-nucleides	Ref. Source
Bioremediation — In Situ Groundwater	Microbial Environmental Services, Inc. (MES) uses bioremediation technologies to treat groundwater aquifers contaminated with petroleum and bioreactor treatment technology for treating industrial effluents containing hazardous organic compounds.	MICROBIAL ENVIRONMENTAL SERVICES (MES) 11270 Aurora Avenue Des Moines, IA 50322-7905 USA	Jack Sheldon	515-276-3434	No	Yes	No	Yes	No	No	No	Yes	Yes	No	No	No	No	No	No	No	No	VISITT
Bioremediation — In Situ Groundwater	In situ bioremediation of groundwater involves enhancing microbial biodegradation of contaminants in the aquifer. The groundwater is continuously or semicontinuously recirculated through the aquifer and is usually amended with nutrients.	OHM REMEDIATION SERVICES CORPORATION 16406 US Route 224 East Findlay, OH 45840-0551 USA	Douglas S. Jerger	419-424-4932	Yes	Yes	No	Yes	No	Yes	Yes	Yes	Yes	No	Yes	Yes	No	No	No	No	No	VISITT
Bioremediation — In Situ Groundwater	In situ bioremediation addresses biodegradable organic contaminants within the saturated zone of a contaminated aquifer. Primary contaminants include nonchlorinated solvents, BTEX, and polyaromatic hydrocarbons.	REMEDIATION TECHNOLOGIES, INC. 127 Kington Drive Chapel Hill, NC 27514 USA	Gaylen Brubaker	919-967-3723	No	Yes	No	Yes	No	Yes	No	Yes	Yes	No	Yes	Yes	No	No	No	No	No	VISITT
Bioremediation — In Situ Groundwater	SBP Technologies, Inc. (SBP) has compiled a collection of patented microorganisms for use in bioremediation applications. These microorganisms utilize organic compounds as sole sources of energy for growth.	SBP TECHNOLOGIES, INC. One Sabine Island Drive Gulf Breeze, FL 32561-3999 USA	James Mueller, Ph.D.	904-934-9282	No	Yes	No	Yes	No	Yes	Yes	Yes	Yes	No	Yes	Yes	Yes	No	No	No	No	VISITT

REMEDIATION TECHNOLOGY MATRIX

Technology	Description	Company	Contact	Phone															Source
Bioremediation — In Situ Groundwater	WST applies microorganisms to remediate soil and water contaminated with organic compounds. Specifically, the polluted matrix is amended with bacteria which augment the rate of degradation of target pollutants species (biostimulation).	WASTE STREAM TECHNOLOGY, INC. 302 Grote Street Buffalo, NY 14207 USA	Jim Hyzy/ Brian S. Schepart, Ph.D.	716-876-5290	No	Yes	No	Yes	No	Yes	Yes	No	Yes	No	No	No	Yes	No	VISITT
Bioremediation — In Situ Groundwater	The process entails control of oxidation-reduction potential, pH, salinity, temperatures, and addition of electron donor(s) and/or electron acceptor(s) to enrich in situ cultures of microorganisms to accomplishing the required biotransformations.	YELLOWSTONE ENVIRONMENTAL SCIENCE, INC. 920 Technology Boulevard Bozeman, MT 59715 USA	Mary M. Hunter	406-586-2002	No	No	No	No	No	Yes	Yes	No	Yes	No	No	No	No	No	VISITT
Bioremediation — In Situ Lagoon	Bio-Genesis Technologies has developed a bioremediation in situ lagoon technology for biodegrading petroleum hydrocarbons. Bio-Genesis' GT 1000HC hydrocarbon bacteria WT1000 and nutrients are blended in 55-gallon drums and introduced into the lagoon.	BIO-GENESIS TECHNOLOGIES 7343 E. Camelback Road #D Scottsdale, AZ 85251 USA	Victor Coukoulis	602-990-0709	Yes	No	No	Yes	No	Yes	Yes	No	No	No	No	No	No	No	VISITT
Bioremediation — In Situ Lagoon	FyreZyme is a multifactoral liquid (aqueous) agent, combining a rich source of bacterial growth-enhancing agents, extracellular enzymes, and bioemulsifiers.	ECOLOGY TECHNOLOGIES INTERNATIONAL, INC. 1225 South 48th Street, Suite 2 Tempe, AZ 85281 USA	Pete Condy	602-985-5524	Yes	No	No	Yes	No	Yes	Yes	No	Yes	No	No	No	No	No	VISITT
Bioremediation — In Situ Lagoon	This process involves adding the required elements directly to the sediments to stimulate indigenous bacterial breakdown of organic contaminants. The chemicals are added to the sediments through an injection boom.	LIMNOFIX INC./ GOLDER ASSOCIATES Suite 213; 2550 Argentina Road Mississauga, Ontario L5N 5R1 Canada	Jay Babin	519-888-0141	Yes	No	No	Yes	No	Yes	No	No	Yes	No	No	No	No	No	VISITT

Technology Type	Description	Vendor	Contact	Phone	Soil, Sed, Sldg	GW	Air	In Situ	Ex Situ	VOC	Chl. VOC	BTEX	Petr, Oil, Lubs	Met	SVOC	PAH	Pest	PCB	Asbst	UXO	Radio-nucleides	Ref. Source
Bioremediation — In Situ Lagoon	Batch biological treatment of contaminants in lagoon sludge, sediments, and/or waters is accomplished by enhancing the indigenous or naturally occurring microbial population. Mineral nutrients, if required, are added to support microbial growth.	OHM REMEDIATION SERVICES CORPORATION 16406 US Route 224 East Findlay, OH 45840-0551 USA	Douglas E. Jerger	419-424-4932	Yes	No	No	No	No	Yes	No	Yes	Yes	No	No	No	No	Yes	No	No	No	VISITT
Bioremediation — In Situ Lagoon	Praxair's MIXFLO System dissolves oxygen into a slurry from a lagoon in a two-stage process. First, pure oxygen is injected into the slurry. Then, the oxygen/slurry mixture is reinjected into the lagoon to assist in bioremediation.	PRAXAIR, INC. 39 Old Ridgebury Road (K-1) Danbury, CT 06810-5113 USA	Gary E. Storms	203-837-2174	Yes	No	No	No	No	Yes	Yes	Yes	Yes	No	Yes	Yes	No	Yes	No	No	No	VISITT
Bioremediation — In Situ Soil	ABB Environmental Services, Inc. (ABB-ES) has developed an in situ soil treatment technology using indigenous bacteria. This treatment technology addresses spills of biodegradable chemicals such as petroleum.	ABB ENVIRONMENTAL SERVICES, INC. Corporate Place 128 107 Audobon Road Wakefield, MA 01880 USA	Jaret Johnson, P.E.	617-245-6606	Yes	No	No	Yes	No	Yes	No	Yes	Yes	No	Yes	Yes	No	No	No	No	No	VISITT
Bioremediation — In Situ Soil	This technology injects air below the water table using positive pressure then withdraws vapor above water table. Stimulation of the microbial community to increase bioremediation of less volatile compounds.	BILLINGS & ASSOCIATES, INC. 3816 Academy Parkway N-N.E. Albuquerque, NM 87109 USA	Dr. Gale K. Billings	505-345-1116	Yes	No	No	Yes	No	Yes	Yes	Yes	Yes	No	Yes	Yes	No	No	No	No	No	VISITT
Bioremediation — In Situ Soil	The aerobic biotreatment system (ABS) is designed to provide access for the incremental delivery of liquid nutrients and enzymatic solutions, and to deliver a continuous source of ambient air throughout the treatment area.	BIO-GENESIS TECHNOLOGIES 7343 E. Camelback Road #D Scottsdale, AZ 85251 USA	Victor Coukoulis	602-990-0709	Yes	No	No	Yes	No	Yes	No	Yes	Yes	No	No	No	No	No	No	No	No	VISITT

REMEDIATION TECHNOLOGY MATRIX

Category	Description	Company	Contact	Phone	C1	C2	C3	C4	C5	C6	C7	C8	C9	C10	C11	Source
Bioremediation — In Situ Soil	Biodegradation is the process in which complex organic compounds are metabolized using enzymatic actions of living microbes. Petroleum hydrocarbons and animal fats, can be reduced to harmless carbon dioxide and water with this process.	BIOGEE INTERNATIONAL, INC. 16300 Katy Freeway Suite 100 Houston, TX 77094-1609 USA	Trey Barber/ Ken Roberts	713-578-3111	No	No	No	No	No	Yes	Yes	No	No	No	No	VISITT
Bioremediation — In Situ Soil	The DETOX Process has the capabilities to destroy chemicals as PCBs, PCP, PAH, BETX, creosote, phenolics, and pesticides. In addition, the process is capable of destroying unrefined and refined petroleum hydrocarbons.	DETOX INDUSTRIES, INC. 10101 Southwest Freeway Suite 400 Houston, TX 77074 USA	Philip Balis	713-240-0892	Yes	No	No	No	Yes	Yes	No	Yes	No	No	No	VISITT
Bioremediation — In Situ Soil	FyreZyme is a multifactoral liquid (aqueous) agent, combining a rich source of bacterial growth-enhancing agents, extracellular enzymes, and bioemulsifiers.	ECOLOGY TECHNOLOGIES INTERNATIONAL, INC. 1225 South 48th Street Suite 2 Tempe, AZ 85281 USA	Pete Condy	602-985-5524	Yes	No	No	Yes	Yes	Yes	No	No	No	No	No	VISITT
Bioremediation — In Situ Soil	The electrokinetic bioremediation (or bio-electrokinetic remediation) technology for continuous treatment of in situ soil or groundwater utilizes either electro-osmosis or electrochemical migration to initiate or enhance in situ bioremediation.	ELECTROKINETICS, INC. LA Business and Technology Center Suite 155, South Stadium Drive Baton Rouge, LA 70803-6100 USA	Robert Marks/ Yalcin Acar/ Robert Gale	504-388-3992	No	No	No	No	Yes	Yes	No	No	No	No	No	VISITT
Bioremediation — In Situ Soil	The technology is based on recovering contaminated groundwater; treating the groundwater in a fixed-film bioreactor with a combination of acclimated indigenous microbes, and then reintroducing the groundwater to the subsurface to flush soils.	ESE ENVIRONMENTAL, INC. 3208 Spring Forest Road Raleigh, NC 27604 USA	M. Tony Lieberman	919-872-9686	Yes	No	Yes	No	Yes	Yes	No	No	No	No	No	VISITT
Bioremediation — In Situ Soil	An in situ biotreatment process has been developed which uses microorganisms to remove the H2S.	GEO-MICROBIAL TECHNOLOGIES, INC. East Main Street P.O. Box 132 Ochelata, OK 74051 USA	Daniel Hitzman	918-535-2281	No	No	No	No	No	Yes	No	No	No	No	No	VISITT

Technology Type	Description	Vendor	Contact	Phone	Soil, Sed, Sldg	GW	Air	In Situ	Ex Situ	VOC	Chl. VOC	BTEX	Petr, Oil, Lubs	Met	SVOC	PAH	Pest	PCB	Asbst	UXO	Radio-nucleides	Ref. Source
Bioremediation — In Situ Soil	Daramend bioremediation consists of adding a solid phase organic soil amendments and distribution of the soil amendments through the target matrix, and the homogenization and aeration of the target matrix using specialized tilling equipment.	GRACE DEARBORN, INC. 3451 Erindale Station Road P.O. Box 3060, Station A Mississauga, Ontario L5A 3T5 Canada	I.J. Marvan	905-279-2222	Yes	No	No	Yes	No	Yes	Yes	Yes	Yes	No	Yes	Yes	Yes	No	No	No	No	VISITT
Bioremediation — In Situ Soil	The nutrients and electron acceptor are delivered to the contaminated area in an aqueous phase. Water is amended with nutrient compounds is percolated into the soil by means of a injection wells. Water is drawn up to form hydrogeological containment zone.	GROUNDWATER TECHNOLOGY, INC. 100 River Ridge Drive Norwood, MA 02062 USA	Peggy Bliss/ Dick Brown	800-635-0053	Yes	No	No	Yes	No	No	No	No	No	No	Yes	Yes	No	No	No	No	No	VISITT
Bioremediation — In Situ Soil	BioInjection is in situ treatment process in which combinations of nutrients, oxygen sources, enzymes and other appropriate materials are injected directly into contaminated soils to enhance the degradation of organic contaminants by microorganisms.	HAYWARD BAKER ENVIRONMENTAL, INC. 1130 Annapolis Road Odenton, MD 21113 USA	George K. Burke, P.E.	410-551-1995	Yes	No	No	Yes	No	Yes	No	No	No	No	No	No	No	No	No	No	No	VISITT
Bioremediation — In Situ Soil	The specialized dual auger equipment system injects site-specific microorganism mixtures, along with the required nutrients, and homogeneously mixes them into the contaminated soils, without requiring any excavation.	IN-SITU FIXATION, INC. P.O. Box 516 Chandler, AZ 85244-0516 USA	Richard P. Murray	602-821-0409	Yes	No	No	Yes	No	Yes	Yes	Yes	Yes	No	Yes	No	Yes	No	No	No	No	VISITT

REMEDIATION TECHNOLOGY MATRIX

Technology	Description	Company	Contact	Phone															Source
Bioremediation — In Situ Soil	A system of infiltration and recovery galleries is installed and connected to a bioreactor and an aeration system. The system is a stack of vertical trenches alternating between infiltration and recovery trenches. Natural groundwater is recirculated.	KEMRON ENVIRONMENTAL SERVICES, INC. 2987 Clairmont Road, Suite 150 Atlanta, GA 30329 USA	James A. Novitsky	404-636-0928	Yes	No	Yes	No	No	No	No	No	No	No	No	No	No	No	No	VISITT
Bioremediation — In Situ Soil	In situ bioremediation of soil is accomplished with batch treatments of a variety of M-1000 microbial consortiums. The principle on which this technology is based is the isolation of bacteria that have special degradation capabilities.	MICRO-BAC INTERNATIONAL, INC. 3200 N. IH 35 Round Rock, TX 78681-2410 USA	Andrew Timmis	512-310-9000	Yes	No	Yes	No	No	No	No	No	No	No	No	No	No	No	No	VISITT
Bioremediation — In Situ Soil	This is a hydraulically controlled biodegradation system designed to prevent migration of the contaminants and nutrients from the treatment site. The results of biodegradation of contaminant are carbon dioxide and water.	MICROBIAL ENVIRONMENTAL SERVICES (MES) 11270 Aurora Avenue Des Moines, IA 50322-7905 USA	Jack Sheldon	515-276-3434	Yes	No	Yes	No	No	Yes	No	No	No	No	No	No	No	No	No	VISITT
Bioremediation — In Situ Soil	Nutrients, oxygen sources, and microbes are added to the subsurface media through a series of injection wells or infiltration trenches or a combination of the two. The additives are circulated through the contaminated media.	QUATERNARY INVESTIGATIONS, INC. 300 W. Olive Street Suite A Colton, CA 92324 USA	Tony Morgan	800-423-0740	Yes	No	Yes	No	No	Yes	No	No	Yes	No	No	No	No	No	No	VISITT
Bioremediation — In Situ Soil	SBP Technologies, Inc. (SBP) has compiled a collection of patented microorganisms for use in bioremediation applications. These microorganisms utilize organic compounds as sole sources of energy for growth.	SBP TECHNOLOGIES, INC. One Sabine Island Drive Gulf Breeze, FL 32561-3999 USA	James Mueller, Ph.D.	904-934-9352	Yes	No	Yes	No	No	Yes	Yes	Yes	Yes	Yes	Yes	No	No	No	No	VISITT
Bioremediation — In Situ Soil	WST applies microorganisms to remediate soil and water contaminated with organic compounds. Specifically, the polluted matrix is amended with bacteria which augment the rate of degradation of target pollutants (bioaugmentation).	WASTE STREAM TECHNOLOGY, INC. 302 Grote Street Buffalo, NY 14207 USA	Jim Hyzy/ Brian S. Schepart, Ph.D.	716-876-5290	Yes	No	Yes	No	No	Yes	Yes	Yes	Yes	Yes	No	No	No	Yes	No	VISITT

Technology Type	Description	Vendor	Contact	Phone	Soil, Sed, Sldg	GW	Air	In Situ	Ex Situ	VOC	Chl. VOC	BTEX	Petr, Oil, Lubs	Met	SVOC	PAH	Pest	PCB	Asbst	UXO	Radio-nucleides	Ref. Source
Bioremediation — In Situ Soil	Biodegradation is the process in which complex organic compounds are metabolized using enzymatic actions of living microbes. This bioremediation process reduces concentrations of halogenated VOCs using indigenous microbes.	WESTINGHOUSE SAVANNAH RIVER CO., P.O. Box 616, Building 773-42A, Aiken, SC 29802 USA		803-725-5178	Yes	No	No	Yes	No	Yes	Yes	No	No	No	No	No	No	No	No	No	No	U.S. EPA
Bioremediation — Not Otherwise Specified	The microbes actively metabolize the mercury compounds. In addition to reducing charged mercury, MERCROBES can disintegrate some organic compounds such as benzene, toluene, etc. to CO2 and water.	AP TECHNOLOGIES, INC. 15800 Southpark Loop Anchorage, AK 99516 USA	Stefan W. Zuckut	907-345-8612	No	No	No	No	No	No	No	No	No	No	No	No	No	No	No	No	No	VISITT
Bioremediation — Not Otherwise Specified	B&S Research combines their microorganisms with emulsifier and nutrients to successfully treat contaminated soil and water. B&S Research effectively degrades hydrocarbons, chlorinated solvents, PCBs, pesticides and other hazardous organics compounds.	B&S RESEARCH, INC. 4345 Highway 21 Embarrass, MN 55732 USA	H.W. Lashmett	218-984-3757	Yes	No	No	Yes	Yes	Yes	Yes	Yes	Yes	No	Yes	Yes	Yes	Yes	No	Yes	No	VISITT
Bioremediation — Not Otherwise Specified	BEAREHAVEN bioremediates contaminants in landfills and soil in situ. TCE, PCB, diesel fuel, and other more complex organic compounds are readily bioremediated. In landfills, organic contaminants are aerobically bioremediated and the volume is reduced.	BEAREHAVEN RECLAMATION, INC. 2410 Chastain Drive Atlanta, GA 30342 USA	Fred Gilliard	404-252-8808	Yes	No	No	Yes	No	Yes	Yes	Yes	Yes	No	Yes	Yes	Yes	Yes	No	No	No	VISITT
Bioremediation — Not Otherwise Specified	BIO-PRO provides enhanced waste, water, soil, and air biotreatment systems, equipment and design services. The main objective is to provide specialized biological control systems for pilot and full-scale biological cleanup technologies.	BIO-PRO CORP. P.O. Box 3064 Butte, MT 59702 USA	Mark J. Shutey	406-494-6823	Yes	No	No	No	Yes	Yes	Yes	Yes	Yes	No	Yes	No	Yes	No	No	No	No	VISITT

REMEDIATION TECHNOLOGY MATRIX

Technology	Description	Company	Contact	Phone																	Source
Bioremediation — Not Otherwise Specified	Fenton's reagent is used in the reaction between hydrogen peroxide and ferrous iron (II) to generate a hydroxyl radical. This radical is second only to fluorine in oxidation potential. This oxidation potential creates nonspecific oxidation with organics.	BIOREMEDIATION SERVICE, INC. 12130 NE Ainsworth Circle Suite 220 Portland, OR 97220-9009 USA	David D. Emery	800-775-9464	Yes	No	No	Yes	Yes	Yes	No	Yes	Yes	Yes	No	Yes	Yes	No	No	VISITT
Bioremediation — Not Otherwise Specified	A consortia of microorganisms is prepared in a humic polymer environment and that are compatible with each other. Among these groups are individuals capable of degrading aliphatic and aromatic hydrocarbons, chlorinated or not.	BIOREMEDIATION TECHNOLOGY SERVICES, INC. P.O. Box 3246 Sonora, CA 95370-3246 USA	Paul Richey	800-865-8808	Yes	No	Yes	Yes	Yes	Yes	No	Yes	Yes	Yes	No	Yes	Yes	No	No	VISITT
Bioremediation — Not Otherwise Specified	ChemPete, Inc. has developed and proven an effective, continuous, and timely cleanup method for transforming gasoline, diesel fuel, fuel oil, kerosene, and chlorinated solvents to nonhazardous organic matter, carbon dioxide, and water.	CHEMPETE, INC. 405 E. Pierce Street Elburn, IL 60119 USA	John Peterson	708-365-2007	Yes	No	Yes	Yes	Yes	Yes	No	Yes	Yes	Yes	No	Yes	No	No	No	VISITT
Bioremediation — Not Otherwise Specified	A bench-scale bioremediation technology using Phanerochaete chrysosporium (white rot fungus) has been developed to degrade chlorinated compounds.	CLYDE ENGINEERING SERVICE P.O. Box 740644 New Orleans, LA 70174 USA	Robert Clyde	504-362-7929	No	No	Yes	Yes	Yes	Yes	No	Yes	Yes	Yes	No	Yes	No	No	No	VISITT
Bioremediation — Not Otherwise Specified	Detox Industries, Inc has adapted naturally occurring microorganisms that can biologically destroy refractory organic chemicals. The process can be applied to soils, sludge, liquids, and sediments.	DETOX INDUSTRIES, INC. 10101 Southwest Freeway Suite 400 Houston, TX 77074 USA	Philip Balis	713-240-0892	Yes	No	Yes	Yes	Yes	Yes	No	Yes	Yes	Yes	Yes	Yes	No	No	No	VISITT
Bioremediation — Not Otherwise Specified	ECO-TEC, Inc. (ETI), utilizes "EnviroMech Gold", (EMG), a proprietary biocatalyst to dramatically accelerate the biodegradation of petroleum hydrocarbons and other biodegradable contaminants.	ECO-TEC, INC./ECOLOGY TECHNOLOGY P.O. Box 1113 Issaquah, WA 98027-1113 USA	Herbert R. Pearse	206-392-0304	Yes	Yes	Yes	Yes	Yes	Yes	No	Yes	Yes	Yes	No	Yes	No	No	No	VISITT

Technology Type	Description	Vendor	Contact	Phone	Soil, Sed, Sldg	GW	Air	In Situ	Ex Situ	VOC	Chl. VOC	BTEX	Petr, Oil, Lubs	Met	SVOC	PAH	Pest	PCB	Asbst	UXO	Radionucleides	Ref. Source
Bioremediation — Not Otherwise Specified	SafeSoil is a rapid, ex situ bioremediation application that involves the excavation of contaminated soil; screening the larger soil aggregates, mixing the soil in the blender with proprietary, nutrient-enriched additives.	ENSITE, INC. 114 TownPark Drive Kennesaw, GA 30144 USA	Dr. Andrew Autry	404-421-3488	Yes	No	No	No	Yes	Yes	Yes	Yes	Yes	No	Yes	Yes	No	No	No	No	No	VISITT
Bioremediation — Not Otherwise Specified	Innovative bioremediation process known as ABR-CIS for in-situ sediment volume reduction and biodegradation in polluted waterways and harbors.	EPG BIOSERVICES INC. 205 Park Avenue Barrington, IL 60010 USA	Jim Olsta, P.E.	708-382-0020	Yes	No	No	Yes	No	Yes	No	Yes	Yes	No	Yes	Yes	No	No	No	No	No	VISITT
Bioremediation — Not Otherwise Specified	The CNP-PLUS is a specialty blended biological activator that stimulates indigenous microbial growth when applied to soils containing organic contaminants.	ETUS, INC. 1511 Kastner Place Sanford, FL 32771 USA	Richard Taylor	407-321-7910	No	No	No	No	No	Yes	Yes	Yes	Yes	No	Yes	Yes	Yes	No	No	No	No	VISITT
Bioremediation — Not Otherwise Specified	MBI has designed and evaluated a dechlorination technology utilizing self-immobilized anaerobic microbial granules to treat PCB-contaminated lake and river sediments. These anaerobic granules are composed of facultative bacteria and acetogenic bacteria.	MICHIGAN BIOTECHNOLOGY INSTITUTE 3900 Collins Road Lansing, MI 48910 USA	Mahendra Jain	517-337-3181	Yes	No	No	No	Yes	No	No	No	No	No	No	No	No	Yes	No	No	No	VISITT
Bioremediation — Not Otherwise Specified	This technique maximizes natural aeration, nutrient availability and moisture availability, to maximize microbial action on contaminants. The speed of degradation of contaminants can be increased by manipulating air, nutrients, and oxygen regimes.	PERINO TECHNICAL SERVICES, INC. 2924 Stanton Street Springfield, IL 62703 USA	Janice V. Perino, Ph.D.	217-529-0090	Yes	No	No	No	Yes	No	No	Yes	Yes	No	No	Yes	No	No	No	No	No	VISITT

REMEDIATION TECHNOLOGY MATRIX

Technology	Description	Company/Address	Contact	Phone														Source
Bioremediation — Not Otherwise Specified	(Biochemical Division) has developed numerous bacterial cultures which are used in a wide range of remedial applications including soil and groundwater decontamination, ex situ as well as in situ.	SYBRON CHEMICALS, INC. Birmingham Road Birmingham, NJ 08011 USA	Mike Scalzi	609-893-1100	Yes	Yes	No	Yes	Yes	Yes	No	Yes	Yes	No	No	No	No	VISITT
Bioremediation — Slurry Phase	Soil slurry-sequencing batch bioreactor capable of degrading explosive organics.	ARGONNE NATIONAL LABORATORY 9700 South Cass Avenue, Argonne, IL 60439 USA	Argonne National Laboratory		Yes	No	No	No	No	No	No	No	No	No	Yes	No	No	U.S. AEC
Bioremediation — Slurry Phase	Bio-Genesis Technologies has developed a bioremediation slurry phase technology for biodegrading petroleum hydrocarbons. For the slurry phase application, Bio-Genesis GT100HC hydrocarbon bacteria are injected into reaction vessel by pump.	BIO-GENESIS TECHNOLOGIES 7343 E. Camelback Road, #D Scottsdale, AZ 85251 USA	Victor Coukoulis	602-990-0709	Yes	No	No	No	Yes	Yes	No	No	No	No	No	No	No	VISITT
Bioremediation — Slurry Phase	The soil slurry-sequencing batch reactor (SS-SBR) system is simply a set of tanks operated on a fill-and-draw basis. Each tank is filled during a discrete period of time and operated as a batch reactor. Reaction times are on the order of days.	BIO SOLUTIONS, INC. P.O. Box 207 Riverdale, NJ 07457 USA	George J. Kehrberger, Ph.D., P.E.	201-616-1158	Yes	No	No	No	Yes	Yes	No	No	No	No	Yes	No	No	VISITT
Bioremediation — Slurry Phase	BioGEE International, Inc., has designed a unique method of removing the oil from the oil-based cuttings which allows the cuttings to be handled as nonhazardous oilfield waste (NoW).	BIOGEE INTERNATIONAL, INC. Two Park Ten Place 16300 Katy Freeway, #100 Houston, TX 77094-1609 USA	Trey Barber	713-578-3111	Yes	No	No	No	Yes	Yes	No	No	No	No	No	No	No	VISITT
Bioremediation — Slurry Phase	Contaminated material is excavated, derocked, pulverized, and slurried with water containing elevated concentrations of acclimated, cultured bacteria. The slurry is then pumped into the biotreatment digested.	BOGART ENVIRONMENTAL SERVICES, INC. 3586 N. MT. Juliet Road P.O. Box 717 Mt. Juliet, TN 37122 USA	Jim League	615-754-2847	Yes	No	No	No	Yes	Yes	Yes	Yes	Yes	Yes	No	No	No	VISITT

Technology Type	Description	Vendor	Contact	Phone	Soil, Sed, Sldg	GW	Air	In Situ	Ex Situ	VOC	Chl. VOC	BTEX	Petr, Oil, Lubs	Met	SVOC	PAH	Pest	PCB	Asbst	UXO	Radionucleides	Ref. Source
Bioremediation — Slurry Phase	FyreZyme is a liquid (aqueous) agent, combining a rich source of bacterial growth-enhancing agents, extracellular enzymes, and bioemulsifiers. The enhanced bacteria utilizes the contaminants as food source for growth.	ECOLOGY TECHNOLOGIES INTERNATIONAL, INC. 1225 South 48th Street, Suite 2 Tempe, AZ 85281 USA	Pete Condy	602-985-5524	Yes	No	No	No	Yes	Yes	Yes	Yes	Yes	No	Yes	Yes	No	No	No	No	No	VISITT
Bioremediation — Slurry Phase	The reactor uses a rake/thickener mechanism to control settled solid particles by raking them to either a central airlift or multiple airlifts at the tank periphery. Internal or external airlifts will effect the recirculation of any settled solids.	EIMCO PROCESS EQUIPMENT CO. 3466 S. Westwood Drive Salt Lake City, UT 84109 USA	Gunter Brox	801-272-2288	Yes	No	No	No	Yes	Yes	Yes	Yes	Yes	No	Yes	Yes	No	No	No	Yes	No	VISITT
Bioremediation — Slurry Phase	EODT Services, Inc., in association with Oak Ridge National Laboratory, has developed several amoebae-bacteria consortia and methods for altering and degrading organic and explosive wastes and contaminants.	EODT SERVICES, INC. 10511 Hardin Valley Road Knoxville, TN 37932 USA	Monirul Haque	615-690-6061	Yes	No	No	No	Yes	Yes	Yes	Yes	No	No	Yes	Yes	No	No	No	Yes	No	VISITT
Bioremediation — Slurry Phase	Technology uses a bioslurry reactor to bioremediate soil contaminated with creosote wood preserving.	EVOCAVA CORP. and WASTE-TECH SERVICES, INC. 800 Jefferson County Parkway Golden, CO 80401 USA			Yes	No	No	No	Yes	Yes	No	Yes	No	No	Yes	Yes	No	No	No	No	No	NRMRL
Bioremediation — Slurry Phase	Genesis' elutriant blend, a biodegradable solution of proprietary surfactants, nutrients, flocculants, pH buffers, and cultured bacteria, is combined into a system to selectively emulsify and degrade the various petroleum hydrocarbon constituents.	GENESIS ECO SYSTEMS, INC. 3341 Fitzgerald Road, Suite D Rancho Cordova, CA 95742 USA	Ken Crabtree	916-638-5733	Yes	No	No	No	Yes	Yes	No	Yes	Yes	No	No	No	No	No	No	No	No	VISITT

REMEDIATION TECHNOLOGY MATRIX

Technology	Description	Contact	Phone												Source
Bioremediation — Slurry Phase	This rapid (within 15 minutes) treatment can reduce the volume of such waste muds by separation of the water and oil phases. The process can also remove the spent drilling mud from the well.	GEO-MICROBIAL TECHNOLOGIES, INC. East Main Street P.O. Box 132 Ochelata, OK 74051 USA	Daniel Hitzman 918-535-2281	Yes	No	No	Yes	No	Yes	No	No	No	No	No	No	VISITT
Bioremediation — Slurry Phase	The main objectives of employing this technology are to oxidize the contaminants of interest and to reduce the volume of impacted material. Bioslurry reactors can provide rapid biodegradation of contaminants due to enhanced mass transfer rates.	IT CORPORATION 312 Directors Drive Knoxville, TN 37923 USA	Kandi Brown 615-690-3211	Yes	No	No	Yes	No	Yes	No	Yes	Yes	No	No	No	VISITT
Bioremediation — Slurry Phase	The SABRE (TM) process employs a bioreactor equipped for monitoring and periodic mixing. Excavation, screening, and homogenization equipment are required to prepare contaminated soils. The reactors are modular and readily adaptable in the field.	J.R. SIMPLOT COMPANY 4122 Yellowstone P.O. Box 912 Pocatello, ID 83204 USA	Russ Kaake, Ph.D. 208-238-2852	Yes	No	No	Yes	No	Yes	No	No	No	No	Yes	No	VISITT
Bioremediation — Slurry Phase	The process treats waste in the form of a slurry or a sludge. Soils are mixed with water to form a slurry. Waste is placed into the bioreactor and air is introduced for mixing and aeration. A nutrient and inoculum solution is added to the bioreactor.	OHM REMEDIATION SERVICES CORPORATION 16406 US Route 224 East Findlay, OH 45840-0551 USA	Douglas E. Jerger 419-424-4932	Yes	No	No	Yes	No	Yes	No	Yes	Yes	No	No	No	VISITT
Bioremediation — Slurry Phase	Praxair's MIXFLO System dissolves oxygen into a slurry from a lagoon in a two-stage process. The resulting mixture passes through a pipeline contactor in which approximately 60% of the injected oxygen dissolves.	PRAXAIR, INC. 39 Old Ridgebury Road (K-1) Danbury, CT 06810-5113 USA	Gary E. Storms 203-837-2174	Yes	No	No	Yes	No	Yes	No	Yes	Yes	No	No	No	VISITT

Technology Type	Description	Vendor	Contact	Phone	Soil, Sed, Sldg	GW	Air	In Situ	Ex Situ	VOC	Chl. VOC	BTEX	Petr, Oil, Lubs	Met	SVOC	PAH	Pest	PCB	Asbst	UXO	Radio-nucleides	Ref. Source
Bioremediation — Slurry Phase	Slurry-phase bioremediation (liquid-slurry treatment) utilizes naturally occurring bacteria to convert hazardous organic chemicals into carbon dioxide and water. The natural degradation process occurs in an engineered reactor.	REMEDIATION TECHNOLOGIES, INC. 7011 N. Chaparral Avenue, Suite 100 Tuscon, AZ 85718 USA	Geoffrey H. Swett	602-577-8323	Yes	No	No	No	Yes	Yes	No	Yes	Yes	No	Yes	Yes	Yes	No	No	Yes	No	VISITT
Bioremediation — Slurry Phase	SBP Technologies, Inc., (SBP) has established a unique collection of patented microorganisms for use in bioremediation applications. The integrated use of these microorganisms, with bioreactor yields an effective and cost-efficient remedial solution.	SBP TECHNOLOGIES, INC. One Sabine Island Drive, Gulf Breeze, FL 32561-3999 USA	James Mueller, Ph.D.	904-934-9282	Yes	No	No	No	Yes	Yes	No	Yes	Yes	No	Yes	Yes	Yes	No	No	No	No	VISITT
Bioremediation — Slurry Phase	In situ bio-slurry technology which utilizes bacteria that uses metal constituents as a nutrient source.	TALLON METAL TECHNOLOGIES, INC. 1961 Cohen Ville St. Laurent, Quebec, Canada	TALLON Metal Technologies, Inc.		Yes	No	No	Yes	No	No	No	No	No	Yes	No	No	Yes	No	No	No	No	Technology Developt. Directorate, Canada
Bioremediation — Slurry Phase	An biological cyanide detoxification reactor designed to remediate heavy metals heaps.	U. S. BUREAU OF MINES, SALT LAKE RESEARCH CENTER, 729 Arapeen Drive, Salt Lake City, UT 84108 USA	Bureau of Mines		Yes	No	No	No	Yes	No	No	No	No	Yes	No	No	No	No	No	No	No	EPA Report
Bioremediation — Slurry Phase	WST applies microorganisms to remediate soil and water contaminated with organic compounds. Specifically, the polluted matrix is amended with bacteria which augment the rate of degradation of target pollutants.	WASTE STREAM TECHNOLOGY, INC. 302 Grote Street, Buffalo, NY 14207 USA	Brian S. Schepart, Ph.D.	716-876-5290	Yes	No	No	No	Yes	Yes	Yes	Yes	Yes	No	Yes	Yes	No	No	No	Yes	No	VISITT

REMEDIATION TECHNOLOGY MATRIX

Technology	Description	Company	Contact	Phone													Source
Bioremediation — Slurry Phase	Biocat (TM) technology is designed to remove organic and inorganic hazardous substances from soil (ex situ) and surface waters. Research to date has focused on removal of acidic metal sulfate solutions produced by mining activities.	YELLOWSTONE ENVIRONMENTAL SCIENCE, INC. 920 Technology Boulevard Bozeman, MT 59715 USA	Mary M. Hunter	406-586-2002	No	No	No	No	No	No	No	No	No	No	No	No	VISITT
Bioremediation — Slurry Phase	The BIOCAT-II (TM) Process for bioremediation of mixed hazardous wastes removes aromatic hydrocarbons, halogenated hydrocarbons, heavy metals, acids and salts from mixed waste streams.	YELLOWSTONE ENVIRONMENTAL SCIENCE, INC. 920 Technology Boulevard Bozeman, MT 59715 USA	Mary M. Hunter	406-586-2002	Yes	No	No	Yes	Yes	No	No	No	No	No	No	No	VISITT
Bioremediation — Solid Phase	Composting technology requires mixing of sludge and other high-strength materials with wood chips or other suitable bulking agents in order to increase permeability to air.	ABB ENVIRONMENTAL SERVICES, INC. Corporate Place 128 107 Audubon Road Wakefield, MA 01880 USA	Jaret Johnson, P.E.	617-245-6606	Yes	No	No	Yes	Yes	No	Yes	Yes	Yes	No	No	No	VISITT
Bioremediation — Solid Phase	ABB-ES has developed a process for treating contaminated soils by landfarming using indigenous soil bacteria. This technology can be used for excavated soil or sludge contaminated with biodegradable chemicals such as petroleum.	ABB ENVIRONMENTAL SERVICES, INC. Corporate Place 128 107 Audubon Road Wakefield, MA 01880 USA	Jaret Johnson, P.E.	617-245-6606	Yes	No	No	Yes	Yes	No	Yes	Yes	Yes	No	No	No	VISITT
Bioremediation — Solid Phase	The bioremediation of soil by static pile treatment using forced aeration amends excavated soil with mineral nutrients.	ABB ENVIRONMENTAL SERVICES, INC. Corporate Place 128 107 Audubon Road Wakefield, MA 01880 USA	Jaret Johnson, P.E.	617-245-6606	Yes	No	No	Yes	Yes	No	Yes	Yes	Yes	No	No	No	VISITT
Bioremediation — Solid Phase	Soils are treated with (Alvarez) solution and remediated to the point of non-detect, i.e., 10 ppm TPH, 0.5 ppm BTEX, and then disposed of on our site. This is a full-scale system.	ALVAREZ BROTHERS, INC. 2004 S. Laurent P.O. Box 2975 Victoria, TX 77901 USA	Bob Alvarez	512-576-0404	Yes	No	No	Yes	Yes	No	No	No	No	No	No	No	VISITT

Technology Type	Description	Vendor	Contact	Phone	Soil, Sed, Sldg	GW	Air	In Situ	Ex Situ	VOC	Chl. VOC	BTEX	Petr, Oil, Lubs	Met	SVOC	PAH	Pest	PCB	Asbst	UXO	Radio-nucleides	Ref. Source
Bioremediation — Solid Phase	This technology is an accelerated biological process for effectively degrading organic compounds in water, soil, and/or sediments.	ARCTECH, INC. 14100 Park Meadow Drive Chantilly, VA 22021 USA	Dr. Daman Walia	703-222-0280	Yes	No	No	No	Yes	No	No	No	No	No	No	No	No	No	No	Yes	No	VISITT
Bioremediation — Solid Phase	Bio-Genesis Technologies has developed a solid phase bioremediation technology to treat hydrocarbon impacted soils. A treatment cell constructed, the contaminated soil is placed in the cell and mixed with Bio-GT's GT1000 bacteria.	BIO-GENESIS TECHNOLOGIES 7343 E. Camelback Road, #D Scottsdale, AZ 85251 USA	Victor Coukoulis	602-990-0709	Yes	No	No	No	Yes	No	No	Yes	Yes	No	No	No	No	No	No	No	No	VISITT
Bioremediation — Solid Phase	BioGEE's HC is composed of specifically selected and adopted bacteria, and is formulated to effectively digest a wide variety of long and short chain hydrocarbons.	BIOGEE INTERNATIONAL, INC. 16300 Katy Freeway Suite 100 Houston, TX 77094-1609 USA	Trey Barber	713-578-3111	Yes	No	No	No	Yes	Yes	No	Yes	Yes	No	Yes	Yes	No	No	No	No	No	VISITT
Bioremediation — Solid Phase	A nonspecific oxidative enzyme system, such as found in the white rot fungus, is the most logical approach to microbial destruction of PCB contamination. This system has been shown to readily degrade PCBs and other complex hydrocarbons.	BIOREMEDIATION SERVICE, INC. 12130 NE Ainsworth Circle Suite 220 Portland, OR 97220-9009 USA	David D. Emery	800-775-9464	Yes	No	No	No	Yes	Yes	Yes	Yes	Yes	No	Yes	Yes	Yes	Yes	No	No	No	VISITT
Bioremediation — Solid Phase	Bioremediation Service recycles hydrocarbon-contaminated soil by applying natural degradation processes in a controlled manner. Naturally occurring microbes exist even in the most heavily contaminated soils, where they gradually metabolize compound.	BIOREMEDIATION SERVICE, INC. 12130 NE Ainsworth Circle Suite 220 Portland, OR 97220-9009 USA	David D. Emery	800-775-9464	Yes										Yes	Yes	Yes	Yes	No	Yes	No	VISITT

REMEDIATION TECHNOLOGY MATRIX

Category	Description	Company	Contact	Phone												Source
Bioremediation — Solid Phase	The process employed by Clean-Up Technology uses naturally occurring microorganisms to convert environmental contaminants into inert, nontoxic substances.	CLEAN-UP TECHNOLOGY, INC. 145 West Walnut Street Gardena, CA 90248 USA	Ron Morris	310-327-8605	Yes	No	No	Yes	Yes	No	No	Yes	No	No	No	VISITT
Bioremediation — Solid Phase	Soil-pile bioremediation employs microorganisms indigenous to the contaminated soils which are metabolically enhanced through optimization of environmental conditions and a supply of nutrients and oxygen.	EARTH TECH 2229 Tomlynn Street Richmond, VA 23230 USA	Michael J. Lamore, P.E.	910-299-9998	Yes	No	No	Yes	Yes	No	No	Yes	No	No	No	VISITT
Bioremediation — Solid Phase	WRF use the same biochemical processes required for lignin degradation to break down a broad variety of carbon-based chemicals, including polycyclic aromatic hydrocarbons, PCBs, coal tars, wood preservatives, chlorinated solvents.	EARTHFAX ENGINEERING, INC. 7324 South Union Park Avenue Suite 100 Midvale, UT 84047 USA	Ray Connors, Larry Dushane	801-561-1555	Yes	No	No	Yes	Yes	No	No	Yes	Yes	Yes	No	VISITT
Bioremediation — Solid Phase	FyreZyme is a liquid (aqueous) agent, combining a rich source of bacterial growth-enhancing agents, extracellular enzymes, and bioemulsifiers. The enhanced bacteria utilizes the contaminants as food source for growth.	ECOLOGY TECHNOLOGIES INTERNATIONAL, INC. 1225 South 48th Street Suite 2 Tempe, AZ 85281 USA	Pete Condy	602-985-5524	Yes	No	No	Yes	Yes	Yes	No	Yes	Yes	No	No	VISITT
Bioremediation — Solid Phase	ENSR Consulting and Engineering has designed, permitted, operated and closed several ex situ full-scale biotreatment facilities. These facilities are capable of treating large quantities of hydrocarbon contaminated soil.	ENSR CONSULTING AND ENGINEERING 35 Nagog Park Acton, MA 01720 USA	Dan Groher	508-635-9500	Yes	No	No	Yes	Yes	Yes	No	Yes	Yes	No	No	VISITT
Bioremediation — Solid Phase	The technology incorporates CNP-PLUS, a biological activator. The enhanced bioremediation technology can treat soils, sludge, and dredged sediment contaminants with organic ranging from diesel fuel to PCBs.	ETUS, INC. 1511 Kastner Place Sanford, FL 32771 USA	Richard Taylor	407-321-7910	Yes	No	No	Yes	Yes	Yes	No	Yes	Yes	No	No	VISITT

Technology Type	Description	Vendor	Contact	Phone	Soil, Sed, Sldg	GW	Air	In Situ	Ex Situ	VOC	Chl. VOC	BTEX	Petr, Oil, Lubs	Met	SVOC	PAH	Pest	PCB	Asbst	UXO	Radio-nucleides	Ref. Source
Bioremediation — Solid Phase	Metal containing catalysts are used in the petrochemical and chemical industries in large volumes for many types of catalytic reactions. Once the catalytic activity has decreased, the spent catalyst becomes a waste product which must be treated.	GEO-MICROBIAL TECHNOLOGIES, INC. East Main Street P.O. Box 132 Ochelata, OK 74051 USA	Daniel Hitzman	918-535-2281	Yes	No	No	No	No	No	No	No	No	Yes	No	No	No	No	No	No	No	VISITT
Bioremediation — Solid Phase	The Daramend products increase the ability of the soil matrix to supply biologically available water and nutrients to target compound degrading microorganisms, and transiently bind pollutants, thereby reducing the acute toxicity.	GRACE DEARBORN, INC. 3451 Erindale Station Road P.O. Box 3060, Station A Mississauga, Ontario L5A 3T5 Canada	I.J. Marvan	905-279-2222	Yes	No	No	No	Yes	Yes	Yes	Yes	Yes	No	Yes	Yes	Yes	No	No	Yes	No	VISITT
Bioremediation — Solid Phase	EBS is a proprietary, patented process for the ex situ bioremediation of soils containing chemical constituents of concern. The treatment system is designed to enhance the natural bioremediation rate of organic constituents.	GROUNDWATER TECHNOLOGY, INC. 100 River Ridge Drive Norwood, MA 02062 USA	Peggy Bliss/ Dick Brown	800-635-0053	Yes	No	No	No	Yes	Yes	Yes	Yes	Yes	No	Yes	Yes	No	No	No	No	No	VISITT
Bioremediation — Solid Phase	The BIOFAST (Biological forced Air Soil Treatment) system is designed to bioremediate soil impacted with biodegradable, volatile, semi- or nonvolatile contaminants.	IT CORPORATION 312 Directors Drive Knoxville, TN 37923 USA	Duane Graves	615-690-3211	Yes	No	No	No	Yes	Yes	Yes	Yes	Yes	No	Yes	Yes	No	No	No	No	No	VISITT
Bioremediation — Solid Phase	Advanced Biological Surface Treatment (ABST) is used to treat excavated contaminated soil on the surface. Nutrients are added to enhance the growth of a targeted population of naturally occurring microorganisms.	MICROBIAL ENVIRONMENTAL SERVICES (MES) 11270 Aurora Avenue Des Moines, IA 50322-7905 USA	Jack Sheldon	515-276-3434	Yes	No	No	No	No	No	No	Yes	Yes	No	No	No	No	No	No	No	No	VISITT

REMEDIATION TECHNOLOGY MATRIX

Technology	Description	Company	Contact	Phone												Source
Bioremediation — Solid Phase	Enzymes produced by various forms of fungi have the ability to degrade many hazardous organic compounds via an oxidation reaction. End products are simple compounds, primarily carbon dioxide and water.	MYCOTECH CORPORATION P.O. Box 4109 Butte, MT 59701 USA	Kevin Harvey	406-782-2386	Yes	No	No	Yes	No	Yes	No	No	Yes	No	Yes	No	VISITT
Bioremediation — Solid Phase	Bioremediation of PCB contaminated soils.	OAK RIDGE NATIONAL LABORATORY, P.O. Box 2008, Oak Ridge, TN 37831 USA	Oak Ridge National Laboratory		Yes	No	No	Yes	No	Yes	No	No	Yes	No	No	No	EPA Report
Bioremediation — Solid Phase	A typical engineered land treatment cell consists of a synthetic liner (if required) covered by a layer of sand. Plastic drain tile is placed in the impoundment within the sand layer.	OHM REMEDIATION SERVICES CORPORATION 16406 US Route 224 East Findlay, OH 45840 USA	Douglas E. Jerger	419-424-4932	Yes	No	No	Yes	No	Yes	No	No	Yes	Yes	No	No	VISITT
Bioremediation — Solid Phase	Perino Technical Services, Inc.'s technique maximizes natural aeration, nutrient availability, and moisture availability, therefore, microbial action on contaminants.	PERINO TECHNICAL SERVICES, INC. 2924 Stanton Street Springfield, IL 62703 USA	Dr. Janice V. Perino, Ph.D.	217-529-0090	No	No	No	No	No	Yes	No	No	Yes	No	No	No	VISITT
Bioremediation — Solid Phase	Prepared bed bioremediation utilizes naturally occurring hazardous organic chemicals into carbon dioxide and water. The natural degradation process occurs in an engineered land treatment cell.	REMEDIATION TECHNOLOGIES, INC. 7011 N. Chaparral Avenue, Suite 100 Tuscon, AZ 85718 USA	Geoffrey H. Swett	602-577-8323	Yes	No	No	Yes	Yes	Yes	Yes	No	Yes	No	Yes	No	VISITT
Bioremediation — Solid Phase	The contaminated soil is processed through a specially designed machine which shreds the soil, adds a compost material, nutrients, and water.	SBP TECHNOLOGIES, INC. One Sabine Island Drive Gulf Breeze, FL 32561 USA	James Mueller, Ph.D.	904-934-9352	Yes	No	No	Yes	Yes	Yes	No	No	Yes	Yes	No	No	VISITT
Bioremediation — Solid Phase	WST applies microorganisms to remediate soil and water contaminated with organic compounds. Specifically, the polluted matrix is amended with bacteria which augment the rate of degradation to target pollutants (bioaugmentation).	WASTE STREAM TECHNOLOGY, INC. 302 Grote Street Buffalo, NY 14207 USA	Brian S. Schepart, Ph.D./Jim Hyzy	716-876-5290	Yes	No	No	Yes	Yes	Yes	No	No	Yes	No	No	No	VISITT

Technology Type	Description	Vendor	Contact	Phone	Soil, Sed, Sldg	GW	Air	In Situ	Ex Situ	VOC	Chl. VOC	BTEX	Petr, Oil, Lubs	Met	SVOC	PAH	Pest	PCB	Asbst	UXO	Radionucleides	Ref. Source
Bioremediation — Solid Phase	Cultured thermophilic organisms show rapid and complete decomposition of explosive compounds. Resulting products are nontoxic. Based on these capabilities, engineered bioreactors have been designed to treat both contaminated soils and liquids.	CONCURRENT TECHNOLOGIES CORPORATION and CENTER FOR HAZARDOUS MATERIALS RESEARCH, 320 William Pitt Way, Pittsburgh, PA 15238 USA	Marvin Scher		Yes	Yes	No	Yes	No	No	No	No	No	No	No	No	No	No	No	No	Yes	No	National Defense Center for Environ. Excellence
Bioremediation — Solid Phase	CTC gas developed a biological regeneration of GAC contaminated with explosives. After GAC is loaded with explosives from contaminated groundwater or processing waste, thermophilic organisms regenerate the GAC in place.	CONCURRENT TECHNOLOGIES CORPORATION and CENTER FOR HAZARDOUS MATERIALS RESEARCH, 320 William Pitt Way, Pittsburgh, PA 15238	Marvin Scher		No	Yes	No	Yes	No	No	No	No	No	No	No	No	No	No	No	No	Yes	No	National Defense Center for Environmental Excellence
Bioventing	Bioventing is a vadose zone remediation technology in which biological degradation of organic compounds occurs in situ. The principle behind bioventing is to deliver oxygen to the zone of contamination to promote biological degradation or organic compound.	ABB ENVIRONMENTAL SERVICES, INC. 107 Audubon Road Corporate Place 128 Wakefield, MA 01880 USA	Jaret Johnson, P.E.	617-245-6606	Yes	No	No	Yes	No	Yes	No	Yes	Yes	No	Yes	Yes	No	No	No	No	No	VISITT
Bioventing	Battelle is currently working on the design of an air injection bioventing demonstration for tight clay soils and an air sparging/bioventing demonstration to simultaneously treat the vadose zone and shallow contaminated groundwater.	BATTELLE PACIFIC NORTHWEST LABORATORIES Battelle Boulevard, MSIN P7-41 P.O. Box 999 Richland, WA 99352 USA	Chris Johnson	509-372-2273	Yes	No	No	Yes	No	No	No	Yes	Yes	No	No	No	No	No	No	No	No	VISITT

REMEDIATION TECHNOLOGY MATRIX

	Description	Company	Contact	Phone															
Bioventing	Bioventing is a method which enhances in situ or ex situ bioremediation of soils contaminated with a variety of organic chemicals. Bioventing incorporates the injection of air through a network of vertical pipes.	DAMES & MOORE 2325 MD Road Willow Grove, PA 19090 USA	Joseph M. Tarsavage	215-657-5000	Yes	No	Yes	Yes	No	Yes	Yes	No	Yes	Yes	No	No	No	No	VISITT
Bioventing	Bioventing is the use of air injection and/or extraction methods to increase soil oxygen levels and enhance the natural biodegradation of petroleum hydrocarbons.	ENGINEERING-SCIENCE, INC. 1700 Broadway Suite 900 Denver, CO 80290 USA	Doug Downey	303-831-8100	Yes	No	Yes	Yes	No	Yes	Yes	No	Yes	Yes	No	No	No	No	VISITT
Bioventing	This technology is applicable to soils with soil gas permeability values of 1 x 10E-8 square cm/sec or greater. Biodegradation of contaminated soil is enhanced through the induction or injection of ambient air into the subsurface via venting wells.	ENSR CONSULTING AND ENGINEERING 35 Nagog Park Acton, MA 01720 USA	Ellen Moyer, Ph.D. P.E.	508-635-9500	Yes	No	Yes	Yes	No	Yes	Yes	No	Yes	Yes	No	No	No	No	VISITT
Bioventing	Bioventing is an in situ technology that is based on the introduction of oxygen into the subsurface to stimulate the growth of microbial activity which results in a reduction of the contaminant.	ENVIROGEN, INC. 480 Neponset Street Canton, MA 02021 USA	Sally Hulsman	617-821-5560	Yes	No	Yes	Yes	No	Yes	Yes	No	Yes	Yes	No	No	No	No	VISITT
Bioventing	Bioventing is the process of degrading a soil contaminant by providing soil bacteria with oxygen and amendments as necessary so that the bacteria can grow and consume the contaminant in the soil. The contaminant must be readily degradable.	ENVIRONEERING 5508 Bent Oak Road Sylvania, OH 43560 USA	Rick Mazur	419-885-3155	Yes	No	Yes	Yes	No	Yes	Yes	No	Yes	Yes	No	No	No	No	VISITT
Bioventing	Bioventing utilizes the natural in situ biodegradation of hydrocarbon contaminants in the vadose zone. Biodegradation is a process in which indigenous bacteria break down the molecular bonds in toxic organic compounds, producing carbon dioxide and water.	H2O SCIENCE, INC. 5500 Bolsa Avenue, Suite 105 Huntington Beach, CA 92649 USA	Mark P. Ausburn	714-379-1157	Yes	No	Yes	Yes	No	Yes	Yes	No	Yes	Yes	No	No	No	No	VISITT

382 STRATEGIES FOR ACCLERATING CLEANUP AT TOXIC WASTE SITES

Technology Type	Description	Vendor	Contact	Phone	Soil, Sed, Sldg	GW	Air	In Situ	Ex Situ	VOC	Chl. VOC	BTEX	Petr, Oil, Lubs	Met	SVOC	PAH	Pest	PCB	Asbst	UXO	Radio-nucleides	Ref. Source
Bioventing	The BioSparge System is a patented closed-loop system of in situ remediation. It utilizes a system of gas injection wells combined with surrounding vapor extraction wells and a mobile surface treatment system to provide injection, capture and cleaning.	HAYWARD BAKER ENVIRONMENTAL, INC. 1130 Annapolis Road Odenton, MD 21113 USA	Derek Rhodes	410-551-1995	Yes	No	No	Yes	No	Yes	Yes	Yes	Yes	No	Yes	No	No	No	No	No	No	VISITT
Bioventing	An air circulation system is established in the unsaturated zone of the given matrix (usually soil). Air movement throughout the contaminated area provides oxygen for the biodegradation process.	IT CORPORATION 312 Directors Drive Knoxville, TN 37923 USA	Maureen Leavitt	615-690-3211	Yes	No	No	Yes	No	Yes	No	No	No	No	Yes	No	No	No	No	No	No	VISITT
Bioventing	Mittelhauser's bioventing technology is remediate soils in situ, primarily in the unsaturated zone. The process is based on the principle that microorganisms have the ability to degrade organic compounds as part of their biological process.	MITTELHAUSER CORPORATION 23272 Mill Creek Drive Suite 100 Laguna Hills, CA 92653 USA	Tim Lester	714-472-2444	Yes	No	No	Yes	No	Yes	Yes	Yes	Yes	No	Yes	No	No	No	No	No	No	VISITT
Bioventing	Bioventing makes use of bioremediation and soil vapor extraction. Airflows are sustained at a level that maintains oxygen in the subsurface soil. Microorganisms in the soil can then biodegrade contaminants under the proper environmental conditions.	OHM REMEDIATION SERVICES CORPORATION 16406 US Route 224 East Findlay, OH 45840-0551 USA	Douglas E. Jerger	419-424-4932	Yes	No	No	Yes	Yes	Yes	No	Yes	Yes	No	Yes	Yes	No	No	No	No	No	VISITT

REMEDIATION TECHNOLOGY MATRIX

Technology	Description	Company	Contact	Phone															VISITT
Bioventing	Quaternary Investigations' (QI) bioventing system enhances the naturally occurring oxygen levels in contaminated subsurface soils to augment the rate of contaminant biodegradation.	QUATERNARY INVESTIGATIONS, INC. 300 W. Olive Street, Suite A Colton, CA 92324 USA	Tony Morgan	800-423-0740	Yes	No	No	Yes	No	Yes	Yes	Yes	No	Yes	No	No	No	No	VISITT
Bioventing	Vapor extraction-enhanced bioremediation is an effective method for remediation of soils in the vadose and saturated zones that are contaminated with volatile or semivolatile compounds.	TERRA VAC, INC. 1555 Williams Drive Suite 102 Marietta, GA 30066-6282 USA	Charles Pineo	404-421-8008	Yes	No	No	Yes	No	Yes	Yes	Yes	No	Yes	No	No	No	No	VISITT
Chemical Treatment — Dechlorination	Halogens are selectively stripped from the halogenated organic material in the soil by the solvated electrons, converting the compounds to calcium halides and hydrocarbon residuals.	COMMODORE ENVIRONMENTAL SERVICES, INC. 1487 Delashmut Avenue Columbus, OH 43212 USA	Albert E. Abel	614-297-0365	Yes	No	No	Yes	No	Yes	Yes	Yes	No	Yes	Yes	No	No	No	VISITT
Chemical Treatment — Dechlorination	SDTX KPEG, (SM) process is a chemical dehalogenation technology for use on soils, sediments, and sludge. The chemistry has been demonstrated at various scales for PCBs, dioxins, PCP, DDT, and chlorinated aliphatic.	SDTX TECHNOLOGIES, INC. 706 Sayre Drive Princeton, NY 08540 USA	Robert Hoch	518-734-4483	Yes	No	No	Yes	No	Yes	Yes	Yes	No	Yes	Yes	No	No	No	VISITT
Chemical Treatment — In Situ Groundwater	The EnviroMetal process uses a reactive metal (usually iron) to degrade dissolved halogenated organic compounds by inducing reduction/oxidation conditions that cause substitution of halogen atoms by hydrogen atoms.	ENVIROMETAL TECHNOLOGIES, INC. 42 Arrow Road Guelph, Ontario N1K 1S6 Canada	John Vogan	519-824-0432	No	Yes	No	Yes	No	Yes	Yes	No	No	No	No	No	No	No	VISITT
Chemical Treatment — In Situ Groundwater	Inorganic contaminants including heavy metals and radionuclides can be removed from the vadose zone and groundwater and fixed in stable, natural geochemical traps.	GEOCHEM DIVISION OF TERRA VAC 12596 West Bayaud Avenue, Suite 205 Lakewood, CO 80228 USA	Charles Pineo	404-421-8008	No	Yes	No	Yes	No	No	No	No	Yes	No	No	No	No	Yes	VISITT

Technology Type	Description	Vendor	Contact	Phone	Soil, Sed, Sldg	GW	Air	In Situ	Ex Situ	VOC	Chl. VOC	BTEX	Petr, Oil, Lubs	Met	SVOC	PAH	Pest	PCB	Asbst	UXO	Radio-nucleides	Ref. Source
Chemical Treatment — In Situ Groundwater	It is a method of surfactant-enhanced aquifer remediation that involves the injection and withdrawal of dilute surfactant solutions into groundwater systems contaminated with non-aqueous phase liquids (NAPL), in particular dense NAPLs (DNAPLs).	INTERA, INC. 6850 Austin Center Blvd., Suite #300 Austin, TX 78731 USA	R.E. Jackson	512-346-2000	No	Yes	No	Yes	No	Yes	Yes	Yes	Yes	No	No	No	No	No	No	No	No	VISITT
Chemical Treatment — Other	The pyrodigestion technology is a thermal and chemical treatment technology. It uses a molten alloy of metals in an anaerobic atmosphere decomposes (pyrolysis) of a variety of organic hazardous or toxic waste streams containing among other elements.	CLEANTECH OF ARKANSAS, INC. 7600 Highway 107 Sherwood, AR 72116 USA	M. Douglas Wood	501-834-7600	Yes	No	No	No	Yes	Yes	Yes	Yes	Yes	No	Yes	Yes	No	Yes	No	No	No	VISITT
Chemical Treatment — Other	CTI has developed and acquired patents on certain innovative chelation chemicals with unique capabilities for control and recovery of radioactive and other types of hazardous metal ions from soils, concrete, steel, and other materials.	CORPEX TECHNOLOGIES, INC. 5400 S. Miami Boulevard Morrisville, NC 27560 USA	John K. Pirotte	919-941-0847	No	No	No	No	Yes	No	No	No	No	Yes	No	No	No	No	No	No	Yes	VISITT
Chemical Treatment — Other	The commercial recovery of gold and uranium from ore is currently undertaken using in-pulp processes. In these processes, the suspension of leached ore particles in the leach solution, called pulp, is directly contacted with a solid adsorbent.	DAVY INTERNATIONAL - ENVIRONMENTAL DIV. Ashmore House Richardson Road Stockton-on-Tees, Cleveland TS183RE England	Mr. George Rowden	164-260-2221	Yes	No	No	No	No	No	No	No	No	Yes	No	No	No	No	No	No	No	VISITT

REMEDIATION TECHNOLOGY MATRIX

Category	Description	Company	Contact	Phone														Source
Chemical Treatment — Other	METRAXT is an aqueous-based product developed specifically for the cleanup of metals on solid surfaces. METRAXT is formulated to extract metals from porous surfaces by bonding with them.	INTEGRATED CHEMISTRIES, INC. 1970 Oakcrest Avenue Suite 215 St. Paul, MN 55113 USA	Cathy Iverson	612-636-2380	No	No	No	No	No	No	No	Yes	No	No	No	No	No	VISITT
Chemical Treatment — Other	RMT process involves the use of a buffering agent with phosphate compounds to render lead- and cadmium-bearing waste, soil, and contaminated media nonhazardous over a wide range of disposal conditions.	RMT, INC. 744 Heartland Trail Madison, WI 53717-1934 USA	Christopher Rehmann	608-831-4444	Yes	No	No	Yes	No	No	No	Yes	No	No	No	No	No	VISITT
Chemical Treatment — Other	Mercon (TM) is a patented liquid mercury vapor suppressant designed to stop and absorb mercury vapors. The chemical process utilized creates a mercuric salt or sulfide. The reagents react with the metal and absorb any ambient vapor.	SOLUCORP INDUSTRIES LTD. 520 Victor Street Saddle Brook, NJ 07663 USA	Noel Spindler	201-368-7902	Yes	No	No	No	Yes	No	No	Yes	No	No	No	No	No	VISITT
Chemical Treatment — Other	A metal contaminated site is excavated. The dirt is piled on a liner. Cyanide is sprayed on the dirt and collected when it reaches the liner. The metal-cyanide solution is stripped of the cyanide and the metals are filtered.	VIKING INDUSTRIES 1015 Old Lascassas Road Murfreesboro, TN 37130 USA	Don T. Pearson	615-890-1018	Yes	No	No	No	Yes	No	No	Yes	No	No	No	No	No	VISITT
Chemical Treatment — Oxidation	The technology uses peroxone, a mixture of ozone and hydrogen peroxide to oxidize organic compounds.	U.S. ARMY, WES 3909 Halls Ferry Road Vicksburg, MS 39180	Mark E. Zappi		No	Yes	Yes	Yes	No	No	No	No	No	No	No	Yes	No	Serdp
Chemical Treatment — Oxidation	The process uses ultra-violet light in combination with ozone to oxidize organic and inorganic contaminants present in air or water.	VM TECHNOLOGY 23901 Remme Ridge Lake Forest, CA 92630			No	Yes	Yes	No	No	Yes	Yes	Yes	Yes	Yes	No	No	Yes	VM Technology, Inc.
Chemical Treatment — Oxidation/reduction	Advanced Recovery Systems, Inc. (ARS) has developed a technology to convert a variety of mercury-bearing hazardous and mixed wastes to final nonhazardous forms meeting disposal criteria.	ADVANCED RECOVERY SYSTEMS, INC. 1219 Banner Hill Road Erwin, TN 37650 USA	Steve Schutt	615-743-6186	Yes	No	No	No	No	No	No	Yes	No	No	No	No	No	VISITT

Technology Type	Description	Vendor	Contact	Phone	Soil, Sed, Sldg	GW	Air	In Situ	Ex Situ	VOC	Chl. VOC	BTEX	Petr, Oil, Lubs	Met	SVOC	PAH	Pest	PCB	Asbst	UXO	Radio-nucleides	Ref. Source
Chemical Treatment—Oxidation/reduction	The process was developed for decontamination of liquid, sludge, sediments, and solid wastes. Halogenated organic compounds can be treated in the presence of other organic or inorganic material as long as sufficient transmission of UV light occurs.	ARCTECH, INC. 14100 Park Meadow Drive Chantilly, VA 22021 USA	Dr. Daman Walia	703-222-0280	No	No	No	No	No	No	No	No	No	No	No	No	No	No	No	No	No	VISITT
Chemical Treatment—Oxidation/reduction	The OZO-DETOX process is a relatively inexpensive novel approach for destroying volatile and nonvolatile organic contaminants in soil and water, on site.	ARCTECH, INC. 14100 Park Meadow Drive Chantilly, VA 22021 USA	Dr. Daman Walia	703-222-0280	Yes	No	No	No	Yes	Yes	No	No	No	No	Yes	Yes	No	No	No	No	No	VISITT
Chemical Treatment—Oxidation/reduction	DETOX(sm) is a patented, catalyzed wet oxidation waste treatment process. Wet oxidation is the nonthermal oxidation of materials, organic materials, with oxygen in a water solution. Wastes and oxygen are fed to a reactor where organic are destroyed.	DELPHI RESEARCH, INC. 701 Haines Avenue, N.W. Albuquerque, NM 87102 USA	Jeffrey Campbell	505-243-3111	Yes	No	Yes	No	Yes	Yes	Yes	Yes	Yes	Yes	Yes	No	No	No	No	Yes	No	VISITT
Chemical Treatment—Oxidation/reduction	The ECO LOGIC process involves the gas-phase reduction of organic compounds by hydrogen at temperatures of 850°C or higher. Chlorinated hydrocarbons, such as PCB, dioxins are chemically reduced to methane, ethylene, and hydrogen chloride (HCl).	ELI ECO LOGIC INTERNATIONAL, INC. 143 Dennis Street Rockwood, Ontario N0B 2K0 Canada	Jim Nash / Martin Hassenbach	519-856-9591	Yes	No	No	No	Yes	Yes	Yes	Yes	Yes	No	Yes	Yes	Yes	Yes	No	No	No	VISITT
Chemical Treatment—Oxidation/reduction	The technology is targeted for battery waste sites. The soil is treated in situ by neutralization using alkali (lime, ammonia, etc.). Equipment designed for soil conditioning and fertilizer injection will be utilized.	EM&C ENGINEERING ASSOCIATES 1665 Scenic Avenue Suite 104 Costa Mesa, CA 92626 USA	Mohamed Elgafi	714-957-6429	No	No	No	No	No	No	No	No	No	No	No	No	No	No	No	No	No	VISITT

REMEDIATION TECHNOLOGY MATRIX

Category	Description	Company	Contact	Phone														Source
Chemical Treatment — Oxidation/reduction	The TR-DETOX technology utilizes the synergetic application of specific inorganic and organic reagents which readily percolate the contaminated soils.	ETUS, INC. 1511 Kastner Place Building #1 Technology Drive Sanford, FL 32771 USA	Richard Dunkel	407-321-7910	Yes	Yes	No	Yes	No	No	No	No	Yes	No	No	No	Yes	VISITT
Chemical Treatment — Oxidation/reduction	G.E.M., Inc.'s chemical treatment technology is a closed system process where hydrocarbons undergo an oxidation process by chemical reaction.	G.E.M., INC. P.O. Box 397, Highway 67 South Malvern, AR 72104 USA	Cleve A. Bond	501-337-9410	Yes	No	No	No	Yes	Yes	No	No	Yes	No	No	No	No	VISITT
Chemical Treatment — Oxidation/reduction	Electrons are accelerated by means of a voltage differential and impact a flowing slurry, producing highly reactive species capable of destroying all toxic organic compounds in aqueous solution.	HIGH VOLTAGE ENVIRONMENTAL APPLICATIONS 9562 Doral Boulevard Miami, FL 33178 USA	Thomas D. Waite, Ph.D.	305-593-5330	Yes	No	No	No	Yes	Yes	No	No	Yes	Yes	No	Yes	No	VISITT
Chemical Treatment — Oxidation/reduction	Organic contaminants in soil are treated in situ by controlled injection of hydrogen peroxide solution into the contaminated zone via a network of injection wells. The hydrogen peroxide oxidizes contaminants to form less toxic compounds.	IT CORPORATION 312 Directors Drive Knoxville, TN 37923 USA	Chuck Parmele	615-690-3211	Yes	No	Yes	No	No	No	No	No	No	No	No	No	No	VISITT
Chemical Treatment — Oxidation/reduction	Organic contaminants in soil are treated in situ by controlled injection of hydrogen peroxide solution into the contaminated zone via a network of injection wells. The hydrogen peroxide oxidizes contaminants to form less toxic compounds.	IT CORPORATION 312 Directors Drive Knoxville, TN 37923 USA	Robert D. Fox	615-690-3211	Yes	No	Yes	No	No	No	No	No	No	No	No	No	No	VISITT
Chemical Treatment — Oxidation/reduction	MMT's Catalytic Extraction Processing is an innovative, proprietary technology that allows organic, organometallic, metallic, and inorganic feeds to be recycled into useful materials of commercial value.	MOLTEN METAL TECHNOLOGY, INC. 51 Sawyer Road Waltham, MA 02154 USA	David Hoey	617-487-7644	Yes	No	No	No	Yes	Yes	Yes	Yes	Yes	No	No	No	Yes	VISITT

Technology Type	Description	Vendor	Contact	Phone	Soil, Sed, Sldg	GW	Air	In Situ	Ex Situ	VOC	Chl. VOC	BTEX	Petr, Oil, Lubs	Met	SVOC	PAH	Pest	PCB	Asbst	UXO	Radio-nucleides	Ref. Source
Chemical Treatment—Oxidation/reduction	The Synthetics Detoxifier uses high-temperature steam and an electrically heated reactor to destroy many hazardous components of a waste using the principle of steam reforming.	SYNTHETICS TECHNOLOGIES, INC. 5327 Jacuzzi Street, Unit 3-0 Richmond, CA 94804 USA	Jisook Park	510-525-3000	Yes	No	No	No	No	Yes	Yes	Yes	Yes	Yes	Yes	Yes	No	No	No	Yes	No	VISITT
Chemical Treatment—Oxidation/reduction	A concentrated H_2O_2 solution is poured or injected into the wells or boreholes and allowed to percolate through the contaminated soils. Once added, the H_2O_2 reacts with any organic material present, yielding carbon dioxide and water.	TERRA VAC, INC. 1555 Williams Drive, Suite 102 Marietta, GA 30066-6282 USA	Charles Pineo	404-421-8008	Yes	Yes	No	Yes	Yes	Yes	No	Yes	Yes	No	Yes	Yes	No	No	No	No	No	VISITT
Chemical Treatment—Oxidation/reduction	The technology involves the injection of the material to be converted along with steam into a fluidized bed to break down the molecular structure of organic materials. The constituents of the organic molecules are then reformed into simple gases.	THERMOCHEM, INC. 5570 Sterrett Place Suite 210 Columbia, MD 21044 USA	Gary Voelker/ Lee Rockvam	410-312-6300	Yes	No	No	No	No	No	No	No	No	No	No	No	No	No	No	No	No	VISITT
Delivery/ Extraction Systems	Drilex Systems, Inc. has developed a horizontal drilling system for use as a remediation tool. The principle involves drilling a borehole by directional or horizontal drilling into contaminated areas that are otherwise difficult or impossible to reach.	DRILEX SYSTEMS, INC. 15151 Sommermeyer (ZIP 77041) P.O. Box 801114 Houston, TX 77280-1114 USA	David S. Bardsley	713-937-8888	Yes	Yes	No	Yes	No	No	No	No	No	No	No	No	No	No	No	No	No	VISITT
Delivery/ Extraction Systems	Horizontal wellbores can be extended beneath surface obstacles such as storage tanks, landfills, buildings and natural obstructions to access contaminated zones. A horizontal screen can be placed through a contaminant plume.	EASTMAN CHERRINGTON ENVIRONMENTAL 1055 Conrad Sauer Street Houston, TX 77043-5201 USA	Glenda Empsall	713-722-7777	Yes	Yes	No	Yes	No	No	No	No	No	No	No	No	No	No	No	No	No	VISITT

REMEDIATION TECHNOLOGY MATRIX

Category	Description	Company	Contact	Phone																	Source
Delivery/ Extraction Systems	This full-scale proprietary technology consists of an array of horizontal wells designed on a site-specific basis to be installed through, above, and below contaminated soils or shallow plumes of contaminated groundwater.	HORIZONTAL TECHNOLOGIES, INC. 2309 Hancock Bridge Parkway (33990) P.O. Box 150820 Cape Coral, FL 33915-0820 USA	Donald R. Justice	813-995-8777	Yes	Yes	No	Yes	No	No	No	No	No	No	No	No	No	No	No	No	VISITT
Delivery/ Extraction Systems	The Detoxifier can inject steam and hot air to strip volatile organic compounds (VOC) and semivolitile organic compounds (SVOC), or inject chemical reagents for stabilization/solidification, plus neutralization, oxidation and bioremediation.	NOVATERRA, INC. 2029 Century Park East Suite #890 Los Angeles, CA 90067 USA	Phil La Mori Ph.D.	310-843-3190	Yes	No	No	No	No	No	No	No	No	No	No	No	No	No	No	No	VISITT
Dual Phase Extraction	The SVVS system injects air below the water table, withdrawn vapor above the water table, and stimulates the microbial community to increase bioremediation of less volatile compounds.	BILLINGS & ASSOCIATES, INC. 3816 Academy Parkway N-N.E. Albuquerque, NM 87109 USA	Dr. Gale K. Billings	505-345-1116	Yes	Yes	No	Yes	Yes	Yes	No	Yes	No	No	No	No	No	No	No	No	VISITT
Dual Phase Extraction	The two-phase vacuum extraction technology was designed to remediate soils contaminated with volatile organic compounds (VOC) in situ.	DAMES & MOORE 2325 MD Road Willow Grove, PA 19090 USA	Joseph Tarsavage	215-657-7134	Yes	Yes	No	Yes	Yes	Yes	No	No	No	No	No	No	No	No	No	No	VISITT
Dual Phase Extraction	NoVOCs combines the concept of an air lift pump with in-well vapor stripping to remove VOCs from the groundwater without the need to remove and discharge a waste water stream.	EG & G ENVIRONMENTAL, INC. Foster Plaza VI, Suite 400 681 Andersen Drive Pittsburgh, PA 15220 USA	Wayne J. Dibartola	412-920-5401	Yes	Yes	No	Yes	Yes	No	No	No	No	No	No	No	No	No	No	No	VISITT
Dual Phase Extraction	FE system is a treatment process by injection and Vapor Extraction in situ used to remove volatile organic contaminants (VOC) that are resistant to removal by traditional groundwater pumping systems.	FIRST ENVIRONMENT, INC. 90 Riverdale Road Riverdale, NJ 07457 USA	Rick Dorrier	201-616-9700	Yes	Yes	No	Yes	No	No	No	No	No	No	No	No	No	No	No	No	VISITT

Technology Type	Description	Vendor	Contact	Phone	Soil, Sed, Sldg	GW	Air	In Situ	Ex Situ	VOC	Chl. VOC	BTEX	Petr, Oil, Lubs	Met	SVOC	PAH	Pest	PCB	Asbst	UXO	Radio-nucleides	Ref. Source
Dual Phase Extraction	IT has developed a technique to remove groundwater from depths of 60 to 70 feet with vacuum suction alone, for deeper aquifer zones, mechanical pumping combined with vacuum can be used. Vacuum pumping systems can be installed for the same or lower costs.	IT CORPORATION 2925 Briar Park Houston, TX 77042 USA	John Mastroianni	713-784-2800	Yes	Yes	No	Yes	No	Yes	No	Yes	Yes	No	Yes	No	No	No	No	No	No	VISITT
Dual Phase Extraction	Entrainment Extraction is a dual vacuum extraction that is effective for remediation of soils with low hydraulic conductivity in the vadose and saturated zones where groundwater and soil are both contaminated with VOCs and SVOCs.	TERRA VAC, INC. 1555 Williams Drive Suite 102 Marietta, GA 30066-6282 USA	Charles Pineo	404-421-8008	Yes	Yes	No	Yes	No	Yes	Yes	Yes	Yes	No	Yes	Yes	No	No	No	No	No	VISITT
Electrical Separation	This process moves water and ions through the soil when direct current is passed through the soil. Electrode systems control both the direction and magnitude of the current through the soil. The contaminants may be extracted through the anode or cathode.	ELECTRO-PETROLEUM, INC. 996 Old Eagle School Road Suite 1118 Wayne, PA 19087 USA	Dr. J. Kenneth Wittle	610-687-9070	Yes	No	No	No	No	No	No	No	No	No	Yes	No	No	No	No	No	No	VISITT
Electrical Separation	Electrokinetic soil processing system is powered by direct current (DC) and generates both electrokinetic and electrochemical (EK/EC) functions which jointly remove the contaminants from the soil.	ELECTROKINETICS, INC. LA Business and Technology Center, Suite 155 South Stadium Drive Baton Rouge, LA 70803-6100 USA	Yalcin Acar	504-388-3992	Yes	No	No	Yes	No	Yes	No	No	No	Yes	Yes	No	No	No	No	No	Yes	VISITT
Electrical Separation	The WASPP Corp.'s electrolysis technology treating heavy metals, acids, bases, and cyanides in sludge and slurries. The technology uses alternating current (AC) electrolysis to electrically separate heavy metals and cyanides from sludge.	WATER AND SLURRY PURIFICATION PROCESS 1264 West 101st Avenue Northglenn, CO 80221 USA	W.E. Wright	303-450-7987	No	No								Yes	No							VISITT

REMEDIATION TECHNOLOGY MATRIX

Technology	Description	Company	Contact	Phone	1	2	3	4	5	6	7	8	9	10	11	12	13	14	15	Source
Electro-thermal Gasification — In Situ	Bio-Electrics has developed an innovative, proprietary electro-thermal process for treating organic contaminants in situ through gasification or electrocarbonization.	BIO-ELECTRICS, INC. 1215 W. 12th Street Kansas City, MO 64101 USA	Dr. Erich Sarapuu	816-474-4895	No	No	No	No	No	No	No	No	No	No	No	No	No	No	No	VISITT
Electrokinetic	The technology uses an induced DC electrical field across electrode pairs to transport contaminants in soil.	GEOKINETICS 74 Muth Dr. Orinda, CA 95563 USA	Stephen Clarke		Yes	No	No	Yes	No	No	No	Yes	No	Yes	No	No	No	No	No	Navy Engin. Field Act. West
Emulsion Recycling	Converts POL, PCB, and metal contaminated soil to a construction commodity, e.g., road base material.	CUNNINGHAM-DAVIS ENVIRONMENTAL, INC. 1691 Jenks Drive, Corona, CA 92720 USA	Gordon Dickson		Yes	No	No	No	No	Yes	Yes	Yes	Yes	No	Yes	No	No	No	No	Marketing Brochure
Magnetic Separation	Magnetic Barrier technology separates and concentrates particles according to magnetic susceptibility.	S.G. FRANTZ CO., INC. 31 East Darrah Lane Lawrence Township, NJ 08648 USA	Thomas D. Wellington	609-882-7100	Yes	No	No	Yes	No	No	Yes	Yes	No	No	No	No	No	No	Yes	VISITT
Materials Handling/ Physical Separation	The technology uses washing, liberation and gravity separation techniques to treat waste from battery wrecking operations and produce recyclable products.	CANONIE ENVIRONMENTAL SERVICES CORP. 94 Inverness Terrace East Suite 100 Englewood, CO 80112 USA	Alistair H. Montgomery	303-790-1747	No	No	No	No	No	No	Yes	Yes	No	No	No	No	No	No	No	VISITT
Materials Handling/ Physical Separation	Microfluidizer equipment employs high-energy mixing, involving sheer impact and cavitation in a patented interaction chamber.	MICROFLUIDICS CORP. 90 Oak Street Newton, MA 02164-9101 USA	Irwin Gruverman	617-969-5452	Yes	No	No	No	No	No	No	No	No	No	No	Yes	No	No	No	VISITT
Materials Handling/ Physical Separation	The Petroleum Separation Technology process separates land-farmed petroleum sludge from Superfund or hazardous waste sites into three phases of oil, water, and solids.	ONSITE • OFSITE INC./BATTELLE PNL P.O. Box 999 Richland, WA 99352 USA	Ed Baker	509-376-1494	Yes	No	No	No	No	Yes	No	No	No	No	No	No	No	No	No	VISITT
Materials Handling/ Physical Separation	Portec/Construction Equipment Division offers predesigned ex situ soil remediation equipment and also offers to manufacture equipment for the soil remediation market.	PORTEC, INC. 700 West 21st Street P.O. Box 20 Yankton, SD 57078-0220 USA	Jim Lincoln	605-665-9311	Yes	No	Yes	No	No	No	No	No	No	No	No	No	No	No	No	VISITT

Technology Type	Description	Vendor	Contact	Phone	Soil, Sed, Sldg	GW	Air	In Situ	Ex Situ	VOC	Chl. VOC	BTEX	Petr, Oil, Lubs	Met	SVOC	PAH	Pest	PCB	Asbst	UXO	Radio-nucleides	Ref. Source
Materials Handling/ Physical Separation	The Mini-Miser Dewatering System mechanically extracts liquids from sludge, sediments, and solids. The unit can operate as a secondary dewatering unit for municipal wastewater treatment plant applications.	RECRA ENVIRONMENTAL, INC. 10 Hazelwood Drive Suite 110 Amherst, NY 14228-2298 USA	James F. Ladue	716-691-2600	Yes	No	No	No	Yes	No	No	No	No	No	No	No	No	No	No	No	No	VISITT
Off-gas Treatment	The Beco "Alka/Sorb" process was developed and refined over the years to provide maximum control efficiency on all typical off-gas contaminants, including: acid gases, fly ash, heavy metals (including mercury and cadmium) and dioxins/furans.	BECO ENGINEERING CO. 800 Third Street P.O. Box 443 Oakmont, PA 15139 USA	B. J. Lerner	412-828-6080	No	No	Yes	No	Yes	Yes	Yes	No	No	Yes	Yes	Yes	Yes	Yes	No	Yes	No	VISITT
Off-gas Treatment	Bohn Biofilter Corp. (BBC) has developed biofilter beds to treat a wide variety of industrial pollutant gases from food and waste processing, petroleum refining, chemical processing, tank vents, and polishing of air after solvent recovery.	BOHN BIOFILTER CORP. P.O. Box 44235 Tucson, AZ 85733-4235 USA	Dr. Hinrich Bohn	602-621-7225	No	No	Yes	No	Yes	Yes	No	Yes	Yes	No	Yes	Yes	No	No	No	No	No	VISITT
Off-gas Treatment	FyreZyme is a multifactural liquid (aqueous) agent, combining a rich source of bacterial growth enhancing agents, extracellular enzymes, and bioemulsifiers. Any off gas generated can be bubbled through a FyreZyme solution.	ECOLOGY TECHNOLOGIES INTERNATIONAL, INC. 1225 South 48th Street Suite 2 Tempe, AZ 85281 USA	Pete Condy	602-985-5524	No	No	Yes	No	Yes	Yes	Yes	Yes	Yes	No	Yes	Yes	No	No	No	No	No	VISITT
Off-gas Treatment	Envirogen, Inc. has developed a biological reactor system that can degrade vapor phase chlorinated contaminants, such as TCE in addition to common aromatic hydrocarbons such as BTEX.	ENVIROGEN, INC. Princeton Research Center 4100 Quakerbridge Road Lawrenceville, NJ 08648 USA	George Skladany	609-936-9300	No	No	Yes	No	Yes	Yes	Yes	Yes	Yes	No	No	Yes	No	No	No	No	No	VISITT

REMEDIATION TECHNOLOGY MATRIX

Category	Description	Company	Address	Contact	Phone															
Off-gas Treatment	The acoustic barrier particulate separator separates particulates in a high-temperature gas flow. The separator produces a high intensity acoustic waveform directed against the gas flow, causing particulates to move opposite to the flow.	GENERAL ATOMICS	3550 General Atomic Court San Diego, CA 92121 USA	Anthony Gattuso	619-455-2910	No	No	Yes	No	No	No	No	Yes	No	No	No	No	No	No	VISITT
Off-gas Treatment	Photocatalytic oxidation uses ultraviolet light and a TiO$_2$ catalyst to destroy VOC in air streams. The oxidation of the VOCs occurs at ambient temperature and pressure and utilizes the oxygen and water already present in the air stream.	IT CORPORATION	312 Directors Drive Knoxville, TN 37923 USA	Richard Miller/ Mike Hoffman	615-690-3211	No	Yes	No	Yes	Yes	Yes	Yes	No	No	No	No	No	No	No	VISITT
Off-gas Treatment	KSE's innovative off-gas emission control technology, employs two stages: a liquid phase absorption process and a photocatalytic oxidation process. The AIR-II process provides low concentration streams of contaminated gas than carbon adsorption.	KSE, INC.	665 Amherst Road Sunderland, MA 01375 USA	Dr. Charles W. Quinlan	413-549-5506	No	Yes	No	Yes	Yes	Yes	Yes	No	No	No	No	No	No	No	VISITT
Off-gas Treatment	The VaporSep (TM) system, developed by Membrane Technology and Research, Inc. (MTR), is an innovative membrane vapor separation technology designed to remove and condense organic vapors from contaminated air streams.	MEMBRANE TECHNOLOGY & RESEARCH, INC.	1360 Willow Road, Suite 103 Menlo Park, CA 94025-1516 USA	Vicki Simmons	415-328-2228	No	Yes	No	Yes	Yes	Yes	Yes	Yes	No	No	No	No	No	No	VISITT
Off-gas Treatment	MBI in conjunction with equipment producer Envirex Incorporated, has available a full-scale biofiltration unit which can be used as either a stand-alone unit for vapor treatment or in conjunction with SVE for contaminated soil cleanup.	MICHIGAN BIOTECHNOLOGY INSTITUTE	3900 Collins Road Lansing, MI 48910 USA	Blaine Severin	517-337-3181	No	Yes	No	Yes	Yes	Yes	Yes	No	No	No	No	No	No	No	VISITT

Technology Type	Description	Vendor	Contact	Phone	Soil, Sed, Sldg	GW	Air	In Situ	Ex Situ	VOC	Chl. VOC	BTEX	Petr, Oil, Lubs	Met	SVOC	PAH	Pest	PCB	Asbst	UXO	Radio-nuclides	Ref. Source
Off-gas Treatment	NUCON International, Inc. has developed and commercialized an innovative off-gas treatment system using a proprietary combination of hot nitrogen stripping of adsorbent beds and a reverse Brayton Cycle turbo expansion system.	NUCON INTERNATIONAL, INC. 7000 Huntley Road Columbus, OH 43229 USA	Jack Jacox	614-846-5710	No	No	Yes	No	Yes	Yes	Yes	Yes	Yes	No	No	No	No	No	No	No	No	VISITT
Off-gas Treatment	Process Technologies, Inc.'s. halocarbon and VOC destruction technology is a process in which vapor phase halogens are photolyzed and the resulting by-products react with a solid liner material to produce a clean air stream.	PROCESS TECHNOLOGIES, INC. 910 Main Street Suite 342 Boise, ID 83701 USA	Michael S. Swan	208-385-0900	No	No	Yes	No	Yes	Yes	Yes	Yes	Yes	No	No	No	No	No	No	No	No	VISITT
Off-gas Treatment	The Product Control Systems (PCS) Multi Venturi Scrubber is specifically designed to treat the off-gas generated by thermal desorption, pyrolysis, and incineration waste treatment processes.	PRODUCT CONTROL La Plaiderie, Guernsey St. Peter Port, Channel Islands GY1 3DQ England	G. Edward Someus	(44)48 1 7264 26	No	No	Yes	No	Yes	Yes	No	No	No	Yes	Yes	No	No	No	No	No	No	VISITT
Off-gas Treatment	The PURUS PADRE (TM) vapor treatment process purifies air streams contaminated with VOC directly from soil vapor extraction wells or from groundwater or wastewater air stripping operations.	PURUS, INC. 2713 North First Street San Jose, CA 95134-2000 USA	Paul Blystone	408-955-1000	No	No	Yes	No	Yes	Yes	Yes	Yes	Yes	No	No	No	No	No	No	No	No	VISITT
Off-gas Treatment	The Thermatrix flameless oxidation process maximizes the destruction of waste hydrocarbon vapors in a safe, durable and low-cost system. It achieves greater than 99.99 percent destruction of hydrocarbons including chlorinated compounds.	THERMATRIX, INC. 3590 N. First Street Suite 310 San Jose, CA 95134 USA	John Schofield	408-944-0220	No	No	Yes	No	Yes	Yes	Yes	Yes	Yes	No	Yes	No	Yes	Yes	No	No	No	VISITT

REMEDIATION TECHNOLOGY MATRIX

Technology	Description	Company	Contact	Phone																Source
Off-gas Treatment	Zapit Technology, Inc. has developed a non-thermal, on-site hazardous waste destruction system for treating contaminated vapor streams that utilizes our proprietary electron beam technology.	ZAPIT TECHNOLOGY, INC. 2350 Mission College Blvd. Suite 400 Santa Clara, CA 95054 USA	Ed Daily, Ph.D.	408-986-1700	No	No	Yes	No	No	Yes	Yes	Yes	Yes	No	No	No	No	No	No	VISITT
Phyto-remediation	Uptake of contaminants from soil into aboveground plant tissue, which is harvested and treated periodically.	PHYTOTECH, INC, One Deer Park Drive, Suite #1 Monmouth Junction, NJ 08852 USA	Burt Ensley		Yes	No	No	No	Yes	No	No	No	No	Yes	No	No	No	No	No	EPA TIO Report
Phyto-remediation	Use of plants to remove, contain, or render harmless environmental contaminants. Plants either uptake the contaminants from soil or produce chemicals to immobilize contaminants.	CENTRAL RESEARCH and DEVELOPMENT Dupont Glasgow, Suite 301 Newark, DE 19714 USA	Dr. S. Cunningham		Yes	No	No	No	Yes	No	No	No	No	Yes	No	No	No	No	No	EPA TIO Report
Phyto-remediation	Remediation of contaminants using lagoons and aquatic plants.	CONCURRENT TECHNOLOGIES CORPORATION and CENTER FOR HAZARDOUS MATERIALS RESEARCH 320 William Pitt Way Pittsburgh, PA 15238	M.A. Qazi		No	Yes	No	No	Yes	Yes	No	No	No	No	No	No	No	Yes	No	National Defense Center for Environmental Excellence
Plasma Arc Treatment	A plasma arc torch is used to treat a wide variety of waste streams in a cost-effective, environmentally safe process. Metals and inorganics exit the system as a nonleachable slag. Organic materials are pyrolyzed and burned.	RETECH, DIV. OF LOCKHEED ENV. SYS.&TECH. 100 Henry Station Road, P.O. Box 997 Ukiah, CA 95482 USA	Ron Womack		Yes	Yes	No	Yes	No	Yes	No	No	No	Yes	No	Yes	Yes	Yes	Yes	National Defense Center for Environmental Excellence
Pneumatic Fracturing	Pneumatic Fracturing Extraction, a process developed jointly by Accutech Remedial Systems, Inc. (ARS) and the NJ Institute of Technology (NJIT), is designed to treat in situ contamination located within geologic formations with low permeability.	ACCUTECH REMEDIAL SYSTEMS, INC. Cass Street & Highway 35 Keyport, NJ 07735 USA	John Liskowitz	908-739-6444	Yes	Yes	No	Yes	No	Yes	Yes	Yes	Yes	No	Yes	Yes	Yes	No	No	VISITT
Pneumatic Fracturing	Injection Vac (TM) is an effective method for supplementing vapor extraction in low permeability soils containing volatile and semivolatile compounds.	TERRA VAC, INC. 1555 Williams Drive Suite 102 Marietta, GA 30066-6282 USA	Charles Pineo	404-421-8008	Yes	Yes	No	Yes	No	Yes	Yes	Yes	Yes	No	Yes	No	No	No	No	VISITT

395

Technology Type	Description	Vendor	Contact	Phone	Soil, Sed, Sldg	GW	Air	In Situ	Ex Situ	VOC	Chl. VOC	BTEX	Petr, Oil, Lubs	Met	SVOC	PAH	Pest	PCB	Asbst	UXO	Radio-nucleides	Ref. Source
Pyrolysis	Bio-Electrics, Inc. has developed an innovative technology for in situ electro-pyrolysis of solid and semisolid hydrocarbons.	BIO-ELECTRICS, INC. 1215 W. 12th Street Kansas City, MO 64101 USA	Dr. Erich Sarapuu	816-474-4895	No	No	No	No	No	No	No	No	No	No	No	No	No	No	No	No	No	VISITT
Pyrolysis	Pyrolytic Waste Reclamation utilizes plasma pyrolysis in a controlled nonoxygen environment to break down the molecular structure of organic liquid wastes into its elemental constituents.	ENERGY RECLAMATION, INC. 114-A River Rd Lyme, NH 03768 USA	Ellen Knights	603-795-2403	No	No	No	No	No	Yes	Yes	No	No	No	Yes	No	No	No	No	No	No	VISITT
Pyrolysis	Plasma Energy Applied Technology has developed the thermal destruction and recovery process to treat materials containing both organic and inorganic compounds. PEAT's TDR system operates in a controlled pyrolysis mode when treating organics.	PLASMA ENERGY APPLIED TECHNOLOGY (PEAT) 4914 Moores Mill Rd. Chase Industrial Park Huntsville, AL 35811 USA	R. D. Dupree, Jr.	205-859-3006	Yes	No	No	No	Yes	No	No	No	No	Yes	Yes	No	No	No	Yes	Yes	No	VISITT
Pyrolysis	The Flash Pyrolysis a reductive, indirect, thermal treatment process of the organic and/or inorganic hazardous and nonhazardous solid waste material.	PRODUCT CONTROL La Plaiderie, Guernsey St. Peter Port, Channel Islands GY1 3DQ England	G. Edward Someus	481-726-426	No	No	No	No	No	Yes	No	Yes	Yes	Yes	Yes	Yes	No	No	No	No	No	VISITT
Pyrolysis	The device is an Incandescently Heated Waste Disposal System by which the process changes the composition of the waste to render it nonhazardous. The process is achieved by exposing the waste to electrically generated intense incandescent heat.	VANCE IDS, INC. 1050 Miller Drive Altamonte Springs, FL 32701 USA	Randy Smith	407-834-2809	No	No	No	No	No	No	No	No	No	Yes	No	No	No	No	No	No	No	VISITT
Slagging	The Horsehead Resource Development Co., Inc. (HRD) Flame Reactor technology is a High Temperature Metal Recovery (HTMR) process for the treatment of metal-bearing wastes.	HORSEHEAD RESOURCE DEVELOPMENT CO., INC. 300 Frankfort Road Monaca, PA 15061 USA	Regis J. Zagrocki	412-773-2289	Yes	No	No	No	Yes	Yes	Yes	No	No	Yes	No	No	No	No	No	No	No	VISITT

REMEDIATION TECHNOLOGY MATRIX

Technology	Description	Company	Contact	Phone															Source
Slagging	Hazardous material in slurry form is pumped to the injector of an entrained bed gasifier and partially oxidized at pressures above 20 atm and temperatures between 2,200°F and 2800°F.	TEXACO, INC. 2000 Westchester Avenue White Plains, NY 10650 USA	Richard B. Zang, P.E.	914-253-4047	Yes	No	No	No	Yes	No	No	No	No	No	No	No	No	No	VISITT
Soil Flushing	Use of water or a chemical reagent to solubilize the contaminants so that the contaminants can be extracted.	HORIZONTAL TECHNOLGIES, INC.	Donald Justice		Yes	No	Yes	No	No	No	No	No	Yes	No	No	No	No	No	EPA TIO Report
Soil Flushing	Soil flushing promotes mobility and migration of metals by solubilizing the contaminants so that the contaminants can be extracted. Leachate is recovered by pump-and-treat methods.	SURTEK INC.	Ken Wyatt		Yes	No	Yes	No	No	No	No	No	Yes	No	No	No	No	No	EPA TIO Report
Soil Flushing—In Situ	This proprietary technology of Horizontal Technology, Inc. consists of a system of trenched horizontal wells designed on a site-specific basis to be installed in plumes of contaminated groundwater and soils.	HORIZONTAL TECHNOLOGIES, INC. 2309 Hancock Bridge Parkway (33990) P.O. Box 150820 Cape Coral, FL 33915-0820 USA	Donald R. Justice/Greg Rawl	813-995-8777	Yes	No	Yes	No	No	No	Yes	Yes	No	Yes	No	No	No	No	VISITT
Soil Vapor Extraction	Vacuum extraction involves withdrawing soil vapors from the subsurface through an extraction well, a well which is screened in soil contaminated with VOC. As the contaminated vapor is extracted from the subsurface soils, a cocentration gradient is created.	DAMES & MOORE 2325 MD Road Willow Grove, PA 19090 USA	Joseph M. Tarsavage	215-657-5000	Yes	No	Yes	Yes	Yes	Yes	Yes	Yes	Yes	Yes	No	No	No	No	VISITT
Soil Vapor Extraction	The integrated AQUADETOX/VES continuous system developed and patented by AWD Technologies, Inc. simultaneously treats groundwater and soil contaminated with volatile organic compounds (VOC).	DOW ENVIRONMENTAL, INC. 15204 Omega Drive Suite 200 Rockville, MD 20850 USA	Carol Bind	301-948-0040	Yes	No	Yes	No	Yes	Yes	Yes	Yes	Yes	Yes	No	No	No	No	VISITT

Technology Type	Description	Vendor	Contact	Phone	Soil, Sed, Sldg	GW	Air	In Situ	Ex Situ	VOC	Chl. VOC	BTEX	Petr, Oil, Lubs	Met	SVOC	PAH	Pest	PCB	Asbst	UXO	Radio-nucleides	Ref. Source
Soil Vapor Extraction	Soil vapor extraction is a nondisruptive, economical solution to soil and groundwater contamination. Soil vapor extraction is a demonstrated successful technology for the in situ remediation of subsurface contamination.	ENVIROGEN, INC. 480 Neponset Street Canton, MA 02021 USA	Scott Drew/ Sally Hulsman	617-821-5560	Yes	No	No	Yes	No	Yes	Yes	Yes	Yes	No	Yes	No	No	No	No	No	No	VISITT
Soil Vapor Extraction	Geo-Con implemented the first full-scale project to use Shallow Soil Mixing/Thermally Enhanced Vapor Extraction and Soil Mixing Extraction to remove VOC contamination from a waste disposal area.	GEO-CON, INC. 4075 Monroeville Boulevard Corporate One, Building II, Suite 400 Monroeville, PA 15146 USA	Linda M. Ward	412-856-7700	Yes	No	No	Yes	No	Yes	Yes	No	No	No	No	No	Yes	No	No	No	No	VISITT
Soil Vapor Extraction	Soil venting or soil vapor extraction is a technology that uses either active or passive equipment to remove air and gases from the subsurface. This technology is applicable for remediation of soil and groundwater contaminated with VOCs.	IT CORPORATION 2925 Briar Park Houston, TX 77042 USA	John Mastroianni	713-784-2800	Yes	No	No	Yes	No	Yes	Yes	Yes	Yes	No	Yes	No	No	No	No	No	No	VISITT
Soil Vapor Extraction	Soil Vapor Extraction (SVE) is the process for removal and treatment of volatile organic compounds (VOC) from the vadose or unsaturated zone. Air streams contaminated with VOCs are extracted from wells and purified in the treatment unit.	KAP & SEPA, LTD. KAP Ltd. Skokanska 80 169 00 Praha 6 Prague, Czech Republic	Peter Kohout (KAP) Kinkor Vlad (SEPA)	(42)2 5200 50	Yes	No	No	Yes	Yes	Yes	Yes	Yes	Yes	No	Yes	No	No	No	No	No	No	VISITT
Soil Vapor Extraction	Soil vapor extraction provides effective economical remediation of volatile and semivolatile contaminants from the vadose zone and capillary fringe soils. Soil vapor extraction provides increased circulation of air through the soil vapor phase.	QUATERNARY INVESTIGATIONS, INC. 300 W. Olive Street Suite A Colton, CA 92324 USA	Tony Morgan	800-423-0740	Yes	No	No	Yes	No	Yes	No	Yes	Yes	No	Yes	No	No	No	No	No	No	VISITT

REMEDIATION TECHNOLOGY MATRIX

Technology	Description	Company	Contact	Phone														Source
Soil Vapor Extraction	Vapor Extraction is an effective method for in situ remediation of vadose zone soils containing VOCs and SVOCs. Vapor extraction wells are installed into the area of contamination with special care taken in the selection of screen depth.	TERRA VAC, INC. 1555 Williams Drive Suite 102 Marietta, GA 30066-6282 USA	Charles Pineo	404-421-8008	Yes	No	Yes	Yes	Yes	Yes	Yes	No	Yes	No	No	No	No	VISITT
Soil Vapor Extraction	DDC system uses air stripping to remove VOCs, including petroleum hydrocarbons, by passing air through the groundwater circulated in the well bore. In addition, the DDC supplies oxygen to promote in situ bioremediation of petroleum hydrocarbons.	WASATCH ENVIRONMENTAL INC., Salt Lake City, UT	Todd W. Schrauf		No	Yes	No	No	No	Yes	Yes	No	No	No	No	No	No	Air force Center for Environmental Excellence
Soil Washing	ARS has developed a technology to treat radioactively contaminated soils and render them suitable for free-release as per NRC Option 1 limits. This process extracts the radionuclides and reduces overall volume of contaminated soil.	ADVANCED RECOVERY SYSTEMS, INC. 1219 Banner Hill Road Erwin, TN 37650 USA	Steve Schutt	615-743-6186	Yes	No	No	Yes	No	No	No	No	No	No	No	No	Yes	VISITT
Soil Washing	AEA Technology's feed preparation and mineral processing techniques for treatment of soil contaminated with metals, petroleum hydrocarbons, and PNAs. Feed preparation processes evaluated include scrubbing, cycloning, and classifying.	AEA TECHNOLOGY Culham Abingdon, Oxfordshire OX14 3DB England	Dr. Peter Allan Wood	123-546-3194	Yes	No	No	Yes	No	No	Yes	No	Yes	No	No	No	No	VISITT
Soil Washing	Soil washing is a physical/chemical process to remove contaminants that reside in specific grain-size domains. It is a batch process and separates the wastestream into "cuts" and focuses on treatment on contaminant/grain size relationship.	ALTERNATIVE REMEDIAL TECHNOLOGIES, INC. 14497 North Dale Mabry Highway Suite 140 Tampa, FL 33618 USA	Michael J. Mann, P.E./ Jill Besch	813-264-3506	Yes	No	No	Yes	No	No	Yes	No	Yes	Yes	No	No	Yes	VISITT

Technology Type	Description	Vendor	Contact	Phone	Soil, Sed, Sldg	GW	Air	In Situ	Ex Situ	VOC	Chl. VOC	BTEX	Petr, Oil, Lubs	Met	SVOC	PAH	Pest	PCB	Asbst	UXO	Radio-nucleides	Ref. Source
Soil Washing	The B&W-NESI ECOSAFE (TM) soil washing process consists of separating contaminated soils into a coarse size fraction and a fine size fraction which represent the clean and contaminated portions, respectively.	B&W NUCLEAR ENVIRONMENTAL SERVICES, INC. 2220 Langhorne Road Lynchburg, VA 24506-0548 USA	T.R. Welch Or Lewis Walton 804-948-4647	804-948-4765	Yes	No	No	No	Yes	No	No	No	No	No	No	No	No	No	No	No	Yes	VISITT
Soil Washing	BenCHEM has developed an aqueous, ambient temperature and pressure process to scrub heavy metals from soil. One reagent is a weak acid solution, all the other reagents are more gentle and biodegradable.	BENCHEM 803 South Negley Avenue, Suite 1 Pittsburgh, PA 15232 USA	Robert Bender	412-361-1426	Yes	No	No	No	Yes	No	No	No	No	Yes	No	No	No	No	No	No	No	VISITT
Soil Washing	Soil/sediment washing is a water-based, volumetric reduction/waste minimization process whereby hazardous contaminants are extracted and concentrated into a small residual portion of the original volume using physical and chemical methods.	BERGMANN USA 1550 Airport Road Gallatin, TN 37066-3739 USA	Richard P. Traver, P.E.	615-452-5500	Yes	No	No	No	Yes	No	No	Yes	Yes	Yes	Yes	No	Yes	Yes	No	Yes	Yes	VISITT
Soil Washing	Soil washing is a vigorous, water-based, scrubbing process for excavated soil which uses mineral processing technology. As an on-site volume reduction process, it generates "clean" washed soil, a contaminated sludge or cake.	BIOTROL, INC. 10300 Valley View Road Eden Prairie, MN 55344 USA	Sandy Clifford	612-942-8032	Yes	No	No	No	Yes	No	No	No	No	Yes	Yes	Yes	Yes	Yes	No	No	No	VISITT
Soil Washing	Soil washing is a process of mixing contaminated soil with water ex situ and mechanically scrubbing and separating the soil fractions to remove the contaminants.	CANONIE ENVIRONMENTAL SERVICES CORP. 94 Inverness Terrace East Suite 100 Englewood, CO 80112 USA	Alistair H. Montgomery	303-790-1747	Yes	No	No	No	Yes	No	No	No	No	Yes	No	No	No	No	No	No	No	VISITT

REMEDIATION TECHNOLOGY MATRIX

Technology	Description	Company	Contact	Phone															
Soil Washing	The SRS-10 is a soil washing system that combines physical separation with a special hydrocarbon mitigation agent that significantly enhances removal of oil, gasoline, organic liquids and solids, and other contaminants.	DIVESCO, INC. 5000 Highway 80 East Jackson, MS 39208 USA	W. L. Strickland	601-932-1934	Yes	No	No	Yes	No	Yes	No	Yes	No	No	No	No	No	No	VISITT
Soil Washing	This technology involves soil washing for the leaching of hazardous metals from soils in a non-acidic, proprietary solution. After mixing the proprietary leaching solution with the soil, solids are separated from the washing liquid.	EARTH DECONTAMINATORS, INC. (EDI) 2803 Barranca Parkway Irvine, CA 92714 USA	Luis Pommier	714-262-2292	Yes	No	No	Yes	No	No	No	No	No	No	No	No	No	No	VISITT
Soil Washing	NuKEM Development has developed a soil cleaning process designed to remove hydrocarbons from contaminated soils.	ENSR CONSULTING AND ENGINEERING 3000 Richmond Avenue Houston, TX 77098 USA	Saeed Darian	713-520-9900	Yes	No	No	Yes	No	Yes	Yes	Yes	Yes	No	No	No	No	No	VISITT
Soil Washing	Genesis' elutriant blend, a biodegradable solution of proprietary surfactants, nutrients, flocculants, pH buffers, and cultured bacteria, is precisely metered into the system to selectively emulsify and degrade the various petroleum constituents.	GENESIS ECO SYSTEMS, INC. 3341 Fitzgerald Road, Suite D Rancho Cordova, CA 95742 USA	Ken Crabtree	916-638-5733	Yes	No	No	Yes	Yes	Yes	Yes	Yes	Yes	No	No	No	No	No	VISITT
Soil Washing	Heap leach technology involves the placing of contaminated soil, sludge, or solid waste on a drainage blanket over an impervious pad and percolating appropriate leach solutions through the waste, under unsaturated flow.	GEOCHEM DIVISION OF TERRA VAC 12596 West Bayaud Avenue, Suite 205 Lakewood, CO 80228 USA	Jim V. Rouse	303-988-8902	No	No	No	No	No	No	No	No	No	No	No	No	No	No	VISITT
Soil Washing	Geocycle Environment has developed a soil washing technology for the treatment of soil contaminated by organic and inorganic compounds.	GEOCYCLE ENVIRONMENT, INC. 2630 Blv. Industrial Chambly, Quebec J3L-4V2 Canada	Pierre Geoffroy/ Francoys Drouin	514-447-5252	Yes	No	No	Yes	No	No	No	Yes	Yes	No	Yes	No	No	No	VISITT

Technology Type	Description	Vendor	Contact	Phone	Soil, Sed, Sldg	GW	Air	In Situ	Ex Situ	VOC	Chl. VOC	BTEX	Petr, Oil, Lubs	Met	SVOC	PAH	Pest	PCB	Asbst	UXO	Radio-nucleides	Ref. Source
Soil Washing	Hydriplex Incorporated has developed a unique soil washing technology that uses its HP-80 compound (a modified sodium silicate) and the hydrocleaner to clean hydrocarbon contaminated soil and hydrocarbon sludge.	HYDRIPLEX, INC. 14730 Sandy Creek Drive Houston, TX 77070 USA	John S. Crowley/ Gary Walter	713-370-2778	Yes	No	No	No	Yes	Yes	No	Yes	Yes	Yes	Yes	Yes	No	No	No	No	No	VISITT
Soil Washing	The ozone-processed, flocculated, filtered water runs through a reverse osmosis system to remove salts. The system will also work for other than petroleum contaminated soils, such as metals, PCBs, SVOCs, VOCs.	KINIT ENTERPRISES 6363 NW 6th Way Suite 210 Ft. Lauderdale, FL 33309 USA	Claus D. Tonn/ Robert Kuhnle	305-776-4829	Yes	No	No	No	Yes	No	No	No	No	Yes	No	No	No	No	No	No	No	VISITT
Soil Washing	The contaminant can be removed by physical segregation. The means utilized will be specific to the mineralogical and morphologic occurrence of the contaminant. Some of the physical separations may be aided by the addition of reagents.	LOCKHEED CORPORATION 980 Kelly Johnson Drive Las Vegas, NV 89119 USA	Ron May	702-897-3313	Yes	No	No	No	Yes	No	No	No	No	Yes	No	No	No	No	No	No	Yes	VISITT
Soil Washing	Soil washing is an ex-situ process which incorporates size classification and vigorous scrubbing of soil particles with water to remove heavy metals or organic contaminants.	OHM REMEDIATION SERVICES CORPORATION 5731 West Las Positas Blvd. Pleasanton, CA 94588 USA	Dwight Gemar	510-227-1105	Yes	No	No	No	Yes	No	No	No	No	Yes	No	Yes	No	No	No	No	No	VISITT
Soil Washing	The "CALOCROMA" soil washing process utilizes mining and enhanced oil recovery techniques economically and efficiently to achieve immediate separation of hydrocarbons and solvents from the soil.	ON-SITE TECHNOLOGIES, INC. 1715 South Bascom Avenue Campbell, CA 95008 USA	Masood Ghassemi, Ph.D., P.E.	408-371-4810	Yes	No	No	No	Yes	Yes	Yes	Yes	Yes	Yes	No	Yes	Yes	No	No	No	No	VISITT

REMEDIATION TECHNOLOGY MATRIX

Technology	Description	Company	Contact	Phone	C1	C2	C3	C4	C5	C6	C7	C8	C9	C10	C11	C12	C13	C14	C15	Source
Soil Washing	The technology is based on the typical distribution of radionuclides in many contaminated soils. Radionuclides are located primarily in specific size fractions, typically the fine silt and clay fractions, as water insoluble substances.	SANFORD COHEN AND ASSOCIATES, INC. 1000 Monticello Court Montgomery, AL 36117 USA	Charles R. Phillips/ William Richardson	334-272-2234	Yes	No	No	Yes	No	No	No	No	No	No	No	No	No	No	Yes	VISITT
Soil Washing	Soil washing is an effective means of volume reduction. Its success is based on the principle that soil contaminants tend to be associated with the fines and organic portions of the soil.	SOIL TECHNOLOGY, INC. 7865 NE Day Road West Bainbridge Island, WA 98110 USA	Richard Sheets	206-842-8977	Yes	No	No	Yes	No	No	No	No	No	No	No	No	No	No	No	VISITT
Soil Washing	Technology Scientific, Ltd. has developed the innovative Flow Consecutor Technology. The technology implements the intensively acting tubular agitator, to replace any traditional agitator, for processing multi-phase mixtures.	TECHNOLOGY SCIENTIFIC, LTD. 152 Ranch Estates Dr. N.W. Calgary, Alberta T3G 1K4 Canada	Dr. Richard Petela	403-239-1239	Yes	No	No	Yes	Yes	Yes	No	No	No	No	No	No	No	No	No	VISITT
Soil Washing	Soils are washed and classified with heated pressurized water in a series of countercurrent extraction augers, screens, cyclones and centrifuges. Chemical treatments for reducing contaminants concentrations are used where appropriate.	TUBOSCOPE VETCO ENVIRONMENTAL SERVICES 2835 Holmes Rd Houston, TX 77051 USA	Dr. Myron I. Kuhlman/ Haridur Karlsson	713-799-5289	Yes	No	No	Yes	No	No	No	No	No	Yes	Yes	Yes	No	No	Yes	VISITT
Soil Washing	Soil washing removes the contaminants from soil such that a large portion of the inlet soil is cleaned and discharged. The extracted contaminants are concentrated in the remaining, smaller portion of the soil for disposal.	WESTINGHOUSE REMEDIATION SERVICES, INC. 675 Park North Boulevard Building F, Suite 100 Clarkston, GA 30021 USA	William E. Norton, P.E.	404-299-4736	Yes	No	No	Yes	No	No	No	No	No	No	No	No	Yes	No	Yes	VISITT
Solvent Extraction	The process, performed in a closed system, consists of a mechanical screening of the contaminated soil, an extraction process, drainage of the extraction liquid, and a stripping process.	A/S PHOENIX CONTRACTORS/ PHOENIX MILJOE Fuglesangsallé 14 DK-6600 Vejen, Ribe Amt Denmark	Poul Thorn	75 3611 11	Yes	No	No	Yes	Yes	Yes	Yes	Yes	Yes	Yes	No	No	No	No	No	VISITT

Technology Type	Description	Vendor	Contact	Phone	Soil, Sed, Sldg	GW	Air	In Situ	Ex Situ	VOC	Chl. VOC	BTEX	Petr, Oil, Lubs	Met	SVOC	PAH	Pest	PCB	Asbst	UXO	Radio-nucleides	Ref. Source
Solvent Extraction	LEEP (Low Energy Extraction Process), a patented process, uses common organic solvent to extract and concentrate organic pollutants from soil, sediments, and sludge.	ART INTERNATIONAL, INC. 100 Ford Road Denville, NY 07834 USA	Werner Steiner/ Genya Mallach	201-627-7601	Yes	No	No	No	Yes	Yes	No	No	No	No	No	Yes	No	No	No	No	No	VISITT
Solvent Extraction	CF Systems' LG-SX solvent extraction technology uses liquefied gas solvents to extract organic components from a feed stream.	CF SYSTEMS CORPORATION 3D Gill Street Woburn, MA 01801 USA	Christopher Shallice	617-937-0800	Yes	No	No	No	Yes	Yes	Yes	Yes	Yes	No	Yes	Yes	Yes	Yes	No	No	No	VISITT
Solvent Extraction	The Carver-Greenfield Process is an innovative, well-proven process which can be applied at Superfund sites for cleanup of hazardous sludge, contaminated solids, or other industrial wastes.	DEHYDRO-TECH CORPORATION 6 Great Meadow Lane East Hanover, NJ 07936 USA	Theodore D. Trowbridge	201-887-2182	Yes	No	No	No	Yes	Yes	No	Yes	Yes	No	Yes	Yes	Yes	Yes	No	No	No	VISITT
Solvent Extraction	This process uses solvent to extract organic contaminants from various wastes. The organics will be separated from the solvent by distillation.	EM&C ENGINEERING ASSOCIATES 1665 Scenic Avenue Suite 104 Costa Mesa, CA 92626 USA	Mohamed Elgafi	714-957-6429	No	No	No	No	No	No	No	No	No	No	No	No	No	No	No	No	No	VISITT
Solvent Extraction	A continuous solvent extraction process to remove hydrocarbons from fine-sized soil particles. The system consists of two well-known chemical processes — aqueous soil washing followed by solvent extraction.	ENSR CONSULTING AND ENGINEERING 3000 Richmond Avenue Houston, TX 77098 USA	Saeed Darian	713-520-9900	Yes	No	No	No	Yes	Yes	No	Yes	Yes	No	Yes	Yes	No	Yes	No	No	No	VISITT
Solvent Extraction	Envirogen has developed and patented an innovative solvent extraction technology (SoPE) that uses a solid phase material (polystyrene beads) as the extractant.	ENVIROGEN, INC. 4100 Quakerbridge Road Lawrenceville, NJ 08648 USA	Michael Shannon, Ph.D.	609-936-9300	Yes	No	No	No	Yes	No	No	No	No	No	No	Yes	No	Yes	No	No	No	VISITT

Technology	Description	Company	Contact	Phone																	
Solvent Extraction	This technology is based on a carbon dioxide (CO_2)-water extraction process which requires neither heat or extreme pressures and will remove the heteroatom containing oil products with no physical modification or destruction of the oil.	GEO-MICROBIAL TECHNOLOGIES, INC. East Main Street P.O. Box 132 Ochelata, OK 74051 USA	Daniel Hitzman	918-535-2281	No	No	No	No	No	No	No	No	No	No	No	No	No	No	No	No	VISITT
Solvent Extraction	CAPSUR(R) is a foam-applied, aqueous based solvent system with emulsifiers developed for the cleanup of polychlorinated biphenyl (PCB) spills on solid surfaces.	INTEGRATED CHEMISTRIES, INC. 1970 Oakcrest Avenue Suite 215 St. Paul, MN 55113 USA	Cathy Iverson	612-636-2380	No	No	No	No	No	No	No	No	No	No	No	No	No	No	No	No	VISITT
Solvent Extraction	Solvent extraction soil remediation (SESR) is an ex situ process which uses an innovative liquid phase agglomeration (LPA) step to separate fine, dispersed particles from the extracting solvent.	NATIONAL RESEARCH COUNCIL OF CANADA Bldg. M12, Montreal Road Ottawa, Ontario K1A OR6 Canada	Terry Kimmel	613-990-8231	Yes	No	No	Yes	No	No	No	No	No	No	No	No	No	No	No	No	VISITT
Solvent Extraction	The B.E.S.T. Process is a patented solvent extraction technology utilizing triethylamine as the solvent. Triethylamine is a biodegradable solvent formed by reacting ammonia and ethyl alcohol.	RESOURCES CONSERVATION CO. 3630 Cornus Lane Ellicott City, MD 21042 USA	Lanny D. Weimer	301-596-6066	Yes	No	No	Yes	No	Yes	Yes	Yes	Yes	Yes	Yes	No	Yes	No	No	No	VISITT
Solvent Extraction	SRE's solvent extraction (SOLV-EX) is an innovative and cost-effective process to treat soils, dredged sludge, and emulsions containing volatile or semivolatile organics, oils, grease, and coal tar compounds.	SRE, INC. 158 Princeton Street Nutley, NJ 07110 USA	Sam Sofer	201-661-5192	Yes	No	No	Yes	No	No	Yes	Yes	Yes	Yes	Yes	Yes	No	Yes	No	No	VISITT
Solvent Extraction	The Terra-Kleen technology for the mobile soil treatment unit removes contaminants from excavated soils, debris, and sediments using non-toxic, environmentally friendly solvents.	TERRA-KLEEN RESPONSE GROUP, INC. 7321 N. Hammond Avenue OK City, OK 73132 USA	Alan B. Cash	405-728-0001	Yes	No	No	Yes	No	Yes	Yes	Yes	Yes	Yes	Yes	Yes	Yes	Yes	No	No	VISITT

Technology Type	Description	Vendor	Contact	Phone	Soil, Sed, Sldg	GW	Air	In Situ	Ex Situ	VOC	Chl. VOC	BTEX	Petr, Oil, Lubs	Met	SVOC	PAH	Pest	PCB	Asbst	UXO	Radio-nucleides	Ref. Source
Solvent Extraction	Innovative process to treat soils containing VOCs, SVOCs, pesticides, and PCBs.	PRECISION WORKS, INC. 126 Wood Road, Suite 112 Camarillo, CA 93010 USA	David Hedman		Yes	No	No	Yes	No	Yes	Yes	No	No	No	Yes	No	Yes	Yes	No	No	No	Vendor Brochure
Sorbent Treatment	Sorbent materials used to treat and stabilize waste streams.	Defense Logistic Agency	Mike Sullivan		No	Yes	No	No	Yes	Yes	Yes	Yes	No	Yes	No	No	No	No	No	No	No	National Defense Center for Environmental Excellence
Surfactant Enhanced Recovery— In Situ	Ecosite applies the basic principles of soil flushing technology to treat saturated and unsaturated soil while improving the distribution of injected flushing solution and underground water control.	ECOSITE, INC. 965, rue Newton, Suite 270 Quebec, Quebec G1P 4M4 Canada	Charles Boulanger	418-872-3600	Yes	Yes	No	Yes	No	Yes	Yes	Yes	Yes	No	Yes	No	No	Yes	No	No	No	VISITT
Surfactant Enhanced Recovery— In Situ	S.S. Papadopulos & Associates, Inc. can extract non-aqueous phase liquids from an aquifer using an innovative technology called Detergent Extraction of Non-Aqueous Phase Liquids in the Subsurface.	S.S. PAPADOPULOS & ASSOCIATES, INC. 7944 WI Avenue Bethesda, MD 20814-3620 USA	Jim Lolcama, M.SC., Remy J.C. Hemett, Ph.D	301-718-8900	No	Yes	No	Yes	No	No	No	Yes	Yes	No	No	No	No	Yes	No	No	No	VISITT
Surfactant Enhanced Recovery— In Situ	Surtek flushing is an in-situ technology that is used to recover nonaqueous phase liquids from groundwater and soils. This method recovers NAPLS by reducing the trapping forces that immobilize them, and solubilizing the residuals.	SURTEK, INC. 1511 Washington Avenue Golden, CO 80401 USA	Kon Wyatt	303-278-0877	Yes	Yes	No	Yes	No	No	No	No	Yes	No	No	Yes	No	No	No	No	No	VISITT
Thermal Desorption	Treatment begins when material enters the counter current flow rotary drier. The vaporization of contaminants is caused by heat. Soil is exposed to a heated air stream elevating soil temperatures to levels of 850-1000°F.	ADVANCED ENVIRONMENTAL SERVICES, INC. Corporate Centre 200, Box 160 200 35th Street Marion, IA 52302-0160 USA	Tad Cooper	800-289-7371	Yes	No	No	No	Yes	Yes	Yes	Yes	Yes	No	Yes	Yes	No	No	No	No	No	VISITT

REMEDIATION TECHNOLOGY MATRIX

Technology	Description	Company	Contact	Phone												Source
Thermal Desorption	AST has thermal treatment units available in two rated capacities, 8 and 30 tons per hour. Contaminated soil is heated in the counter flow rotary kiln, the process gas combines with evaporated soil moisture and volatilized hydrocarbon.	ADVANCED SOIL TECHNOLOGIES 4570 Churchill Street Suite 3000 St. Paul, MN 55126-2222 USA	Kirk D. Shellum	612-486-7000	Yes	No	No	Yes	Yes	Yes	No	Yes	Yes	No	No	No	VISITT
Thermal Desorption	Ariel Industries, Inc. designs and manufactures low temperature thermal desorption systems for soil remediation. The system is designed specifically for processing soils contaminated with VOC and hydrocarbons at 15-60 tons per hour.	ARIEL INDUSTRIES, INC. 2204 Industrial South Road Dalton, GA 30721 USA	Timothy L. Boyd	706-277-7070	Yes	No	No	Yes	Yes	Yes	No	Yes	Yes	No	No	No	VISITT
Thermal Desorption	This patented thermal desorption process utilizes high temperatures to volatilize organics from a waste matrix in a nonoxidizing atmosphere, while pulverizing the waste material.	BIRD ENVIRONMENTAL GULF COAST, INC. 2700 Avenue S San Leon, TX 77539 USA	Brad Hogan	713-339-1352	Yes	No	No	Yes	Yes	Yes	No	Yes	Yes	No	No	No	VISITT
Thermal Desorption	The low-temperature thermal aeration technology thermally separates contaminants like chlorinated and nonchlorinated hydrocarbons, solvents and VOC, chlorinated and nonchlorinated pesticides, and low levels of PAHs from excavated soils.	CANONIE ENVIRONMENTAL SERVICES CORP. 800 Canonie Drive Porter, IN 46304 USA	Douglas Anderson	219-926-8651	Yes	No	No	Yes	Yes	Yes	No	Yes	Yes	No	No	No	VISITT
Thermal Desorption	The medium temperature thermal desorption (MTTD) process operated by Carlo Environmental Technologies (CET) can be a highly advantageous choice of remediation for many hydrocarbon contaminated sites.	CARLO ENVIRONMENTAL TECHNOLOGIES, INC. 44907 Trinity Drive P.O. Box 744 Clinton Township, MI 48038-0744 USA	Keith Flemingloss	810-468-9580	Yes	No	No	Yes	Yes	No	No	Yes	Yes	No	No	No	VISITT

408   STRATEGIES FOR ACCLERATING CLEANUP AT TOXIC WASTE SITES

Technology Type	Description	Vendor	Contact	Phone	Soil, Sed, Sldg	GW	Air	In Situ	Ex Situ	VOC	Chl. VOC	BTEX	Petr, Oil, Lubs	Met	SVOC	PAH	Pest	PCB	Asbst	UXO	Radio-nucleides	Ref. Source
Thermal Desorption	Carson Environmental has developed a transportable thermal desorption off-gas treatment method for remediating particulate media utilizing the oxidative capabilities of ozone hydrogen peroxide and ultraviolet light.	CARSON ENVIRONMENTAL 11734 Gateway Boulevard Los Angeles, CA 90064 USA	R.Carson Later	310-478-0792	No	No	No	No	No	Yes	Yes	Yes	Yes	No	Yes	No	No	No	No	No	No	VISITT
Thermal Desorption	Caswan Environmental Services Ltd.'s Thermal Distillation and Recovery treatment process uses the principles of volatilization and condensation to remove hydrocarbon contaminants from the waste matrix.	CASWAN ENVIRONMENTAL SERVICES LTD. Bay 1, 2916 5th Avenue, N.E. Calgary, Alberta T2A 6K9 Canada	R. Welke	403-235-9333	Yes	No	No	No	Yes	Yes	No	Yes	Yes	No	Yes	Yes	Yes	No	No	No	No	VISITT
Thermal Desorption	Thermal Desorption volatilizes contaminants from soil at relatively low temperatures and at conditions below the lower explosive limit (L.E.L.).	CLEAN-UP TECHNOLOGY, INC. 145 West Walnut Street, Gardena, CA 90248 USA	Ron Morris	310-327-8605	Yes	No	No	No	Yes	Yes	No	Yes	Yes	No	Yes	Yes	Yes	No	No	No	No	VISITT
Thermal Desorption	The low temperature thermal absorber uses a rotary kiln combining proven equipment and design with innovative aerospace technology.	CONTAMINATION TECHNOLOGIES, INC. 348 Turnpike Street Canton, MA 02021 USA	David Hodgdon or David Lesky	617-575-8920	Yes	No	No	No	Yes	Yes	Yes	Yes	Yes	No	Yes	No	Yes	Yes	No	Yes	No	VISITT
Thermal Desorption	ConTeck's thermal desorption system heats soil from 500 to 1,000°F, depending upon the temperature needed to volatilize the organic constituent(s) from the soil particles.	CONTECK ENVIRONMENTAL SERVICES, INC. 22460 Highway 169 Northwest Elk River, MN 55330-9235 USA	Chris Kreger	612-441-4965	Yes	No	No	No	Yes	Yes	No	Yes	Yes	No	Yes	Yes	No	No	No	No	No	VISITT
Thermal Desorption	Covenant Environmental Technologies, Inc. has developed a Mobile Retort Unit (MRU) which, in the absence of air, can remove hydrocarbons from contaminated soil or other media.	COVENANT ENVIRONMENTAL TECHNOLOGIES, INC. 65 Germantown Court Suite 210 Memphis, TN 38018 USA	Rick P. Newman	901-759-5874	Yes	No	No	No	Yes	Yes	No	Yes	Yes	No	Yes	No	No	Yes	No	No	No	VISITT

REMEDIATION TECHNOLOGY MATRIX

Technology	Description	Company	Contact	Phone														Source
Thermal Desorption	A fully transportable system consists essentially of three semi-trailer loads; it includes a primary unit (rotary kiln), a secondary unit (thermal oxidizer), and ancillary components such as control house and discharge augers/conveyors.	DBA, INC. 773 Lido Drive Livermore, CA 94550 USA	Hal Miller	510-447-4711	Yes	No	Yes	No	Yes	Yes	No	No	No	No	No	No	No	VISITT
Thermal Desorption	ETTS is a rotary kiln based thermal treatment system for soil contaminated with various chemicals. The rotary kiln operates at a high throughput and low temperature. In an afterburner the evaporated contaminants are oxidized at a high temperature.	ECOTECHNIEK B.V. Het Kwadrant 1 Maarssen, 3606 AZ Netherlands	J. Bouman	346-557-700	Yes	No	Yes	No	Yes	Yes	No	Yes	Yes	Yes	Yes	Yes	No	VISITT
Thermal Desorption	Enviro-Klean Soils, Inc. is a highly mobile system mounted on a 25-foot trailer that can be moved to a small or crowded contaminated soils site to remove petroleum contaminates with a carbon chain length of 45 or less.	ENVIRO-KLEAN SOILS, INC. P.O. Box 2003 Snoqualmie, WA 98065 USA	R.T. Cokewell	206-888-9388	Yes	No	Yes	No	Yes	Yes	No	Yes	No	No	No	No	No	VISITT
Thermal Desorption	Soil is placed into a hopper which screens out debris. It is then conveyed in a continuous stream into the desorber unit. Using a propane heating source, the unit indirectly heats soil to over 500°F, desorbing any petroleum contaminants.	ENVIRO-SOIL REMEDIATION, INC. P.O. Box 2212 Matthews, NC 28106 USA	Dmitri Wilkinson, Tom Evans	704-821-6000	Yes	No	Yes	No	Yes	Yes	No	No	No	No	No	No	No	VISITT
Thermal Desorption	The process, developed jointly by Hazen Research, Inc., and the Chlorine Institute, Inc., thermally treats mercury-containing wastes to produce a nontoxic residue, elemental mercury for recycling, and a clean off-gas.	HAZEN RESEARCH, INC. 4601 Indiana Street Golden, CO 80403 USA	Charles W. (Rick) Kenney	303-279-4501	Yes	No	Yes	No	No	No	Yes	No	No	No	No	No	No	VISITT

Technology Type	Description	Vendor	Contact	Phone	Soil, Sed, Sldg	GW	Air	In Situ	Ex Situ	VOC	Chl. VOC	BTEX	Petr, Oil, Lubs	Met	SVOC	PAH	Pest	PCB	Asbst	UXO	Radio-nucleides	Ref. Source
Thermal Desorption	The HRUBOUT process is a mobile thermal treatment batch process that removes VOCs and SVOCs from contaminated soils. In the process, heated air is injected into the soil below the zone of contamination, removing VOCs and SVOCs.	HRUBETZ ENVIRONMENTAL SERVICES, INC. 5949 Sherry Lane, Suite 525 Dallas, TX 75225 USA	Barbara Hrubetz	214-363-7833	Yes	No	No	No	Yes	Yes	No	Yes	Yes	No	Yes	No	No	No	No	No	No	VISITT
Thermal Desorption	IT Corporation's Thermal Desorption System is a high-capacity, continuous, transportable, cost-effective alternative for soil or sludge decontamination. The process involves heating soils to a temperature high enough to volatilize organic contaminants.	IT CORPORATION 312 Directors Drive Knoxville, TN 37923 USA	Edward Alperin/ Stuart Shealy	615-690-3211	Yes	No	No	No	Yes	Yes	Yes	Yes	Yes	Yes	Yes	Yes	Yes	Yes	No	No	Yes	VISITT
Thermal Desorption	The batch steam distillation/ metal extraction treatment process is a two-stage process to remove organics and heavy metals from contaminated soils.	IT CORPORATION 312 Directors Drive Knoxville, TN 37923 USA	Edward Alperin/ Stuart Shealy	615-690-3211	Yes	No	No	No	Yes	Yes	Yes	Yes	Yes	Yes	No	Yes	No	No	No	No	No	VISITT
Thermal Desorption	This aboveground treatment is a continuous operation. Kal Con's full-scale system can operate up to 24 hours a day. Low-temperature thermal treatment is employed to remove the hydrocarbon contamination from soil.	KALKASKA CONSTRUCTION SERVICE, INC. 500 S. Maple P.O. Box 427 Kalkaska, MI 49646 USA	David Hogerheide/ Justin Straksis	616-258-9134	Yes	No	No	No	Yes	Yes	Yes	Yes	Yes	No	Yes	Yes	No	No	No	No	No	VISITT
Thermal Desorption	Use of two-stage process to remove a wide variety of contaminants from soils, sediments, and sludge. System components consist of rotary kiln, soil discharge/ dust suppression system, cyclone, quench tower, baghouse, ID fan, and stack.	MAXYMILLIAN TECHNOLOGIES, INC. 1801 East Street Pittsfield, MA 01201 USA	Neal Maxymillian	617-557-6077	Yes	No	No	No	Yes	Yes	Yes	Yes	Yes	No	Yes	Yes	No	No	No	No	No	VISITT

Technology	Description	Company	Contact	Phone																			Source
Thermal Desorption	Mercury Recovery Services, Inc. (MRS) has developed and commercialized a medium-temperature thermal desorption process for removing mercury from soils and industrial wastes.	MERCURY RECOVERY SERVICES, INC. 700 Fifth Avenue New Brighton, PA 15066 USA	William F. Sutton	412-843-5000	Yes	No	No	Yes	No	No	No	No	No	Yes	No	Yes	No	No	No	No	No	No	VISITT
Thermal Desorption	The processes are planned and designed to avoid combustion by using low temperatures, 300 to 1,200°F, in the primary unit. Typical off-gas systems include thermal or catalytic oxidation.	MIDWEST SOIL REMEDIATION, INC. 27W010 St. Charles Road Wheaton, IL 60188 USA	John Sweeney	708-231-5115	Yes	No	No	Yes	Yes	Yes	Yes	Yes	No	Yes	Yes	Yes	No	No	No	No	No	No	VISITT
Thermal Desorption	Low temperature thermal desorption is a process based on mass transfer principles in which contaminants are removed from the soil through volatilization from the liquid to the vapor phase. Volatilization is accomplished through agitation of soil.	O'BRIEN & GERE TECHNICAL SERVICES, INC. 5000 Brittonfield Parkway East Syracuse, NY 13057 USA	James A. Fox, P.E.	315-437-6400	Yes	No	No	Yes	Yes	Yes	Yes	No	No	Yes	No	No	No	No	No	No	No	No	VISITT
Thermal Desorption	Low Temperature Thermal Desorption-Off-Gas Treatment Technology uses heat in a controlled environment to volatilize various organic compounds and therefore removes them from the soil.	PET-CON SOIL REMEDIATION, INC. P.O. Box 205 Spring Green, WI 53588 USA	Tom Labudde	608-588-7365	Yes	No	No	Yes	Yes	Yes	Yes	Yes	No	Yes	Yes	Yes	No	No	No	No	No	No	VISITT
Thermal Desorption	Thermal desorption of contaminants from soil has been proven effective at many remediation sites. Thermal desorption is preferable because it requires less energy.	PHILIP ENVIRONMENTAL SERVICES CORP. 10 Duff Road, Suite 500 Pittsburgh, PA 15235 USA	Teresa Sabol Spezio	412-244-9000	Yes	No	No	Yes	No	No	No	No	No	Yes	No	Yes	No	No	No	No	No	No	VISITT
Thermal Desorption	Product Control has developed two main alternatives of the Product Control Systems (PCS) Thermal Desorbers: Indirectly Fired PCS Thermal Desorbers, and Directly Fired PCS Thermal Desorbers.	PRODUCT CONTROL La Plaiderie, Guernsey St. Peter Port, Channel Islands GY1 3DQ England	G. Edward Someus	481-726-426	No	No	No	No	No	No	No	No	No	Yes	No	No	No	No	No	No	No	No	VISITT

Technology Type	Description	Vendor	Contact	Phone	Soil, Sed, Sldg	GW	Air	In Situ	Ex Situ	VOC	Chl. VOC	BTEX	Petr, Oil, Lubs	Met	SVOC	PAH	Pest	PCB	Asbst	UXO	Radio-nucleides	Ref. Source
Thermal Desorption	RSI's low-temperature, high-volume Desorption and Vapor Extraction (DAVE) system separates toxic and hazardous waste from soils, sludge, and sediments. The process is designed to separate and capture both volatile and semi-volatile organic contaminants.	RECYCLING SCIENCE INTERNATIONAL, INC. 175 West Jackson Blvd. Suite A1910 Chicago, IL 60604-2601 USA	William C. Meenan or Neil Ryan	312-663-4242	Yes	No	No	No	Yes	No	No	Yes	Yes	No	No	No	No	Yes	No	No	No	VISITT
Thermal Desorption	The process utilizes high temperature to volatilize organics from contaminated sludge and soils. It can remediate contaminated soils and sediments containing heavy hydrocarbons such as PNAs and PCBs.	REMEDIATION TECHNOLOGIES, INC. 9 Pond Lane Damonmill Square Concord, MA 01742 USA	Mark Mccabe	508-371-1422	Yes	No	No	No	Yes	Yes	Yes	Yes	Yes	Yes	Yes	Yes	Yes	Yes	No	No	No	VISITT
Thermal Desorption	Low temperature thermal desorption is a two-stage process where hydrocarbon contaminants are removed from the soil and then destroyed in a secondary thermal oxidizer.	REMTECH, INC. 9109 West Electric Avenue Spokane, WA 99204-9035 USA	Keith G. Carpenter	509-624-0210	Yes	No	No	No	Yes	Yes	Yes	Yes	Yes	No	Yes	No	No	No	No	No	No	VISITT
Thermal Desorption	The LT3 is a continuous operation that utilizes a hollow flight screw conveyor to indirectly heat the soil to approximately 560°F. A high-temperature fluid such as steam or another heat transfer fluid is circulated.	ROY F. WESTON, INC. 1 WESTON Way West Chester, PA 19380 USA	Michael G. Cosmos, P.E.	610-701-7423	Yes	No	No	No	Yes	Yes	Yes	Yes	Yes	No	Yes	Yes	No	No	No	No	No	VISITT
Thermal Desorption	The X*TRAX process uses an indirectly heated rotary dryer to continuously volatilize water and organic contaminants from soil sludge and sediment in a sealed system. Mercury wastes can also be treated.	RUST INTERNATIONAL, INC. Clemson Technology Center 100 Technology Drive Anderson, SC 29625 USA	Carl Palmer/ Chetan Trivedi	803-646-2413	Yes	No	No	No	Yes	Yes	Yes	Yes	Yes	No	Yes	Yes	Yes	Yes	No	No	Yes	VISITT

REMEDIATION TECHNOLOGY MATRIX

Technology	Description	Company	Contact	Phone													Source
Thermal Desorption	High Temperature Thermal Distillation can operate up to 2,200 degrees Fahrenheit. The desired material temperature is reached gradually as it progresses through the system.	SEAVIEW THERMAL SYSTEMS P.O. Box 3015 1767 Sentry Parkway West, Suite 250 Blue Bell, PA 19422 USA	Tim Carlsen	215-654-9800	Yes	No	No	Yes	Yes	Yes	No	Yes	Yes	Yes	No	No	VISITT
Thermal Desorption	The MX-2000 thermally desorbs the volatile organic compounds (VOC) from the centrifuge cake, and the MX-2500 thermally desorbs the semivolatile organic compounds (SVOC) from the MX-2000 "dried" solids material.	SEPARATION AND RECOVERY SYSTEMS, INC. 1762 McGaw Avenue Irvine, CA 92714-4962 USA	Brad Miller	714-261-8860	Yes	No	No	Yes	No	Yes	No	Yes	No	No	No	No	VISITT
Thermal Desorption	The remediation plant is designed to store and treat contaminated soils in a safe and effective manner. The untreated soil is stored within an enclosed storage building until it is conveyed into the rotary kiln. The soil is heated to cause desorption.	SOIL REMEDIATION OF PHILADELPHIA, INC. 3201 South 61st Street Philadelphia, PA 19153 USA	Matthew Paolino	215-724-5520	Yes	No	No	Yes	Yes	Yes	No	Yes	No	No	No	No	VISITT
Thermal Desorption	The SoilTech ATP System is a thermal treatment for remediation of contaminated soils. The SoilTech ATP System removes organic contaminants from feed material through anaerobic thermal desorption in an indirectly fired rotary kiln.	SOILTECH ATP SYSTEMS, INC. 800 Canonie Drive Porter, IN 46304 USA	Joe Hutton	219-926-8651	Yes	No	No	Yes	Yes	Yes	No	Yes	Yes	Yes	No	No	VISITT
Thermal Desorption	The LTTD plant is a mobile thermal processing unit for removing petroleum products and chlorinated hydrocarbons from excavated soil. Contaminants are destroyed during processing, thus eliminating future liability concerns.	SOUTHWEST SOIL REMEDIATION, INC. 3951 East Columbia Street Tucson, AZ 85714 USA	Trevor Johansen	602-571-7174	Yes	No	No	Yes	Yes	Yes	No	Yes	No	No	No	No	VISITT
Thermal Desorption	Thermal desorption is the process of contaminant removal from soils by transferring contaminants from one phase to another.	SPI/ASTEC P.O. Box 72515 Chattanooga, TN 37407 USA	Wendell R. Feltman, P.E.	706-861-0069	Yes	No	No	Yes	Yes	Yes	No	Yes	Yes	No	No	No	VISITT

Technology Type	Description	Vendor	Contact	Phone	Soil, Sed, Sldg	GW	Air	In Situ	Ex Situ	VOC	Chl. VOC	BTEX	Petr, Oil, Lubs	Met	SVOC	PAH	Pest	PCB	Asbst	UXO	Radio-nucleides	Ref. Source
Thermal Desorption	Texarome's ex-situ continuous desorption process uses highly superheated steam as a conveying and stripping gas in a "pneumatic" type conveying system to treat contaminated wastes, and difficult emulsions containing VOCs and SVOCs.	TEXAROME, INC. 1.5 Miles East Highway 337 P.O. Box 157 Leakey, TX 78873 USA	Gueric R. Boucard	210-232-6079	Yes	No	No	No	Yes	Yes	No	No	No	No	Yes	No	No	No	No	No	No	VISITT
Thermal Desorption	The 1420 HW is a transportable unit that operates at 1,400°F and is designed to remediate soil contaminated with PCB and other toxic chlorinated hydrocarbons.	THERMOTECH SYSTEMS CORPORATION 5201 N. Orange Blossom Trail Orlando, FL 32810 USA	A. A. Dishian	407-290-6000	Yes	No	No	No	Yes	Yes	No	Yes	Yes	No	Yes	Yes	No	Yes	No	No	No	VISITT
Thermal Desorption	The process is heating the soil to approximately 600 to 700°F. In the same manner that water turns into steam when heated to 212 degrees F, liquid petroleum will turn into vapor when heated to higher temperatures.	TPS TECHNOLOGIES, INC. 1964 S. Orange Blossom Trail Apopka, FL 32703 USA	James Lousararian	407-886-2000	Yes	No	No	No	Yes	Yes	No	Yes	Yes	No	Yes	Yes	No	No	No	No	No	VISITT
Thermal Desorption	WRI has developed a thermal reactor where solids are moved through various temperature regimes using screws. In the process volatiles or liquids are pulled off using a sweep gas.	WESTERN RESEARCH INSTITUTE 365 North 9th Street Laramine, WY 82070 USA	John Nordin/ Alan Bland	307-721-2443	Yes	No	No	No	Yes	Yes	No	Yes	Yes	No	Yes	Yes	No	No	No	No	No	VISITT
Thermal Desorption	Soil is fed into the Westinghouse TDU, in which the soil temperature is raised above the boiling point of the contaminants present, causing the contaminants to volatilize. The vaporized contaminants are condensed and captured.	WESTINGHOUSE REMEDIATION SERVICES, INC. 675 Park North Boulevard Building F, Suite 100 Clarkston, GA 30021-1962 USA	Jeff Rouleau	404-299-4698	Yes	No	No	No	Yes	Yes	Yes	No	No	No	Yes	No	No	Yes	No	No	No	VISITT

REMEDIATION TECHNOLOGY MATRIX

Technology	Description	Contact	Phone														Source
Thermal Desorption	Thermal desorption of DDT, dieldrin, toxaphene, and lindame contaminated soil.	No specific name		Yes	No	Yes	No	No	No	No	No	No	No	Yes	No	No	No	WILLIAMS ENVIRONMENTAL SERVICES, INC. 2076 West Park Stone Mountain, GA 30087 USA	U.S. EPA
Thermal Desorption	The Nitrem Process is used to destroy nitroaromatic energetic materials present in wastewater from ammunition plants where energetic materials are produced, loaded, or packaged.	Paul Loeffler		No	Yes	No	No	No	No	No	No	No	No	No	No	Yes	No	BATTELLE PACIFIC NORTHWEST	National Defense Center for Environmental Excellence
Thermal Treatment	The Adams Process is a patented high temperature process that uses the reaction between liquid and gaseous sulfur and the energetic or nonenergetic materials that needs to be destroyed.	Jim Hendricks		Yes	Yes	No	No	No	Yes	No	No	No	No	Yes	Yes	Yes	No	BURNS AND ROE INDUSTRIAL SERVICES, CO. 800 Kinderkamack Rd. Oradell, NJ, 07049 USA	National Defense Center for Environ. Excellence
Thermally Enhanced Recovery — In Situ	In situ heating is a process that uses common AC electricity to heat soils as a means of significantly improving the performance of conventional soil-venting techniques.	Theresa Bergsman	509-376-3638	Yes	No	No	Yes	Yes	No	No	Yes	No	No	No	No	No	No	BATTELLE PACIFIC NORTHWEST LABORATORIES Battelle Boulevard, P.O. Box 999 Mailstop B1-40 Richland, WA 99352 USA	VISITT
Thermally Enhanced Recovery — In Situ	Bio-Electrics has developed an innovative, proprietary electro-thermal process for volatilizing organic compounds and recovering the emission products.	Dr. Erich Sarapuu	816-474-4895	No	No	No	No	No	No	No	No	No	No	No	No	No	No	BIO-ELECTRICS, INC. 1215 W. 12th Street Kansas City, MO 64101 USA	VISITT
Thermally Enhanced Recovery — In Situ	Steam is injected through the injection wells. The pressure gradient between the injection wells and the grid of recovery wells provide the driving force for steam to flow. The steam will vaporize VOCs and drive out nonvolatile similar to steam stripping.	Mohamed Elgafi	714-957-6429	No	No	No	No	No	No	No	No	No	No	No	No	No	No	EM&C ENGINEERING ASSOCIATES 1665 Scenic Avenue Suite 104 Costa Mesa, CA 92626 USA	VISITT
Thermally Enhanced Recovery — In Situ	The process utilizes downhole burners for in situ soil treatment. Fossil burners are used typically, but the burner type and design depend on applications.	Mohamed Elgafi	714-957-6429	No	No	No	No	No	No	No	No	No	No	No	No	No	No	EM&C ENGINEERING ASSOCIATES 1665 Scenic Avenue Suite 104 Costa Mesa, CA 92626 USA	VISITT

416 STRATEGIES FOR ACCLERATING CLEANUP AT TOXIC WASTE SITES

Technology Type	Description	Vendor	Contact	Phone	Soil, Sed, Sldg	GW	Air	In Situ	Ex Situ	VOC	Chl. VOC	BTEX	Petr, Oil, Lubs	Met	SVOC	PAH	Pest	PCB	Asbst	UXO	Radio-nucleides	Ref. Source
Thermally Enhanced Recovery — In Situ	The HRUBOUT (TM) process is a mobile thermal treatment batch process that removes VOCs and SVOCs from contaminated soils. In the in situ process, heated air is injected into the soil below the zone of contamination.	HRUBETZ ENVIRONMENTAL SERVICES, INC. 5949 Sherry Lane, Suite 525 Dallas, TX 75225 USA	Barbara Hrubetz	214-363-7833	Yes	No	No	Yes	No	Yes	No	Yes	Yes	No	Yes	No	No	No	No	No	No	VISITT
Thermally Enhanced Recovery — In Situ	The in situ radio frequency (RF) heating process utilizes electromagnetic energy in the RF band to heat soil rapidly without injection of heat transfer media or on-site combustion. The process can be used to heat soil to a temperature range of 150-200°C.	IIT RESEARCH INSTITUTE 10 West 35th Street Chicago, IL 60616 USA	Harsh Dev	312-567-4257	Yes	No	No	Yes	No	Yes	Yes	Yes	Yes	No	Yes	No	No	No	No	No	No	VISITT
Thermally Enhanced Recovery — In Situ	The KAI in situ radio frequency heating technology imparts heat to nonconducting materials through the application of carefully controlled radio frequency transmissions. The technology can be used to heat variety of contaminants.	KAI TECHNOLOGIES, INC. 170 West Road #7 Portsmouth, NH 03801 USA	Raymond S. Kasevich	603-431-2266	Yes	No	No	Yes	No	Yes	Yes	No	No	No	Yes	No	No	No	No	No	No	VISITT
Thermally Enhanced Recovery — In Situ	The Detoxifier is a patented, full-scale, transportable treatment unit that uses in situ physical and chemical treatment to remediate contaminated soil.	NOVATERRA, INC. 2029 Century Park East Suite #890 Los Angeles, CA 90067 USA	Phil La Mori Ph.D.	310-843-3190	Yes	No	No	Yes	No	Yes	Yes	Yes	Yes	No	Yes	Yes	No	No	No	No	No	VISITT
Thermally Enhanced Recovery — In Situ	The in situ steam enhanced extraction process is designed to remove volatile and semivolatile organic compounds from an area of contaminated soil without the need for excavation.	PRAXIS ENVIRONMENTAL TECHNOLOGIES, INC. 1440 Rollins Road Burlingame, CA 94010 USA	Dr. Lloyd Stewart	415-548-9288	Yes	No	No	Yes	No	Yes	Yes	Yes	Yes	No	Yes	No	No	No	No	No	No	VISITT

REMEDIATION TECHNOLOGY MATRIX

Technology	Description	Company	Contact	Phone																Source
Thermally Enhanced Recovery — In Situ	R.E Wright Environmental, Inc. has implemented a steam extraction/free product recovery system to remove and recover petroleum from the unsaturated soil zone. The injection of steam into the subsurface decreases the viscosity of the contaminants.	R.E. WRIGHT ENVIRONMENTAL, INC. (REWEI) 3240 Schoolhouse Road Middletown, PA 17057-3595 USA	Richard C. Cronce, Ph.D.	717-944-5501	Yes	Yes	No	Yes	No	No	Yes	Yes	No	Yes	Yes	No	No	No	No	VISITT
Thermally Enhanced Recovery — In Situ	Steam Injection and Vacuum Extraction (SIVE) is an innovative in situ hazardous waste site remediation technology that uses underground steam injection to enhance the soil vapor extraction and groundwater pump-and-treat remedial methods.	SIVE SERVICES 555 Rossi Drive Dixon, CA 95620 USA	Douglas K. Dieter	916-678-8358	Yes	No	No	Yes	No	No	Yes	Yes	No	No	No	No	No	No	No	VISITT
Thermally Enhanced Recovery — In Situ	Steam injection-soil vapor extraction (SI-SVE) has been demonstrated at a number of sites. The process enhances the extraction of organics which are tough to remove from soil and groundwater by volatilizing them with hot steam.	THERMATRIX, INC. 3590 N. First Street Suite 310 San Jose, CA 95134 USA	John Schofield	408-944-0220	No	No	No	No	No	No	Yes	Yes	No	Yes	No	No	No	No	No	VISITT
Vitrification	B&W cyclone furnace is an innovative thermal technology that may be used in treating liquids, sludge, or soils containing organics, heavy metals, and/or radionuclide contaminants.	B&W NUCLEAR ENVIRONMENTAL SERVICES, INC. 2220 Langhorne Road Lynchburg, VA 24506-0548 USA	T.R. Welch	804-948-4765	Yes	No	No	No	Yes	No	No	No	Yes	Yes	Yes	No	No	No	No	VISITT
Vitrification	Terra-Vit is a continuous, above ground melted technology that oxidizes, melts, and transforms a broad spectrum of wastes into a glass- or rock-like material. The melting energy is derived from direct electrical heating of the molten material.	BATTELLE PACIFIC NORTHWEST LABORATORIES Battelle Boulevard, P.O. Box 999 Mail Stop P7-41 Richland, WA 99352 USA	Chris Chapman	509-376-6576	Yes	No	No	No	Yes	No	No	No	No	Yes	Yes	Yes	No	No	Yes	VISITT

418 STRATEGIES FOR ACCLERATING CLEANUP AT TOXIC WASTE SITES

Technology Type	Description	Vendor	Contact	Phone	Soil, Sed, Sldg	GW	Air	In Situ	Ex Situ	VOC	Chl. VOC	BTEX	Petr, Oil, Lubs	Met	SVOC	PAH	Pest	PCB	Asbst	UXO	Radio-nucleides	Ref. Source
Vitrification	The technology uses electrodes placed in situ to apply a high electric current to the contaminated area. The electrical energy heats up the contaminated soil. Once the heat is high enough, the soil begins to melt into a molten glass form.	BIO-ELECTRICS, INC. 1215 W. 12th Street Kansas City, MO 64101 USA	Dr. Erich Sarapuu	816-474-4895	No	No	No	No	No	No	No	No	No	No	No	No	No	No	No	No	No	VISITT
Vitrification	Dredged sludge is "sintered" in a rotating oven. The heat and oxygen in the oven result in burning off all organic compounds, including the harmful ones. Inorganic compounds, particularly metals are converted into baked ceramic product.	ECOTECHNIEK B.V. Het Kwadrant 1 Maarssen, 3606 AZ Netherlands	J. Bouman	346-557-700	Yes	No	No	No	Yes	Yes	Yes	Yes	Yes	Yes	Yes	Yes	Yes	Yes	No	Yes	No	VISITT
Vitrification	EET Corporation has developed a thermal process to solidify inorganic wastes using microwave energy. The process produces a synthetic mineral matrix, incorporating hazardous or radioactive components of the waste in the crystal structure.	EET CORPORATION 11217 Outlet Drive Knoxville, TN 37932 USA	Robert D. Petersen	615-671-7800	Yes	No	No	No	No	No	No	No	No	Yes	No	No	No	No	No	No	Yes	VISITT
Vitrification	Electro-Pyrolysis, Inc. (EPI) pioneered the use of DC arc furnace technology for the destruction and treatment of hazardous waste resulting in a gas phase and melt phase.	ELECTRO-PYROLYSIS, INC. 996 Old Eagle School Road Suite 1118 Wayne, PA 19087 USA	Dr. J. Kenneth Wittle	610-687-9070	Yes	No	No	No	Yes	No	No	No	No	Yes	No	No	No	No	No	No	No	VISITT
Vitrification	The unique feature of the process is the addition of fluxing material to achieve vitrification at a relatively low temperature. This low temperature is possible since the fluxing material behaves like impure material.	EM&C ENGINEERING ASSOCIATES 1665 Scenic Avenue Suite 104 Costa Mesa, CA 92626 USA	Mohamed Elgafi	714-957-6429	No	No	No	No	No	No	No	No	No	No	No	No	No	No	No	No	No	VISITT

REMEDIATION TECHNOLOGY MATRIX

Technology	Description	Company	Contact	Phone	C1	C2	C3	C4	C5	C6	C7	C8	C9	C10	C11	C12	C13	C14	Source
Vitrification	In situ vitrification involves the electric melting of contaminated solids for purposes of destroying/ removing hazardous organics and immobilizing/ removing hazardous inorganic contaminants in a glass and microcrystalline residual product form.	GEOSAFE CORPORATION 2950 George Washington Way Richland, WA 99352 USA	James E. Hansen or Matt Haass	509-375-0710	Yes	No	No	Yes	Yes	Yes	Yes	Yes	Yes	Yes	Yes	Yes	Yes	Yes	VISITT
Vitrification	Xtalite is a patented process for converting inorganic waste concentrates derived from smelting and incineration into a non-toxic synthetic mineral. The mineral form is chosen on the basis of established geological principles.	MULTIPLEX XTALITE-TEXILLA ENVIRONMENTAL 4570 Westgrove Dr. Suite 255 Addison, TX 75248 USA	Irfan A. Toor, Ph.D.	214-733-3378	No	No	No	No	No	No	No	No	No	No	No	No	No	No	VISITT
Vitrification	The Plasma ARC Centrifugal Treatment System (PACT) uses electric energy from an arc, which operates between a plasma torch and a rotating tub, to detoxify the material feed. The tub rotates on a vertical axis inside a sealed chamber.	RETECH, DIV. OF LOCKHEED ENV. SYS. & TECH. 100 Henry Station Road, P.O. Box 997 Ukiah, CA 95482 USA	Ronald K. Womack	707-462-6522	Yes	No	No	Yes	Yes	Yes	No	No	No	Yes	No	No	No	No	VISITT
Vitrification	Seiler offers a vitrification process for the treatment of various solids and sludge containing inorganic and organic contaminants. The system is designed to use the organic components of the waste to provide fuel value to heat the melter system.	SEILER POLLUTION CONTROL SYSTEMS, INC. 5555 Metro Place North Suite 100 Dublin, OH 43017 USA	Alan Sarko	614-791-3272	Yes	No	No	No	No	No	No	No	Yes	Yes	No	No	Yes	No	VISITT
Vitrification	Stir-Melter, Inc. is offering compact electric resistance glass melter for vitrifying inorganic hazardous waste residues. Toxic heavy metal oxides go into solution in glass and become nonleachable per TCLP standards.	STIR-MELTER, INC. (SUBSID/ GLASSTECH, INC.) Ampoint Industrial Park 995 Fourth Street Perrysburg, OH 43552 USA	Ken Kormanyos	419-661-9500	Yes	No	No	No	No	No	No	No	No	No	No	No	No	No	VISITT
Vitrification	Vortec has created an aboveground oxidation and vitrification process for remediation of soils, sludge, ashes, and sediments that have organic, inorganic, and heavy metal contamination.	VORTEC CORPORATION 3770 Ridge Pike Collegeville, PA 19426-3158 USA	James G. Hnat	610-489-2255	Yes	No	No	Yes	No	No	No	No	Yes	No	No	No	No	No	VISITT

Technology Type	Description	Vendor	Contact	Phone	Soil, Sed, Sldg	GW	Air	In Situ	Ex Situ	VOC	Chl. VOC	BTEX	Petr, Oil, Lubs	Met	SVOC	PAH	Pest	PCB	Asbst	UXO	Radio-nucleides	Ref. Source
Vitrification	Waste Destruction Technologies, Inc. has developed an innovative in situ waste destruction and vitrification process. The process uses a graphite arc melter system which is lowered to a previously installed borehole which penetrate through the waste zone.	WASTE DESTRUCTION TECHNOLOGIES, INC. 1850 Four Mile VUE Butte, MT 59701 USA	Lawrence C. Farrar	406-494-5555	Yes	No	No	Yes	No	No	No	No	No	No	No	No	No	No	No	No	No	VISITT
Vitrification	The CCBA process converts heavy metals in soils, sediments, and sludge to nonleaching silicates. The process is also capable of oxidizing organics in the waste stream and incorporating the ash into the ceramic pellet matrix.	WESTERN PRODUCT RECOVERY GROUP, INC. 10626 Cervaza Drive Escondido, CA 92026 USA	Bert Elkins		Yes	No	No	No	Yes	No	No	No	Yes	Yes	Yes	Yes	No	Yes	No	No	No	Vendor Brochure

Note: Version 4.0 Of VISITT.Matrix Source: Matrix Source: this remediation technology matrix was adapted from a file the Navy Environmental Leadership Program developed.

INDEX

A

Abandoned mine reclamation programs, 198–201
Academic research, 174–178
Accelerated Cleanup Model, Superfund, See Superfund Accelerated Cleanup Model
Accelerated Response Centers, 34
Accelerated Site Characterization for Confirmed or Suspected Petroleum Releases (ASTM), 220–221
Acid extraction, 356–357
Acoustic geophysics, 232
Action memorandum, 30
Administrative support, 298–299
　cost reduction, 158
Adsorption/absorption technology, 357
AFCEE, See Air Force Center for Environmental Excellence
Air Force, See U.S. Air Force
Air Force Center for Environmental Excellence (AFCEE), 113–128, 257, 335
　innovative technologies, 125–128
　Installation Restoration Program Information Management System, 122–125
　intrinsic remediation protocol, 115–122, 127–128
　list of services, 114
　proactive communication, 126–128
　Quality Assurance Project Plan (QAPP), 128
Air sparging, 357–358
Alessandra, Tony, 266
Alternative approaches, See Innovative approaches
Alternative technologies, See Innovative technologies
Alternative Treatment Technology Information Center (ATTIC), 19
Ambient levels, 84–85
American Society for Testing and Materials (ASTM)

Accelerated Site Characterization Guide, 220–221
Risk Based Corrective Action (RBCA) model, 188, 191, 217–220
Analytical level, 297–298
Analytical models, 225–226
Applicable or relevant and appropriate requirements (ARARs)
　DoD ROD process, 104
　waivers, 52, 290–291
Aquifer remediation, determining technical impracticality of, 74–76
ARAR, See Applicable or relevant and appropriate requirements
Areas of Concern (AOC), 40
Army sites, 142–143
Arsenic, 84–85, 162
Asbestos, 73–74
As-generated waste, 61
ASTM, See American Society for Testing and Materials

B

Balko Plant, 313
Baseline Environmental Management Report (BEMR), 148
Base Realignment and Closure (BRAC), 82, 86–96
　California program, 202–204
　Cleanup Plan (BCP), 89
　personnel, 88
　public involvement, 91
　land use determination, 293
Battelle Columbus Laboratories, 160
Behavior styles, 265–266
Bioplume II, 117
Bioremediation, 162–163, See also Natural attenuation
　California program, 189
　defined, 161–162

421

DOE NABIR Program, 161–171
interstate technology cooperation, 196
oxygenation, 175
vendor information matrix, 359–380
    groundwater, 359–363
    lagoon, 363–364
    slurry phase, 371–375
    soil, 364–368
    solid phase, 375–380
Bioslurping, 125, 126–127
Bioventing, 163, 380–383
Block, Peter, 270
Borehole logging, 235
BRAC, See Base Realignment and Closure
Brownfields Initiative, 4, 250, 302–314
    assessment demonstration pilots, 304–306
    environmental program integration, 304, 306
    state voluntary cleanup programs, 306–314
    strategic planning, 314–316
Brownfields redevelopment
    liability issues, 302–303
    uncertainty, 304
Brownfields Remediation, LLC, 314
Buber, Martin, 264–265
Budgetary considerations, See Cost-reduction; Funding

C

California Base Closure Environmental Committee, 202–204
California UST programs, 189–194
C&M Plating, 312
CERCLA, 4, 24–58, See also Removal actions; Superfund Accelerated Cleanup Model; U.S. Environmental Protection Agency
    abandoned mine reclamation programs, 198–201
    Air Forces's Installation Restoration Program, 106
    ARAR waivers, 290–291
    cross-program integration, 73
    EPA Brownfields Initiative, 306
    federal facilities, 25, 82, See also U.S. Department of Defense; U.S. Department of Energy
    DoD's "Road to ROD," 102–106
    flexibility and strategic planning, 289–291
    interim remedial actions, 33, 101
    land use determination, 292
    managing uncertainty, 209
    progress report, 100
    record of decision (ROD), See Record of decision
    response action matrix, 28–29
    Superfund case study for accelerated cleanup, 55–58
    Superfund process flow chart, 26
    voluntary cleanup programs, 178–179
CERCLIS, 15, 309
Chaotic conditions, 257–258
Chapple, Tom, 264–265
Chemical treatment, vendor information matrix, 383–388
CLEAN contract, 139–140
Cleanup Information Bulletin Board System (CLU-IN), 19
CLU-IN, 21
Co-disposal mixtures, 145–146
Communication, 259–267
    basic dialogue, 261–263
    behavior styles, 265–266
    DoD'S ROD process, 105
    DOE research programs, 171
    gender and cultural differences, 267
    I and Thou concepts, 264–265
    learning process, 260–261
    military programs, 105, 107, 126–128, 131–132
    outdated models, 257
    publishing case studies, 263
    skills training, 258
    state-to-state level, 195–196
    storytelling, 263
    success and failure, 264
    trust, 279–280
    verbal skills, 261–263
    writing skills, 259–260
Community Environmental Response Facilitation Action (CERFA), 88
Community relations
    Department of Defense policy, 81
    EPA removal actions, 30–31
    military environmental programs, 91, 98, 110
    Superfund programs, 42
Comprehensive contingency plans, 66–67
Comprehensive Environmental Response, Compensation and Liability Act, See CERCLA
Comprehensive Environmental Response, Compensation and Liability Information System (CERCLIS), 15, 309
Comprehensive Long-Term Environmental Action Navy (CLEAN) contract, 139–140
Computational performance indicators, 323
Computer applications, 248
    administrative and organizational tools, 298
    cost estimation software, 112
    cost-saving approaches, 317
    modeling, 221–231, 248, See also Modeling
Conceptual design, 138

INDEX 423

Conceptual model
  analytical models, 225–226
  intrinsic remediation protocol, 120, 121
  observational method, 211
Cone penetrometers, 125
Conflict of interest, 110
Construction risk, contracting issues, 133
Contamination Cleanup Grant Progrm, 309
Contingency planning
  comprehensive (under RCRA), 66–67
  observational approach, 206–214
    sequential risk mitigation, 214–217
  schedules, 296
Contingency remedy record of decision, 55
Contracting
  Air Force program, 110, 114
  Comprehensive Long-Term Environmental Action Navy (CLEAN) contract, 139–140
  cost considerations, 134, 318
  DOE Environmental Restoration Program, 156
  Navy integrated project schedule approach, 133–134
  observational method and, 214
  project design level and, 135
  remediation technologies and suitability matrix, 136
Contract Laboratory Program (CLP), 40
Cooperation, 5, 6, See also Teamwork
Corrective action management units (CAMU), 61–64, 289
  cross-program integration, 73
Corrective actions, 59–69, See also Resource Conservation and Recovery Act
Corrective Actions Plan, CAMU provision, 64
Cost-effective technologies, DOE application, 157
Cost considerations
  caps for EPA removal actions, 31
  contracting issues, 134, 318
  practicality, 319
  remedial performance evaluations, 322–324
  under Superfund Accelerated Cleanup Model, 35, 37
Cost estimation software, 112
Cost of project (COP) contracts, 133
Cost reduction, 283, 316–327
  acceleration and streamlining approaches, 317
  contracting issues, 134, 318
  DOE Environmental Restoration Program, 158
  efficiency and motivational factors, 327
  EPA Brownfields Initiative, 314–315
  interstate cooperation program, 194–195
  property tax abatement, 324–326
  remediation costs, 319–322
  site investigation, 320–322
  strategic planning considerations, 295

technology-based approaches, 317
waste minimization motivation, 71
Covey, Stephen, 276, 277
Cross-program integration, See Integrated environmental programs
Cynics, 277–278

D

Data collection, See Investigation; Site characterization
Databases, 298
  Department of Energy, 342–343
  EPA, 19, 21, 337–339
  Installation Restoration Program Information Management System, 122–125
  military, 340
  miscellaneous government agencies, 344–345
  modeling information, 224
  private and state, 346–349
Data validity, 297
Dechlorination, 383
Decision-making process, incorporating into strategic plans, 291–292, 299
Defense Environmental Restoration Account (DERA), 82
Definitive design, 138
Delivery/extraction systems, 388–389
Dense non-aqueous phase liquids (DNAPLs), 74–76, 210
Design
  AFCEE services, 114
  level of, 134–137
  Navy integrated project schedule approach, 134–139
Detroit Rockwell, 313
Directed Business Concepts (DBC), 260
Dual phase extraction, 389–390

E

Early actions, 25–30
  integrated assessments, 41
  under SACM, 35–37
Early remedial action, 35
  interim remedial actions vs., 55
Electrical separation, 390
Electromagnetics, 234
Electro-thermal gasification, 391
Emergency removal actions, 27
Emerging Technology Program, 17, 18
Empowerment, 107, 272–273
Emulsion recycling, 391
Energy dispersive X-ray fluorescence, 22
Energy Technology Engineering Center, 158

Engineering evaluation/cost analysis (EE/CA), 27, 30, 132
   design considerations, 137, 138
Engineering Field Activity (EFA), 130
Environmental Baseline Survey (EBS), 89
Environmental Impact Statement (EIS), 89
Environmental justice, 304
Environmental Leadership Program, 129–130
Environmental Protection Agency (EPA), See U.S. Environmental Protection Agency
Environmental regulations, intent consistent with accelerated cleanup/decision-making, 3
Environmental resource guide, 329–349, See also Information resources
Environmental Restoration Program, 148–161, See also U.S. Department of Energy
   acceleration strategy, 149–158
   budgetary linkages to compliance milestones, 156
   budget review and funding priorities, 152
   contracting strategies, 156
   cost reduction, 158
   land use, 157
   lead regulator concept, 155–156
   privatization, 157
   resource allocation, cleanup vs. studies, 152
   summary table, 151
   "Workouts" for achieving stakeholder consensus, 154–155
Environmental Restoration Strategic Plan, 149
Environmental security program, 80–81
EPA, See U.S. Environmental Protection Agency
Excavation
   investigation by, 252–253
   over excavation, 185, 253
Explosive materials, 142–143, 145, 163
Exposure pathways analysis, 118, 121–122
   observational method application, 210

F

Facility, defined, 73
Fate and transport modeling, 117–118, 121
Feasibility studies, 320–322, See also Remedial investigation/feasibility study
   design considerations, 139
   ROD time frame, 51
Federal Facility Agreement (FFA), 103
Federal Facility Site Remediation Agreement (FFSRA), 103
Federal sites, See U.S. Department of Defense; U.S. Department of Energy
Final remedy, See also Record of decision
   interim remedial actions, 33
   removal actions, 25

Firm fixed price (FFP) contracts, 133
Firm price remedial action contract (FRAC), 133, 136
Fixed fee contracts, 156
Fixed-price contract, 318
Formerly Utilized Sites Remedial Action Program (FUSRAP), 150
Funding, See also Cost considerations
   Air Force's Installation Restoration Program (IRP), 108
   budgetary linkages to compliance milestones, 156
   DOE Environmental Restoration Program priorities, 152
   HRS score and, 38
   military programs, 83–85, 131
   1990s reductions, 2
   strategic planning considerations, 294–296
   Superfund, 24
   underground storage tank reimbursement, 183–184
Future beneficial land use, See Land use

G

Gehm Corporation, 239
Gender differences, 267
Geographical information systems (GIS), 142
Geophysical tools, 230–245
   case study (electromagnetic offset logging), 239–245
   selecting and planning, 231, 238
   table of methods and applications, 232–237
Global positioning system (GPS), 298
Goals and objectives, 6–7, 268–269, See also Land use
Graphical presentation, 229, 323
Gravity surveys, 236
Ground penetrating radar, 233
Groundwater remediation
   AFCEE intrinsic remediation protocol, 115–122
   observational method application, 209–211
   technical impracticability determination, 74–76
   vendor information matrix
      bioremediation, 359–363
      chemical treatment, 383–384

H

Hazardous and Solid Waste Amendments, 59
Hazardous Substance Spill Law, 310
Hazardous Waste Consultant, 285
Hazardous Waste Identification Rule for Contaminated Media (HWIR-media), 61
Hazard Ranking System (HRS), 24, 38, 309
   integrated assessment, 40–41

INDEX

High explosive materials, 142–143
High-yield aquifers, 210
Horizontal wells, 127
Howard, Philip, 277
HRS, See Hazard Ranking System
Human factors, 5–6, 12, 255–281, See also Communication; Leadership; Teamwork; Trust
   cost benefits, 258–259
   efficiency and motivational factors, 327
   environmental industry need to value people, 256–259
   science-based information, 259
   strategic planning, 291–292, 299
   training and intervention, 258, 280–281

I

I and Thou concepts, 264–265
Idaho National Engineering Laboratory (INEL), 157
Illinois voluntary cleanup program, 307, 311
Immunoassay techniques, 22
Implementation work plan, 138
In situ approaches, 145, See also Natural attenuation; specific approaches
In situ bioremediation, 163, 167, 196, See also Bioremediation; Natural Attenuation
   vendor information matrix, 359–380
Incineration, 47
Indiana voluntary cleanup program, 307–308, 312
Induced polarization, 237
Information Management System, 122–125
Information resources, 329–349
   Department of Energy, 341, 342–343
   EPA, 330–334, 337–339
   military, 114, 335–336, 340
   miscellaneous government agencies, 344–345
   private and state, 346–349
Innovative approaches, 7–8, 246, 250–251, See also Brownfields Initiative; Superfund Accelerated Cleanup Model (SACM)
   Defense Department proactive nonlinear approach, 85–86, 141
   real-time data collection maximization, 251–252
   strategic planning, 297
Innovative technologies, 246, See also Computer applications
   commercialization, 194
   comparing with established technologies, 249
   concepts of, 248–250
   cost-reduction approaches, 317
   developing, 249
   environmental innovation concepts, 246–248

EPA evaluation projects, 351–354
EPA information transfer/demonstration programs, 17–19
implementation levels, 247
military programs, 81, 111
   AFCEE, 125–128
   Navy Environmental Leadership Program (NELP), 129–130, 239
NCP requirements, 16
private sector, 205, 246–253
RCRA, 60
remediation technology matrix, 355–420
selecting, 249–250
slowness in application of, 2
soil sample analytical methods, 22–23
state and academic research programs, 174–178
state and interstate cooperation programs, 194–197
strategic planning, 297
Superfund Innovative Technology Evaluation (SITE) program, 17–19
Installation Restoration Program (IRP), 82, 106–113
   Information Management System (IRPIMS), 122–125
   Management Action Plan Guidebook, 112–113
Integrated environmental programs, 71–74, See also Partnerships
   DOE's bioremediation research program, 168–171
   DOE Environmental Restoration Program, 154–156
   EPA Brownfields Initiative, 304, 306
   lead regulator concept, 155–156
   Navy integrated project schedules, 130–139
Integrated assessment, 35, 38–42
Integrated project schedules, 130–139, See also under U.S. Navy
Interagency Agreement (IAG), 103, 109–110
Interim measures, 108
   RCRA corrective actions, 65–66
Interim remedial actions, 33, 101
   ASTM RBCA standard, 221
   military programs, 108
International Groundwater Modeling Center (IGWMC), 224
International leadership, 4
Internet addresses, See On-line resources
Interstate Technology and Regulatory Cooperation (ITRC) Working Group, 194–197
Intrinsic remediation, See Natural attenuation
Investigation, See also Geophysical tools; Site characterization
   accelerated, EPA programs, 20–24
   AFCEE services, 114

analytical level of data, 297–298
data validity, 297
DOE Environmental Restoration Program resource allocation, 152
innovative methods, See Innovative technologies
intrinsic remediation protocol, 118
Monitoring and Measurement Technologies Program (MMTP), 17, 20
observational method and, 206, 209–211
potential conflicts of interest, 110
real-time data collection, 251–252
simultaneous cleanup, 141–142
Investigation by excavation, 252–253
Iowa, UST fund, 197
IRPTools, 124

J

Job satisfaction, 278–279
John Deere Plow Works, 311

K

Kessler Products, 312

L

Lagoon bioremediation, 363–364
Land disposal restrictions (LDR), 59, 289
  corrective action management units, 62–63
  corrective action management units and, 73
Landfarming, 48, 163, 181–182
Landfill sites, presumptive remedies, 46
Land Recycling Program, 311
Land use, 7, 292–293
  DOE Environmental Restoration Program, 157
  military programs, 98–99, 107
    Base Realignment and Closure program, 89, 92
  record of decision considerations, 51
  strategic planning, 292–293
Lawrence Livermore National Laboratory (LLNL) LUST research, 189–192
Leadership, 273–279
  cost benefits, 258–259
  job satisfaction concerns, 278–279
  management and, 276–277
  power of cynics, 277–278
  skills, 275–279
Leadregulator concept, 155–156
Leaking underground storage tanks, See Underground storage tank (UST) programs

Liability issues, greenfields vs. brownfields development, 302–303
Light non-aqueous phase liquids (LNAPLs), electromagnetic offset logging, 239–245
Listening skills, 262
Local programs, 173, See also State and local programs; Underground storage tank (UST) programs
Long-term monitoring, AFCEE intrinsic remediation protocol, 122
Low Risk Groundwater Case, 191–193
Low-yield aquifers, 210

M

Magnetic separation, 391
Magnetometer surveys, 233
Management, leadership and, 276–277
Management Action Plan Guidebook, 112–113
Marston, William, 265
Materials handling/physical separation technology, 391–392
Measurement, See Investigation; Site characterization
Meetings and negotiations, 261, See also Communication
Mercury, 162
Methyl tertiary butyl ether (MTBE), 193
Michigan voluntary cleanup program, 308, 313–314
Microwave-assisted digestion, 22
Military bases, See Base Realignment and Closure; U.S. Department of Defense
Mine reclamation programs, 198–201
Minimum technology requirements (MTR), 59
  corrective action management units, 62–63
Minnesota voluntary cleanup program, 309–310, 312
Mission and vision, 267–268
Mixing zones, 69
Mixtures of contaminants, 145–146
MMTP, 17, 20
Modeling, 221–231
  analytical approaches, 225–227
  case studies, 227
  evaluating need for, 223–224
  information resources, 224
  innovative approach, 248
  intrinsic remediation, 117–118, 121
  model selection and assessment, 224–225
  numerical approaches, 227–229
  results presentation, 229
MODFLOW, 248
Money management tools, 298

INDEX
427

Monitoring and Measurement Technologies Program (MMTP), 17, 20
Montana
  abandoned mine reclamation program, 199–201
  Streamside Tailings project, 201–202
Mound Plant, 160
Multi-tasking, 140

N

NABIR, See Natural and Accelerated Bioremediation Research Program
NAPLs, See Non-aqueous phase liquids
National Oil and Hazardous Substances Contingency Plan (NCP), 3, 16, 285, See also Removal actions
National Priorities List (NPL) sites, 15, 24, See also CERCLA
  military facilities, 79
  risk management under Superfund Accelerated Cleanup Model, 38
National Zinc Company site, 55–58
Natural and Accelerated Bioremediation Research Program (NABIR), 161–171
  goals, 164–165
  participants, 167–168
  program elements, 165
  rationale for DOE research, 165–167
  scientific integration strategies, 168–171
Natural attenuation (intrinsic remediation), See also Bioremediation
  advantages, 116
  Air Force protocol, 115–122, 127–128
    demonstration of intrinsic bioremediation, 116–117
    implementation protocol, 118–122
    long-term monitoring, 122
    modeling, 117–118, 121
    site characterization, 120–121
  bioremediation processes, 162–163
  California LUST program, 189, 192
  conceptual model, 120, 121
  DOE's NABIR Program, 161–171, See also Natural and Accelerated Bioremediation Research Program
  no-action RODs, 53
Naval Explosive Ordinance Disposal Technology Division (NAVEODTECHDIV), 143
Navy, See U.S. Navy
Navy Environmental Leadership Program (NELP), 129–130, 239
Nevada, one-stop permit application, 197
Nevada Test Site, 164

No Action decisions, 52–53, 111
No Further Action decisions, 53, 111
Non-aqueous phase liquids (NAPLs), 74–76, 117, 210
  electromagnetic offset logging, 239–245
Nonlinear cleanup approaches, 7–8, 85–86, 141
Non-time-critical removal actions, 30
North Carolina UST reimbursement program, 184
Northeast Geophysical Services, 231
Nuclear facilities, 145, 149, 157, See also U.S. Department of Energy
Numerical models, 227–229

O

Observational approach, 206–214
  contingency plan, 212–214
  limitations, 207
  sequential risk mitigation, 214–217
Off-gas treatment, 392–395
Office of Environmental Management (OEM), 147–148, 160–161, See also U.S. Department of Energy
  acceleration strategy, 149–158
  case studies, 158–160
  Environmental Restoration Program, 148–161, See Environmental Restoration Program
Office of Health and Environmental Research, 161–171, See also Natural and Accelerated Bioremediation Research Program
Ohio voluntary cleanup program, 310, 312
Oklahoma case study, 55–58
On-line resources
  Department of Energy, 342–343
  EPA, 337–339
  military, 340
  miscellaneous government agencies, 344–345
  private and state, 346–349
  SITE programs, 18–19
On site, defined, 73
Operable units (OU), 5
  Air Force program, 109
  Base Realignment and Closure program, 92
Operational units
Organizational skills, 298
Outcomes monitoring, See Progress and performance evaluation
Over excavation, 185, 253
Owen, Harrison, 273
Oxidation, 385
Oxidation/reduction, 385–388
Oxygenation, 175

P

Partnerships, 269–272, See also Integrated environmental programs
 abandoned mine reclamation programs, 198
 defense programs, 81, 269
 DOE's bioremediation research program, 168–171
 DOE Environmental Restoration Program, 154–155
 military programs, 96–97, 107, 128, 130
 in strategic plans, 291
PCBs, 30, 163, 166
Penetrometers, 196
People training and intervention, 280–281
Performance tracking, See Progress and performance evaluation
Performance-based specifications
 DOE Environmental Restoration Program, 156–157
 Navy integrated scheduling approach, 137
Pesticides, 145
Peters, Tom, 257, 277
Petroleum wastes
 ASTM Risk Based Corrective Action (RBCA) model, 188, 191, 217–220
 ASTM site characterization standard, 220–221
 DOE bioremediation research program, 162–171, See also Natural and Accelerated Bioremediation Research Program
 electromagnetic offset logging, 239–245
Phytoremediation, 395
Pinellas Plant, 159
Planning, See also Strategic planning
 DoD's ROD process, 104
 uncertainty, 208–209
 Workouts for achieving stakeholder consensus, 154–155
Plasma arc treatment, 395
Plug-in remedies, 42–45
Pneumatic fracturing, 395
Point-of-compliance, 68, 122
Pollution prevention, 176
 AFCEE services, 114
 EPA Brownfields Initiative, 304
 military programs, 142
 RCRA waste minimization, 70–72
 Resource Conservation and Recovery Act, 59
 UST releases, 183
Polychlorinated biphenyl (PCB), 30, 163, 166
Pre-Notice Program, 307, 311
Presumptive remedies, 42–45
 benefits of, 46–48
 integrated assessments, 41

 landfill, 46
 military environmental programs, 108
 RCRA, 60
 VOCs, 45–46
Private sector, 205, See also Geophysical tools; Modeling; Observational approach; Professional organizations
 information resources, 346–349
 innovative technologies and approaches, 205, 246–253
Privatization, 157
Proactive cleanup approaches, 85–86, 126–128
Problem identification, 320
Process Action Teams (PAT), 203
Professional organization, 205
 ASTM
  Accelerated Site Characterization Guide, 220–221
  RBCA model, 188, 191, 217–220
 information resources, 346–349
Progress and performance evaluation, 322–324
 budget linkages under DOE program, 156
 DOE Environmental Restoration program, 150
 military environmental programs, 99–101
 remedial performance evaluations, 322–324
 strategic plan implementation, 287
Property tax abatement, 314, 324–326
Pump and treat, 175
 relative uncertainty levels, 210
Pyrolysis, 396

Q

Quality, AFCEE services, 114
Quality Assurance Project Plan (QAPP), 128
Quality assurance/Quality control (QA/QC) procedures, 40

R

Radar, 233
Radioactive waste, 145, 149, 157, See also U.S. Department of Energy
Radionuclides, 145, 162
RCRA, See Resource Conservation and Recovery Act
Real-time data collection, 16, 251–252
Record of Decision (ROD)
 communication and coordination, 105
 contingency remedy, 55
 Department of Defense process, 102–106
 flexibility under state programs, 201–202
 improving planning process, 104
 Interagency Agreement, 103
 interim remedial action, 33, 53–55

INDEX

429

no-action, 52–53
slowness of process, 2
Recycling, 71
Regulatory program, cross-program integration, See Integrated environmental programs
Relative risk
  evaluation, 84–85
  sequential risk mitigation, 214–217
Remedial action, interim, See Interim remedial actions
Remedial action contract (RAC), 136
Remedial Action Cost Engineering and Requirements (RACER) system, 112
Remedial design/remedial action (RD/RA), 100
Remedial investigation/feasibility study (RI/FS), 16
  design considerations, 139
  modeling, 229
  observational method application, 209
  ROD time frame, 51
Remedial performance evaluations, 322–324
Remedial technology deployment strategy, 156–157
Remediation cost reduction, 319–322, See also Cost reduction
Remediation technology contracting suitability matrix, 136
Remediation technology vendor matrix, 355–420
Removal actions, 25–32
  application of, 31–32
  case study, 32
  emergency, 27
  integrated assessments, 41
  non-time-critical, 30
  time-critical, 27, 30
Research
  DOE's bioremediation program (NABIR), 161–171, See Natural and Accelerated Bioremediation Research Program
  state programs, 174–178
Resistivity, 235
Resource Conservation and Recovery Act (RCRA), 4, 58–71, 289
  Air Forces's Installation Restoration Program, 106
  comprehensive contingency plans, 66–67
  corrective actions program, 59–69
    corrective action management units, 61–64
  cross-program integration, 73
  innovative investigation and remedial technologies, 60
  interim measures, 65–66
  military installations and, 82
  mixing zones, 69

point of compliance, 68
presumptive remedies, 60
site stabilization, 68–69
state underground storage tank programs, 182
temporary units, 64–65
underground storage tank programs, 69–70
waste minimization, 70–72
Rhode Island Brownfields Pilot, 305
Risk, accelerated cleanup-related, 9
Risk assessment, land use determination, 293
Risk-based corrective actions (RBCA), 188, 191, 217–220
Risk management
  Department of Defense relative risk evaluation, 84–85
  Superfund Accelerated Cleanup Model, 37–38
Rockwell Hot Lab, 158
Rocky Flats Environmental Technology Site, 159
Rolling milestone approach, 156

S

Sandia National Laboratories, 159
Scheduling
  Navy accelerated scheduling, 141–142
  strategic planning considerations, 296, 301
Scientific method, 9–10
Scope of work, 139–141
Seismic reflection, 232
Seismic refraction, 232
Selenium, 162, 163
Sequential risk mitigation, 214–217
Shoal Test Area, 159
Site characterization, 16, See also Geophysical tools; Investigation
  ASTM Risk Based Corrective Action (RBCA) model, 218
  ASTM standard for petroleum releases, 220–222
  California Base Closure Environmental Committee, 203
  California LUST remediation program, 191
  cost reduction, 320–322
  defining desired end results, 7
  demonstrating impracticality of groundwater cleanup, 74–76
  documenting intrinsic remediation, 117
  EPA technology evaluation projects, 351–354
  integrated assessment, 39
  intrinsic remediation protocol, 120–121
  investigation by excavation, 252–253
  modeling considerations, 223
  observational method and, 206, 209–211
  screening under SACM, 34–35

Site Characterization and Analysis Penetrometer System Laser-Induced Fluorescence (SCAPS-LIF), 196–197, 249
Site inspection/remedial investigation (SI/RI) stage, 34–35
SITE program, 17–19
Site stabilization, 68–69
Slagging, 396–397
Slurry phase bioremediation, 371–375
Snake River Aquifer, 164
Software, See Computer applications
Soil bioremediation, vendor information matrix, 364–368
Soil flushing, 397
Soil sample analytical methods, 22–23
Soil sampling and analysis, innovative approaches, 252
Soil vapor extraction (SVE), 31, 69, 109, 126, 182, 186
    presumptive remedies, 45
    vendor information matrix, 397–399
Soil washing, 399–403
Solid-phase bioremediation, 375–380
Solid-phase microextraction, 23
Solid Waste Management Units (SWMU), 58, 59
Solvent extraction, 403–406
Source control, Base Realignment and Closure program, 92
Source reduction, 71
Standardized cleanup, See Presumptive remedies
Standard plans and specifications, 109
State and local programs
    abandoned mine reclamation, 198–201
    ASTM Risk Based Corrective Action (RBCA) model, 220
    California Base Closure Environmental Committee, 202–204
    flexibility in, 179–180
    flexible RODs under CERCLA, 201–202
    interstate programs, 194–197
    new technology facilitation, 194–197
    research programs, 174–178
    underground storage tanks, See Underground storage tank (UST) programs
    voluntary cleanup and negotiating actions, 178–182, 306–314, See also Brownfields Initiative
    Western Governors Association, 194–197
Statistical methods, 323
Story boards, 299
Storytelling, 263
Strategic planning, 269, 283–316
    administrative and organizational elements, 298–299
    cost saving, 295

EPA Brownfields Initiative, 314–316
EPA information resource, 284
flexibility, 285
fundamental phases, 286–287
funding requirements and allocation, 294–296
future land use considerations, 292–293
implementation process, 286–287
innovative technologies and approaches, 297
large-scale elements for project acceleration and streamlining, 287–297
regulatory process and flexibility, 289–291
scheduling considerations, 296, 301
science and technology elements, 294, 297–298
small-scale elements for project acceleration and streamlining, 297–301
story boards, 299
teamwork, communication, and decision making, 291–292, 299
technical strategy development, 284, 294
uncertainty components, 296
Streamlined Approach for Environmental Restoration (SAFER), 153–154
Supercritical fluid extraction, 23
Superfund, 24–26, See CERCLA
Superfund Accelerated Cleanup Model (SACM), 33–38, 85, 216, 250, 290
    case study, 55–58
    cost factor, 35, 37
    early actions, 35–37
    integrated assessment, 38–42
    presumptive and/or plug-in remedies, 42–45
    risk management, 37–38
    site screening, 34–35
Superfund Amendments and Reauthorization Act (SARA), 4, 24, 82
Superfund Innovative Technology Evaluation (SITE) program, 17–19
Superfund Record of Decision (ROD), See Record of Decision
Surfactant-enhanced recovery, 406

T

Tax abatement, 314, 324–326
Teamwork, 5, 6, 267–273
    common goals and objectives, 268–269
    efficiency and motivational factors, 327
    empowerment, 107, 272–273
    leadership, 273–279
    military programs, 107, 128
    mission and vision, 267–268
    outdated models, 257
    strategic planning, 286, 291–292, 299
    training, 258

INDEX 431

Technical impracticability determination, 74–76
Technical Review Committee (TRC), 105
Technical strategy, 284, 294
Technical support cost reduction, 158
Technical tunnel vision, 7
Technology-based cost reduction approaches, 317
Technology development, 249
Technology evaluation projects (EPA), 351–354
Technology Innovation Office (TIO), 19, 21
Technology matrix (including vendors), 355–420
Technology screening, 157
Technology transfer program, 17, 18–19
  Air Force Technology Transfer Division, 125–128
Temporary units (TU), 61, 64–65
Texas, innovative technology program, 197
Thermal desorption, 406–415
Thermally enhanced recovery, 415–417
Third party reviews, 111
Time-critical removal actions, 27
TIO, 19, 21
Toxic Substances Control Act (TSCA), 74, 82
Training and intervention, 258, 280–281
Treatment, storage, and disposal (TSD) facilities, 58, 59
  interim measures, 65
  RCRA comprehensive contingency plans, 66–67
Trichloroethylene (TCE), 162, 163
Trip, Dennis, 319
Trust, 279–280, 292

U

Uncertainty
  Brownfields redevelopment, 304
  field personnel empowerment, 272–273
  groundwater remediation modalities, 210
  human relations aspects, 257–258
  managing and planning for, 208–209
  observational method and, 206, 211
  strategic planning, 285, 296
Underground storage tank (UST) programs, 4, 173, 182–194
  ASTM Risk Based Corrective Action (RBCA) model, 220
  California program, 189–194
    containment zone designation, 193
    Low Risk Groundwater Case, 191–193
  Illinois program, 311
  investigation by excavation, 252–253
  Iowa program, 197
  new USTs, requirements for, 70
  pilot cleanup technologies, 185–188
  presumptive remedies, 48
  RCRA and, 69–70
  reimbursement programs, 183–184
  risk-based corrective actions, 188, 191
University research, 174–178
Uranium Mill Tailings Remedial Action (UMTRA), 150
Uranium tailings, 147, 150
U.S. Air Force, 106–128, See also U.S. Department of Defense
  AFCEE, 113–128, See Air Force Center for Environmental Excellence
  innovative technologies program, 125–128
  Installation Restoration Program, 106–113
    Information Management System, 122–125
  proactive communication, 126–128
U.S. Army, 142–143
U.S. Army Environmental Center (USAEC), 142–143
U.S. Department of Defense (DOD), 25, 79–143, See also Base Realignment and Closure
  AFCEE, See Air Force Center for Environmental Excellence
  Air Force, 106–128, See U.S. Air Force
  Army, 142–143
  BRAC, See Base Realignment and Closure
  budgeting priorities, 83–85, 131
  CERCLA and, 82
  common facility contaminants, 80
  community relations, 81, 91, 98, 110
  contracting considerations, 110, 133–134, 139–140
  Environmental Leadership Program, 129–130
  environmental security program, 80–81
  Information Management System, 122–125
  information resources, 335–336, 340
  innovative technologies, 81, 111, 125–128
  Installation Restoration Program (IRP), 82, 112–113
  Interagency Agreement, 109–110
  land use determination, 92, 98–99, 107
  lead agency responsibility, 82
  Management Action Plan Guidebook, 112–113
  Navy, 128–139, See U.S. Navy
  partnering, 81, 96–97, 107, 127, 269
  proactive site cleanup approach, 85–86, 126–128
  progress measurement, 99–101
  relative risk evaluation, 84–85
  ROD process ("Road to ROD"), 102–106
    communication and coordination, 105
    improving planning process, 104
    Interagency Agreement, 103
U.S. Department of Energy (DOE), 145–171, See also Environmental Restoration Program

bioremediation research program, 161–171, See also Natural and Accelerated Bioremediation Research Program
Environmental Restoration Strategic Plan, 149
field research centers, 168
information resources, 341, 342–343
intractable problems, 164
land use determination, 292–293
Office of Environmental Management, 147–148, 160–161, See also specific programs
   case studies, 158–160
partnerships, 168–171
privatization, 157
R&D remediation and waste management focus areas, 147–148
release sites and facilities, 153
U.S. Environmental Protection Agency, 15–76, See also Brownfields Initiative; CERCLA; specific programs
analytical level of data, 297–298
Brownfields Initiative, See Brownfields Initiative
completed monitoring and measurement technologies evaluation projects, 351–354
databases, 19, 21
DoD Federal Facility Agreement, 103
early actions, 25–30
   SACM, 35–37
enforcement policy, 37
evaluating groundwater remediation impracticability, 74–76
information resources, 18–19, 76, 330–334, 337–339
information transfer and technology demonstration programs, 17, See also specific programs
   accelerated investigation, 20–24
   remedial technologies (SITE), 17–19
   VISITT database, 19, 356–420
integrated assessment, 38–42
integration of environmental programs, 71–74
land use determination, 292–293
modeling information, 224
presumptive and/or plug-in remedies, 42–48
RCRA, 58–71, See Resource Conservation and Recovery Act
regional alignments, 66–67
removal actions, 25–32
strategic planning resource, 284
voluntary corrective actions, 178–179
U.S. Navy
accelerated scheduling, 141–142
Comprehensive Long-Term Environmental Action Navy (CLEAN) contract, 139–140
integrated project schedules, 130–139
   contracting issues, 133–134
   design considerations, 137–139
   design level, 134–137
   funding prioritization, 131
   project coordination, 132
Navy Environmental Leadership Program, 129–130, 239
scope of work, 139–141
UXO, 142–143

V

Validity of data, 297
Value added for project completion, 136
Vendor FACTS, 20, 21
Vendor Information System for Innovative Treatment Technologies (VISITT), 19, 356–420
Verbal skills, 261–263
VISITT, 19, 356–420
Visual MODFLOW, 248
Vitrification, 417–420
Volatile organic compounds, presumptive remedies, 45–46
Voluntary Action Program, 310, 312
Voluntary cleanup, 178–182, 295, See also Brownfields Initiative
   case study, 180–182
   program flexibility, 179–180
   state programs, 306–314

W

Waivers, 52, 290–291
Waste minimization, 70–72
Weather patterns, 108
Western Governors Association, 194–197
Wisconsin voluntary cleanup program, 310–311, 313
World economy, 4
Writing skills, 259–260